AF364665

Good Laboratory Practices and Compliance Monitoring

Good Laboratory Practices and Compliance Monitoring

Prof. Trupti Patil-Dongare

Master in Industrial Pharmacy, PhD

PharmaMed Press

An imprint of Pharma Book Syndicate

A unit of BSP Books Pvt. Ltd.

4-4-309/316, Giriraj Lane,

Sultan Bazar, Hyderabad - 500 095.

Published by

PharmaMed Press

An imprint of Pharma Book Syndicate

A unit of BSP Books Pvt. Ltd.

4-4-309/316, Giriraj Lane, Sultan Bazar, Hyderabad - 500 095.

Phone: 040-23445688; Fax: 91+40-23445611

E-mail: info@pharmamedpress.com

www.pharmamedpress.com/pharmamedpress.net

ISBN: 978-93-89974-26-3

FOREWORD

In the recent past, pharmaceutical manufacturing organizations worldwide have devoted a great deal of resources to ensure compliance with cGMP guidelines as issued by the licensing authorities. The Pharmaceutical industry, in particular, has demonstrated remarkably its ability to comply with the stringent requirements of regulatory authorities like the USFDA, European, MHRA and other regulatory bodies, a prerequisite to enter the lucrative generic market in the regulatory.

While many manufacturing units in India are already in compliance with the requirements of these stringent guidelines, a large number of medium and small-scale units still need to upgrade their facilities and procedures to comply with these requirements. Apart from facilities and systems involved in the manufacturer of pharmaceuticals, the current guideline lay great emphasis on the documentation of the procedures used to manufacture and store pharmaceutical products. This upgradation of the systems and procedures are essential to ensure a uniform quality product reducing the possibility of inter-batch and multiple locations.

Change in the regulatory focus it is very important for sustain successes. A illustrative of this book is the component essential to achieve sustained success of the laboratory management.

In view of these changes, implemented by drug licensing authorities all over the world, I am happy these will be easily understood for greater benefit of the pharmaceutical industry, through this book. I am confident that this book will become a comprehensive reference guide for all technical personnel in the entire drug industry, especially personnel involved with the manufacturer of pharmaceutical formulation. More efforts such as these will go on long way in creating great awareness of these revised guidelines and will help to develop better preparedness among our technical personnel to face audits from different regulatory authorities in developed and developing countries.

Laboratory management, in its numerous arrangements, is a cost-effective means of improving the innovativeness and meets the regulatory requirement. The approaches marked in this book are both old and new. They may be difficult to accept and even more difficult to implement. They do, however, work. This book will basis you to think. "It is an opportunity to knocks, but it needs yours address.

Enjoy Reading!

Sudhir M Dongare

PREFACE

You are about to be familiarized to some old and new tactics to the application of laboratory management program. The pharmaceutical quality control laboratory serves one of the most important functions in pharmaceutical production and control. Forceful universal rivalry and extraordinary changes in machineries -technologies necessitate us to challenge past practices. Old antagonistic methods of seek, point, and blame will no longer work. Using some of the basic laboratory management principles, we can observe the practicality and enactment of all controls as they apply to core and peripheral operations. Laboratory, along with other forms of evaluation, can help us to determine if our own controls and our supplier controls work effectively.

This book aids quality laboratories for the laboratory Innovativeness, reduced costs, improved efficiency, Improved transparency to lab works, new regulatory requirements and Improved quality and compliance.

The main task of this book is to maintain the laboratory as per cGLP standards, to understand the regulatory requirement and meet the regulatory requirements.

This book gives a guideline for enhancing an organization ability to achieve regulatory sustained success. This document provides a self-assessment tool to review the extent to which the laboratory management has adopted the concepts in this book. This book provides knowledge of fundamental good quality laboratory practice and skill. It is intended to provide background information on auditing practice and the evaluation of the requirements from a regulatory point of view, with an overview of the applicable regulations.

This book applicable to any organization regardless of its type, size and activity.

Drugs being a very important component of healthcare, these need special attention especially with respect to Quality, Efficacy and Safety. This has always been a matter of concern from the consumer point of view.

In India, initially drugs were imported. After Nineteen Hundred fifties, the Indian Pharmaceutical Industry has made phenomenal progress and reached the present state where now India exports both bulk drugs and formulations even to the developed countries as against a situation till Nineteen Hundred Forties when India was importing formulation.

Such growth and acceptance of products in the international market could happen only if Indian products could meet the quality standards as set by the International Regulator's. The Quality Control concept is basically meant that the quality of the product conformed to pre-determined standards as given in the Pharmacopeia.

If the drugs are to qualify for export to non–regulatory market, then it can be only with assurance that products are manufactured as per the GMP guidelines defined in Schedule M or as per International regulatory guideline.

The pharmaceutical products must only be produced by licensed manufacturers, holders of the Manufacturing Authorization and whose activities are regularly inspected by the relevant National Sanitary Authorities. This Regulation of Good Manufacturing Practices (GMP) must be taken as a reference during the inspection of the plant facilities, of the production processes and quality control, and as training material for the inspectors in pharmaceutical products area, as well as, for the training of professionals responsible for the production process and Quality Control in the industries.

The GMP are applicable to all operations related to pharmaceutical products manufacturing, including those medicines being developed for clinical assays.

The Good Laboratory Practices (GLP) described in this document are subject to continuous updating, keeping up with the evolution of new technologies, where alternative actions might be adapted to attend to necessities of a determined product, provided that alternative actions are validated to guarantee the product quality.

The GMP do not contain aspects linked to the safety of the staff involved in the manufacturing process; those aspects are ruled by a specific legislation. Nevertheless, the manufacturer must guarantee the safety of his workers.

To compile this book, a self-effacing effort has been made to bring the various GMP/GLP guidelines under one reference book.

It is sincerely hoped that the compilation of various cGLP guideline in this book in a simple language will help in bringing about harmonization of the quality systems in all big and small Pharma companies.

The implementation of cGLP and harmonized quality systems will ensure that there are control on out of specification and other non-conformance through continual improvement in Quality Systems which can be achieved through self-assessment. It will also ensure Active Pharmaceutical Ingredients and Formulations meeting standards of quality, efficacy and safety, thereby meeting customer satisfaction and the demand for export in International Markets.

Dr. Trupti Patil-Dongare

ACKNOWLEDGEMENTS

First and foremost, I would like to thank my mentor, Mr. Rajkumar Mungarwadi, Mr. Sukhvinder Banga and Mr. Shirish Ambulgekar, for having introduced me to this exciting field of pharmaceutical, for his inspiring guidance and constant encouragement during the course of this work. I owe my heartfelt gratitude to him for providing me an opportunity to pursue research with excellent laboratory facilities. The successful completion of this thesis is entirely due to his innovative ideas, critical analysis, excellent advice and detailed discussions during the entire study period. I am most fortunate and to have a great researcher as my research supervisor.

Last not the lease I want to thank my parents, book publisher –Mr. Naresh, Mr. Santosh and my friend who were also a backbone to this research work.

I have been working on laboratory management for a long time and different hurdles has been crossed by many to achieve this result which is the reason that this book here.

Some major forces have influenced me in the past 16 years. As an Pharmacist, I first learned about to provide quality assured drug in affordable price, quality management and leadership principles. Upon leaving the diverse pharmaceutical organization, I exposed the quality concepts while employed by a pharmaceutical industry. I also developed my pharmaceutical-regulatory knowledge and skill, although I had some help from the existing agreement standards, it was mainly a trial-and-error process. I made several mistakes, but I learned from them. I started developing, training to form sustained system and meets the stringent regulatory requirements. I collide with out on my own as a mentor. Since then, I continue to learn from each company I visit/work and each class I teach.

Warm Regards,

Dr. Trupti Patil-Dongare

CONTENTS

Foreword... (v)

Preface .... (vii)

Acknowledgements .... (ix)

CHAPTER 1

OVERVIEW OF GOOD LABORATORY
PRACTICES IN PHARMACEUTICAL INDUSTRY

1.1 Fundamental of GLP 2

1.2 GLP Organization and Personnel Management..2

1.3 Organization Chart, Job Description
    and Specimen Signature in GLP 3

1.4 Personnel and Personnel Hygiene...3

1.5 Instruments and Equipment..3

1.6 Laboratory Glassware 4

1.7 Chemicals, Reagents and Analytical Standards 5

1.8 Laboratory Reference Standards and
    Culture Media in Good Laboratory Practice 5

1.9 Good Housekeeping and Safety 6

1.10 Sampling 6

1.11 Testing and Review 7

1.12 Documentation...8

1.13 Incident, OOS and OOT 8

1.14 Self-Inspection (Internal Audits) 9

1.15 Validation/Verification...9

1.16 Measures in Good Laboratory Practice..9

1.17 Technical Transfer of Testing Methods in Good Laboratory Practice.....10

1.18 On-going Stability Programme in Good Laboratory Practice 10

1.19 Documentation in Good Laboratory Practice 11

1.20 Sampling in Good Laboratory Practice...11

1.21 Testing in Good Laboratory Practice 12

1.22 Annexure...13

CHAPTER 2

OVERVIEW OF GOOD MICROBIOLOGY PRACTICES IN PHARMACEUTICAL INDUSTRY

2.1 Premises, Layout and Zone 18

2.2 Laboratory Equipment 19

2.3 Personnel 19

2.4 Media and Preparation 20

2.5 Reference Culture 21

2.6 Sampling 22

2.7 Laboratory Testing 22

2.8 Disposal of Contaminated Waste 23

2.9 Laboratory Data Management 23

CHAPTER 3

QUALITY ASSURANCE MANAGEMENT IN GLP ENVIRONMENT

3.1 Requirement of Quality Assurance Management in GLP 25

3.2 Responsibilities of the Quality Assurance Personnel 26

3.3 Qualifications of QA Personnel 26

3.4 Quality Assurance Activity in GLP 26

3.5 QA and Non-Regulatory Test 27

3.6 The QA Statement 28

3.7 Regulatory Expectations in GLP 28

CHAPTER 4

LABORATORY QUALITY MANUAL

4.1 Contents of the Laboratory Quality Manual 29

4.2 Topics that should be Covered in the Final Section on
     Quality Policies for Specific Regulation Elements 31

4.3 Format for Laboratory Quality Manual 33

CHAPTER 5

LABORATORY QUALITY POLICY

5.1 Scope of GLP Quality Policy 35

5.2 GLP General Requirements 35

5.3 Risk Management 36

5.4 Documentation, Records, and Data Controls 36

5.5 Management Responsibility 36

5.6 Resource Management 37

5.7 Product Realization 39

5.8 Product Sourcing... 40

5.9 Measurement, Analysis and Improvement................................. 40

CHAPTER 6

LABORATORY SITE MASTER FILES

6.1 Content of Site Master File.. 42

6.2 Appendix 48

6.3 Specimen Copy of Laboratory Site Master File.......................... 48

CHAPTER 7

VALIDATION MASTER PLAN

7.1 Types of Methods.. 60

7.2 Method Validation Requirements 61

7.3 Content of Analytical Procedure 61

7.4 Validation Characteristics... 62

7.5 Analytical Method Transfer 63

7.6 Analytical Revalidation... 64

7.7 Analytical Procedure Life Cycle Management 65

7.8 Definition 66

CHAPTER 8

LABORATORY COMPUTER-SOFTWARE VALIDATION AND QUALIFICATION

8.1 GAMP 5: Good Automated Manufacturing Practice.................... 69

8.2 Computer and Software Validation Master Plans 70

8.3 Computer Qualification 70

8.4 Software Validation 73

8.5 Validation Protocol/Report... 75

8.6 Validation of Hardware and Software....................................... 75

8.7 Computer System and Software Life Cycle Management 76

8.8 Security Management... 77

8.9 Electronic Records and Electronic Signature 77

8.10 Handling of Audit Trial 78

8.11 System Periodic Review 80

8.12 Data Back-up, Recovery, and Archiving 80

8.13 Electronic Data Integrity for Long Term in GLP 80

8.14 System Retirements Management 81

8.15 Business Continuity Plan (BCP) and
Disaster Recovery Plan (DRP).. 82

8.16 Approach to Managing Regulatory Risks and
Achieving in GPL Compliance.. 84

CHAPTER 9

CONTROL OF SPREADSHEETS IN GLP

9.1 Classification of Spreadsheet... 85

9.2 Control of the Spreadsheet 85

9.3 Validation of the Spreadsheet 86

9.4 Live use of Spreadsheet.. 87

9.5 Spreadsheet Inventory Management 87

9.6 Data Storage and Security.. 88

9.7 Case Study for Master Spreadsheet Validation 88

CHAPTER 10

ANALYTICAL INSTRUMENT AND EQUIPMENT QUALIFICATION

10.1 GLP Qualification and Validation 92

10.2 Components of Data Quality in GLP 92

10.3 Analytical Instrument and Equipment Design.................................. 93

10.4 Analytical Instrument and Equipment Qualification Phase............. 93

10.5 Steps involved in Qualification.. 94

10.6 Requalification Frequency 97

10.7 Instrument and Equipment Operating Ranges 97

10.8   Categories of Analytical Instrument 98

10.9   Analytical Instrument and Equipment
Standard Operating Procedure 99

10.10 Calibration Management 99

10.11 Planned Preventive Maintenance Program 100

10.12 Usage Log of Analytical Instrument and Equipment 100

10.13 Handling Defective and Nonqualified
Instruments and Equipment 101

10.14 Analytical Instrument and Equipment Audits 101

10.15 Definition 101

CHAPTER 11

ELECTRONIC AND PAPER-BASED DATA MANAGEMENT IN GLP

11.1   ALCOA 104

11.2   Data Integrity GLP Regulatory Expectations 104

11.3   When does Electronic Data become a
cGMP and GLP Record? 105

11.4   Confidentiality, Integrity and Availability Module (CIA) 105

11.5   GLP Data Governance System 106

11.6   Designing GLP Systems to Assure
Data Quality and Integrity 106

11.7   GLP Triggers of Data Integrity Loss 107

11.8   Data Integrity Metrics in GLP 107

11.9   Electronic Data Management in GLP as
per USFDA Part 11 and EU GMP Annex 11 108

CHAPTER 12

LABORATORY TRAINING AND QUALIFICATION MANAGEMENT PROGRAM

12.1 Training Program Management 113

12.2 Star Bugs of Training Program Management 114

12.3 Identification and Evaluation of Training Needs 115

12.4 Identification of Trainer 115

12.5 Training Program Material, Assets, Methods and Tools 116

12.6 Training Program Management Plan 116

(xvi) Contents

<hr>

12.7 Evaluation of Effectiveness of Training	116
12.8 Documentation of Training Management	117
12.9 Automated Training Management	118
12.10 Analyst Trainee Training Qualification	118
12.11 Reviewer Qualification	120
12.12 Format used for Training Qualification Program Management	121
12.13 Format Used for Analyst Trainee Qualification Program Management	125
12.14 Format Used for Reviewer Qualification Program Management	130

CHAPTER 13

REVIEWER QUALIFICATION MANAGEMENT

13.1 Documentation Requirements for Reviewer Qualification	132
13.2 Qualification Procedure	133
13.3 Evaluation and Approval	134
13.4 Acceptance Criteria for Qualification	134
13.5 Requalification	135
13.6 Loss of Qualification	135
13.7 Planning, Reviewing and Documenting Qualification Activities	135
13.8 Reviewer Qualification Management	136
13.9 Definition	136

CHAPTER 14

LABEL MANAGEMENT IN GLP ENVIRONMENT

14.1 Master Specimen Label Copy	137
14.2 Procurement, Receipt, Storage, Issuance, Reconciliation and Destruction of Label	137
14.3 Format Used for Label Management	138

CHAPTER 15

GLASSWARE MANAGEMENT IN GLP

15.1 Classification of Glass as per United State of Pharmacopeia 146

15.2 Types of Glassware used in Laboratory Management 148

15.3 Common Glassware used in the Laboratory Management 148

15.4 Glassware Classification by Accuracy 149

15.5 Laboratory Glassware Procurement 149

15.6 Laboratory Glassware Verification 149

15.7 Glassware Calibration 150

15.8 Glassware Use and Cleaning 150

15.9 Glassware Care and Storage 153

15.10 Verification of Laboratory Glassware Procedure 154

15.11 Definitions 158

15.12 Pictorial Diagram of Glassware used in
Laboratory Management 158

CHAPTER 16

GLP COLUMN MANAGEMENT

16.1 High Liquid Chromatographic Column 162

16.2 Gas Chromatography Column 164

16.3 Receipt, Issuance, Maintenance, Regeneration
and Destruction of Column 165

16.4 Flow Chart of the Column Management 166

16.5 General Maintenance and Care 166

16.6 Records Maintained for Column Management 167

CHAPTER 17

GLP STANDARD OPERATING PROCEDURE (SOP) MANAGEMENT

17.1 SOP Process 170

17.2 Writing Styles 170

17.3 Format and Structure 170

17.4 SOP Review and Approval 172

17.5 Frequency of Revisions and Reviews 172

17.6 Checklists 173

17.7   Document Control 173

17.8   SOP Document Tracking and Archival 173

17.9   SOP General Format 174

17.10 Definitions... 175

CHAPTER 18

GLP SPECIFICATION MANAGEMENT

18.1   Procedures and Requirements 179

18.2   Types of the Laboratory Specifications................................... 179

18.3   Specification Template 181

CHAPTER 19

GLP CERTIFICATE OF ANALYSIS (COA) MANAGEMENT

19.1   Process of Certificate of Analysis (COA)................................ 184

19.2   Contents of COA 185

19.3   Control on COA 186

19.4   Format of COA 186

CHAPTER 20

ANALYTICAL ROUNDING OF RESULTS

20.1   Rules Involved in the Rounding of Results............................. 187

20.2   Definitions... 188

CHAPTER 21

HANDLING OF RESIDUAL SOLVENT

21.1   Classification of Residual Solvents by Risk Assessment........ 189

21.2   Analytical Procedures 190

21.3   Residual Solvent Level Reporting 190

21.4   Limits of Residual Solvents 190

21.5   Diagram Relating to the Identification of
       Residual Solvents and the Application of Limit Tests........... 193

21.6   Glossary... 194

CHAPTER 22

RESERVE SAMPLE MANAGEMENT

22.1 Reserve Sample Collection Management 195

22.2 Reserve Sample Size and Retention Time 195

22.3 Storage Management for Reserve Sample 196

22.4 Periodic Visual Inspection of Reserve Sample 197

22.5 Reserve Sample Issuance and Retrieval Management 197

22.6 Destruction of Reserve Sample 197

22.7 Format Used in Reserve Sample Management 197

CHAPTER 23

PHARMACEUTICAL PRODUCT STABILITY MANAGEMENT

23.1 Stability Testing Condition and Testing Condition 201

23.2 Testing of Stability Sample 201

23.3 Selection of Batches for Stability Sample 203

23.4 Stability Sample Quantity 203

23.5 Stability Protocol and Report 204

23.6 Analysis of Stability Sample 204

23.7 Evaluation of Stability Data 204

23.8 Extrapolation of Stability Data 206

23.9 Significant Change at Accelerated Condition 206

23.10 Destruction of Stability Samples 207

23.11 Stability Storage Chamber Management 207

23.12 Mean Kinetic Temperatures 208

23.13 Format used for Stability Study Management 209

23.14 Definition 211

CHAPTER 24

GOOD CHROMATOGRAPHIC INTEGRATION PRACTICE IN GLP

24.1 Steps Involved Chromatographic Integration 215

24.2 Integration of Well Resolved Peaks and
Properly Integrated Single Peak 215

24.3 Acceptable Practice of Peak Integration for a
Single Peak during Assay Test 216

24.4 Chromatographic Integration of Unresolved Peaks 218

24.5 Noisy Baseline 223

24.6 Integration for Tests Involving Qualification of
       Minor Peaks in Combination with Principle Peak(s)..............224

24.7 Chromatographic Inhibit Integration....................................225

24.8 Chromatographic Manual Integrations 225

24.9 Flow Chart of Good Chromatographic Practice in GLP.........227

24.10 Definition 228

CHAPTER 25

MANAGEMENT OF UNKNOWN AND
EXTRANEOUS PEAKS IN CHROMATOGRAPHIC TESTS

25.1 Impurities as per ICH Q3A...230

25.2 Classification of Impurities...230

25.3 Unknown or Extraneous Peaks
       Observed during a Chromatographic Test 231

25.4 Reporting, Identification and Qualification
       Thresholds for Unknown Peaks.. 233

25.5 Investigation Trigger for Extraneous Peaks...........................238

25.6 Investigating "Unknown" or "Extraneous" Peaks.................238

25.7 Definition 241

25.8 Annexure...241

CHAPTER 26

QUALITY AGREEMENTS IN LABORATORY

26.1 Elements of a Quality Agreement 244

26.2 Activities in Quality Agreements 245

26.3 Disqualification 246

26.4 Template used for Quality Agreement of Contract Laboratory.............247

CHAPTER 27

SAMPLING AND TESTING IN LABORATORY MANAGEMENT

27.1 Sampling Responsibility 261

27.2 Different Types of Sampling in the
       Pharmaceutical Industry..262

27.3 Different Sampling Category 264

27.4 Sampling Process 265

27.5 Sampling Plan... 265

27.6 Sampling Facility used in Pharmaceutical Industry 265

27.7 Sampling Tools... 266

27.8 Sampling Container and Labeling.. 269

27.9 Storage for Sampled Material... 269

27.10 Golden Rules of Sampling 269

27.11 Definition 273

27.12 Quality Control Sample Management Flow 275

CHAPTER 28

LABORATORY STANDARD MANAGEMENT

28.1 Source Material for Laboratory Standard.................................... 276

28.2 Traceability of Laboratory Standard 277

28.3 Reference Standards 277

28.4 Steps Followed in Reference Standard Project Design 278

28.5 Receipt and Storage Standard Reference Standard..................... 278

28.6 Working Standard.. 279

28.7 Receipt and Storage Standard Working Standard 279

28.8 Chemical and Physical Methods used
in Evaluating Substances 280

28.9 Definition 280

CHAPTER 29

ANALYTICAL REAGENT, INDICATORS AND VOLUMETRIC SOLUTION MANAGEMENT

29.1 Requirement of Reagent, Indicators and Volumetric Solution............... 283

29.2 Reagents Management... 283

29.3 Indicators Management 284

29.4 Volumetric Standard Management 285

29.5 Batch Number System for Reagent,
Indicators and Volumetric Solution 287

29.6 Label for Reagent, Indicators and Volumetric Solution 287

(xxii) *Contents*

CHAPTER 30

STATISTICAL TOOLS FOR PHARMACEUTICAL INDUSTRY

30.1 Concept of Six Sigma and Lean Manufacturing 289

30.2 Common Statistical Tools Pharmaceutical Industry can use to meet Regulatory Requirements 289

30.3 Other Statistical Tools used in Pharmaceutical Industry 291

30.4 Qualitative and Quantitative Analysis 294

30.5 Patterns and Trends 294

30.6 Summary 295

CHAPTER 31

FAILURE INVESTIGATION: TO PREVENT REOCCURRENCE

31.1 Failure Investigation Regulatory Expectation 296

31.2 Failure Investigation Approach 297

31.3 Why do we need Investigation? 297

31.4 Why do we have Failures? 300

31.5 Failure Investigation Methodology: A Key to Success in a World Failure 300

31.6 Failure Investigation Data Review and Evaluation 305

31.7 What is Corrective & Preventive Actions? 306

31.8 Failure Investigation Report 307

CHAPTER 32

LABORATORY FAILURE INVESTIGATION MANAGEMENT REPORT WRITING

32.1 Investigation Management 312

32.2 Investigation Star Bugs 312

32.3 Challenges in Investigation 317

CHAPTER 33

DEVIATION MANAGEMENT IN GOOD LABORATORY PRACTICES

33.1 Deviation Categorization 318

33.2 Management of Deviation 319

33.3 Steps Involved in Deviation Investigation 320

33.4 Most Common Deviation in the Laboratory, but not Limited to 320

33.5 Deviation Trend Analysis 320

33.6 Deviation Case Study 320

CHAPTER 34

HANDLING OF OUT OF TREND (OOT) RESULT IN LABORATORY MANAGEMENT

34.1 Reporting and Login of the OOT 327

34.2 Laboratory Investigation Phase I 328

34.3 Phase II: Full Scale Investigation 330

34.4 OOT Trend Analysis 331

34.5 OOT Methodology Calculation 332

34.6 Challenges in Implementation of OOT 332

34.7 Definition 333

34.8 Annexes 334

CHAPTER 35

HANDLING OF OUT OF SPECIFICATION LABORATORY RESULTS

35.1 Reporting and Login of the OOS 345

35.2 Phase I: Laboratory Investigation 345

35.3 Phase II: Full Scale Investigation 348

35.4 OOS Trend Analysis 350

35.5 Field Alter Report (FAR) 350

35.6 Definition 350

35.7 Annexes 353

CHAPTER 36

PHARMACEUTICAL CHANGE CONTROL MANAGEMENT

36.1 Regulatory Expectation in Change Control Managements 365

36.2 Steps Involved in Change Control System 365

36.3 ICH Q8, Q9 and Q10 Change Management and Application of Changes 367

36.4 E-Change Control System 368

36.5 Format Used in Change Control 368

36.6 Definition 369

CHAPTER 37

DATA INTEGRITY IN PHARMACEUTICAL QUALITY CONTROL LABORATORIES

37.1 Procedural versus Technical Controls 371

37.2 Certificates of Software Validation or Capability: Do they Provide Value? 372

37.3 Your Responsibilities: Audit, Assess, and Validate 372

37.4 Assessing the Software's (and the Vendor's) Compliance Support...373

37.5 Laboratory Audit Trail 373

37.6 Audit Trails and Design..373

37.7 Audit Trail Beginning...374

37.8 System Audit Trail Features 374

37.9 Audit Trail Content 375

37.10 Audit Strategies...375

37.11 Regulatory Main Elements of Data Integrity and Audit Trail Reviews..376

37.12 Audit Trail Review...376

37.13 Beyond the Audit Trail 377

37.14 Data Integrity from a Regulators' Point of View 377

CHAPTER 38

ALARM MANAGEMENT IN GOOD LABORATORY PRACTICES

38.1 Key Aspects of Alarm Management Program..380

38.2 Regulators Expectation of Alarm Management System...........................380

38.3 Element of a Good Alarm on Risk Assessment in Pharmaceutical Industry 382

38.4 Categorization of Alarm 383

38.5 Principle of Alarm Management System..383

CHAPTER 39

HANDLING OF OBJECTIONABLE ORGANISMS – THE REGULATORY PERSPECTIVE

39.1 Regulatory Expectations 385

39.2 Objectionable Microorganism Needs...385

39.3 Microbiological Requirements 385

39.4 Detection of Objectionable Organism 386

39.5 What the Laboratories Can Do on Identification
of Objectionable Organism? 386

39.6 Risk Assessment 387

39.7 Resources for Assessing the Potential Harm 389

39.8 FDA Product Recalls due to Objectionable
Microorganism given in the Table below 390

39.9 Summary 390

CHAPTER 40

QUALITY BY DESIGN (QBD) APPROACH IN THE PRODUCT LIFE CYCLE

40.1 Elements of QbD 391

40.2 Steps Involved in QbD Development Process 392

40.3 The Principle Steps in QbD are 392

40.4 QbD Documents 393

40.5 Process Analytical Technology (PAT) Frame Work 394

40.6 Use of PAT Tools in Continuous Manufacturing Process 394

40.7 PAT Regulatory Approach 395

40.8 Product Life Cycle 395

40.9 Definition 397

CHAPTER 41

VALIDATION MASTER PLAN (VMP)

41.1 Relation between Validation and Qualification 398

41.2 Validation Master Plan 399

41.3 Validation Protocol/Report 401

41.4 Time Plans of Each Validation Project and Sub-Project 402

41.5 Validation, Sampling Schedule 402

41.6 Definitions 402

CHAPTER 42

AUDIT IN LABORATORY MANAGEMENT

42.1 Traditional Approach to IA 409

42.2 Forecasting Approach to IA 410

42.3 Benefits of QRM to IA Program 411

42.4 IA Approach to Organization 411

42.5 Planning 415

42.6 Audit Sampling 417

42.7 Audit Tools 418

42.8 Audit Evidence and Analysis Phase 418

42.9 Audit CLOSE-OUT and FOLLOW-UP 418

42.10 Good Laboratory Checklist for Internal Audit 419

CHAPTER 43

LABORATORY ENVIRONMENT CONDITION AND MONITORING

43.1 Laboratory Environmental Conditions.....................................442

43.2 Laboratory Temperature and Humidity Recording Program...............442

43.3 Monitoring of Differential Pressure in Different Areas...........443

43.4 Microbiological Environmental Monitoring 443

43.5 Air and Surface Environment Monitoring 445

43.6 Difference between Viable Particle and Non-Viable Particle 446

CHAPTER 44

LABORATORY QUALITY STANDARDS MANAGEMENT

44.1 International Laboratory Management Quality Standards 448

44.2 Steps Involved in Laboratory Quality Management System...................449

44.3 ISO 15189 Laboratory Standards 450

44.4 Steps Involved in Implementing the
Quality Standard in Laboratory...450

44.5 Laboratory Management Error 451

44.6 Laboratory Management Phases 451

44.7 Checklist for Laboratory Quality Standards Management.....................454

CHAPTER 45

LABORATORY INSTRUMENT CALIBRATION PROGRAM

45.1 Laboratory Instrument Calibration Authorized Agencies.....................464

45.2 Method used for the Laboratory Instrument Calibration 465

45.3 Setting of Laboratory Instrument Calibration Limits 466

45.4 Calibration Certification Requirement......................................466

45.5 Instrument Calibration Procedure...466

45.6 Instrument Calibration Responsibility..467

45.7 Calibration Frequency Determination 467

45.8 Laboratory Calibration Schedules 468

45.9 Instrument Labelling and Documentation......................................468

45.10 Relocation of Laboratory Instruments 469

45.11 External Calibration Contractors 469

45.12 Definitions..469

CHAPTER 46

LABORATORY SAFETY MANAGEMENT PROGRAM

46.1 Laboratory Safety Standard - Roles and Responsibilities 472

46.2 Requirement of Safety Department..473

46.3 Safety Rules for Laboratories 474

46.4 Laboratory Personnel Information and Training 475

46.5 Material Safety Data Sheets (MSDSs)...477

46.6 Safety Hazards 478

CHAPTER 47

ROLL BACK GOOD ANALYTICAL PRACTICES

47.1 Analytical Technique Procedure...479

47.2 Laboratory Instrument Qualification..480

47.3 Statistical Concepts 481

47.4 Analysis Platinum Rules 481

47.5 Laboratory Data Integrity 481

47.6 Laboratory Management Practice..482

47.7 Risk-Based Approach 483

47.8 Important Element in Quality Control...483

47.9 Laboratories Contents and its Verification....................................485

CHAPTER 48

LABORATORY ENTRY-EXIT PROCEDURE

48.1 Entry Procedure in Pharmaceutical Laboratory 500

48.2 Exit Procedure in Pharmaceutical Laboratory 501

References 503

Chapter - 1

Overview of Good Laboratory Practices in Pharmaceutical Industry

Introduction

GLP refers to a quality system to ensure the uniformity, consistency, reliability, reproducibility, quality and integrity of the data. GLP gives true reflection of tested results.

Good Laboratory Practices - Good laboratory practices embody a set of principles that provide framework within which laboratory work is planned, performed, monitored, recorded, reported and archived.

Good laboratory practice must be planned, reliable, accurate, recorded, reported, monitor and archive all data generated during analysis or testing.

GLP helps in providing confidence to regulatory authorities and customers that the data submitted are a true reflection of the results obtained during the testing and can therefore be relied upon when making risk/safety assessment. Below important aspects shall be taken into account to achieve GLP.

The purpose of testing items is to obtain information on their safety with respect to human health and environment. GLP is also required for registration purpose and licensing of pharmaceuticals, pesticides, food additives, veterinary drug products and some bio-products.

Personnel working in laboratory must have education, training, and experience, or combination thereof, to enable that individual to perform the assigned functions. Facilities must be adequate and environment control.

1.1 Fundamental of GLP

Good Laboratory Practice is defined as "a quality system concerned with the organisational process and the conditions under which non-clinical health and environmental safety studies are planned, performed, monitored, recorded, archived and reported."

Fundamental of GLP depends on:

(a) Resources: Organization, personnel, facilities and equipment;

(b) Characterization: Test items and test systems;

(c) Rules: Protocols, standard operating procedures (SOPs);

(d) Results: Raw data, final report and archives;

(e) Quality Assurance: Independent monitoring of research processes

(f) Compliance across the Pharmaceutical laboratory

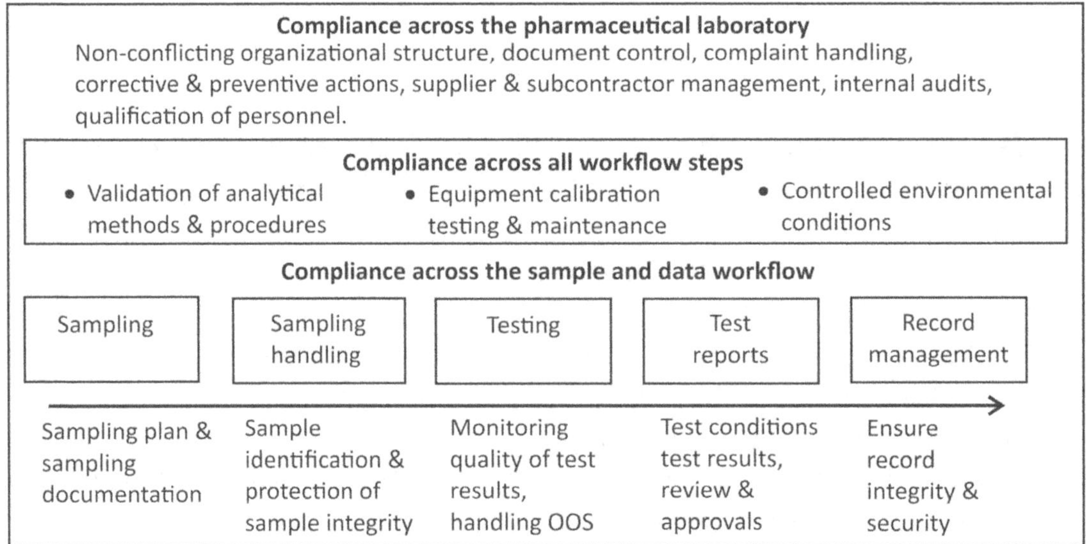

1.2 GLP Organisation and Personnel Management

GLP regulations require clear definitions of the structure of the research organisation and the responsibilities of the research personnel. This means that the organisational chart should reflect the reality of the institution and should be kept up to date. Organisational charts and job descriptions give an immediate idea of the way in which the laboratory functions and the relationships between the different departments and posts.

GLP also stresses that the number of personnel available must be sufficient to perform the tasks required in a timely and GLP-compliant way. The responsibilities of all personnel should be defined and recorded in job descriptions and their qualifications and competence defined in education and training records. To maintain adequate levels of competence, GLP attaches considerable importance to the qualifications of staff, and to both internal and external training given to personnel.

1.3   Organization Chart, Job Description and Specimen Signature in GLP

- The laboratory should have an organization chart depicting key positions and the names of responsible persons. The organization chart should be dated, authorized and kept up to date.
- There should be job descriptions for all personnel, including a description of their responsibilities. Every job description should be signed and dated by the staff member to whom it applies.
- There should be a list of signatures of the authorized personnel performing tasks during each study.

1.4   Personnel and Personnel Hygiene

There should be adequate number of personnel qualified in terms of education/training/ experience. All personnel must be provided training in respective working areas and training record of all the personnel should be maintained as per applicable SOP. All personnel prior to employment should be medically examined and periodic re-examination to be carried out for medical fitness. There should be written job descriptions for all persons working in Quality control.

Smoking, eating, drinking, chewing or keeping plants, food, drinks and personal medicines shall not be permitted in laboratory areas. Person suffering from any infectious disease or having any open lesions should not engage in activities which could affect the quality of analysis. All Personnel shall wear the company's uniform or aprons as applicable in the laboratory premises. Entry/exit and gowning procedure shall be followed wherever applicable.

Adhere to the procedure for personnel clothing for entering into microbiological testing laboratories. Personnel shall practice good sanitation and health habits. All personnel should follow the time schedule with respect to shift and should be punctual in attendance.

1.5   Instruments and Equipment

Laboratory should be furnished with all types of Instruments/Equipment which are necessary to carry different activities. Qualification of all instruments/equipment shall be ensured prior to routine usage. Allot identification number to all analytical instruments/ equipment. Calibrations and preventive maintenance shall be done as per schedule. Relevant SOP shall be displayed preferably near the instrument/equipment or arranged in such a way that it shall be readily available to user for reference.

The activities performed on instrument/equipment shall be updated and displayed as status board near instrument/equipment as per annexure T03. Usage Log shall be maintained for instruments/equipment as per procedure mentioned in applicable SOP. At the end of day's operation or after use, instrument/equipment shall be switched off and maintained properly.

List of 'Authorized Users' for instruments/equipment/laboratory areas shall be maintained or displayed based on requirement as per respective SOP. Temperature & humidity of stability walk in chambers or incubators shall be maintained and data shall be recorded as per requirement. Desiccant replacement from desiccator/instruments shall be done as per requirement and status shall be displayed as per annexure T04. Water from instruments/equipment shall be changed as per requirement and status shall be displayed as per annexure T05.

1.6   Laboratory Glassware

Class `A' Glassware shall be used for preparation and its certificate of compliance shall be maintained.

Clean all the laboratory glassware following applicable procedure of glassware cleaning. Cleaned and dried glassware shall be stored in dust free storage area.

Examine glassware prior to use for damage i.e. cracked, chipped or any other defective glassware.

Do not use such defective glassware and dispose it in glass bin. Do not dispose defective glassware in general waste bin. Always use clean and dry glassware for analysis. Use guard for volumetric flasks and measuring cylinder to avoid breakages.

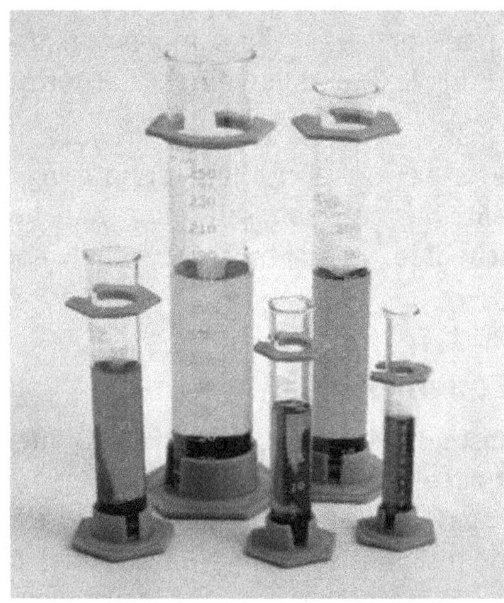

Figure 1 Illustrative example of measuring cylinders with guard

1.7   Chemicals, Reagents and Analytical Standards

Use appropriate grade and highly purified chemicals. After receipt, label all chemicals and readymade reagents as per applicable SOP. Reagents/solutions should be prepared, standardized and stored according to respective reagents/solution specification and label appropriately.

Storage and handling of chemicals and reagent should be done in a manner considering the physicochemical properties of these substances and hazards involved in their use. In-compatibility chart/MSDS of chemicals and reagents, should be available and should be displayed preferably near the chemical storage area.

All the solutions, dispensed solvent and waste solvents collected in beaker/vessel should be covered entirely with appropriate cover by using aluminum foil or glass lid to avoid probable contamination.

Ensure the validity of the chemicals before usage and if any chemical is beyond validity period, discard the same from laboratory.

All solid and liquid chemical bottles shall be kept in the places intended for them, immediately after use. They must not be left on the work benches. Chemical stock should be checked periodically and based on the requirement if required purchase request for new chemicals should be raised. Examine chemicals for any change in appearance or nature (Colour, clarity etc.) and shall discard if any abnormalities are observed Safe disposal of chemicals shall be always ensured. Follow the appropriate chemical disposal procedure strictly.

1.8   Laboratory Reference Standards and Culture Media in Good Laboratory Practice

Qualified Reference standards must be suitable for their intended use. Qualification and certification should be clearly stated and documented. Whenever compendial reference standards from an officially recognized source exist, these should preferably be used as primary reference standards unless fully justified (the use of secondary standards is permitted once their traceability to primary standards has been demonstrated and is documented). These compendial materials should be used for the purpose described in the appropriate monograph unless otherwise authorized by the National Competent Authority.

Culture media should be prepared in accordance with the media manufacturer's requirements unless scientifically justified. The performance of all culture media should be verified prior to use.

Used microbiological media and strains should be decontaminated according to a standard procedure and disposed of in a manner to prevent the cross-contamination and retention of residues. The in-use shelf life of microbiological media should be established, documented and scientifically justified.

Animals used for testing components, materials or products, should, where appropriate, be quarantined before use. They should be maintained and controlled in a manner that assures their suitability for the intended use. They should be identified, and adequate records should be maintained, showing the history of their use.

1.9   Good Housekeeping and Safety

The laboratory shall have adequate first aid kit and fire extinguishers located at right places and staff must be familiar and trained.

Safety data sheets must be made available to staff before testing is carried out for respective material.

All staff must wear safety apparels or other protective clothing including PPE wherever required.

While handling any chemicals, cytotoxic materials, steroids etc.; use personnel protective equipment and follow safety instructions and precaution.

Never handle any chemicals, raw materials, intermediate or bulk finished with bare hands.

Do not taste any chemical or solvents and do not touch face, mouth or eyes during analysis.

Operators carrying out sterility tests shall wear sterilized garments including headgear, face masks and shoes.

The staff must be educated in the first aid techniques and other emergency care. Water showers shall be installed at appropriate place in the laboratory. Appropriate facilities for the collection, storage and disposal of waters shall be made available. Avoid spillage of sample, chemicals and reagents, in case any spillage occurs then clean the area as quickly as possible by following appropriate safety precautions.

Staff must be aware of methods for safe disposal of corrosive/hazardous materials by using neutralization or deactivation method. Always do a visual check on electrical equipment before use looking for obvious wear or defects.

1.10 Sampling

Sampling shall be done by trained personnel in accordance with approved written procedure.

Samples shall be stored in tightly closed bags/containers at appropriate storage conditions, until their destruction. Reseal/close the sample bags/bottles immediately after use. Sampling personnel shall have knowledge of the nature of the samples to be handled and should refer respective specification for the same.

1.11 Testing and Review

Daily work plan as per analysis requirements shall be designed by supervisor in consultation with analyst and status shall be updated as per annexure T06.

The analysis shall be done as per approved version of standard testing procedure or method of analysis.

Label all test and standard preparations for all tests with at least details such as Analytical Reference No/Batch No. Solution name and appropriate replicate preparation number wherever applicable (in a suitable short form in case of space constraint but in an identifiable manner), sign and date with legible marker pen.

```
A. R. No./Batch No. :  ____________________
Solution Name       :  ____________________
Sign & Date         :  ____________________
```

Figure 2 Illustrative example of glassware label (standard and test solutions)

Use clean spatula/clean pipette/glass weighing boats/butter papers or other suitable receivers for weighing and transferring the samples.

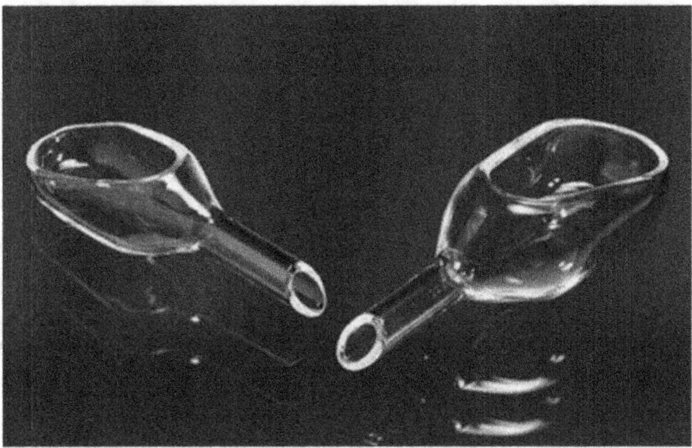

Figure 3 Illustrative example of weighing receivers (glass weighing boats)

Balance print shall be taken for all the weighments. This print shall be stamped by using stamp for weight print as per annexure T07 and details shall be mentioned. If columns are provided to mention details on weight print itself then stamp is not required.

Wipe the tip of the burette and pipette before dispensing the withdrawn liquid. Use approved Record of analysis to enter raw data. The status of analysis preparation shall be indicated as per annexure T08.

The results shall be checked for compliance against specification. The completed ROA and analytical data shall be submitted to reviewer for review as per applicable SOP.

1.12 Documentation

Always follow good documentation practices. Clearly written documentation prevents errors from spoken communication. Pharmacopoeia & its supplement or addendum, technical books etc. as required, should be available in Quality Control Laboratory.

Ensure the availability of current versions of controlled documents in the laboratory.

SOPs, Specifications, Testing procedures and logbooks must be maintained securely, at the workplace where it can be easily accessed. All analytical work shall be carried out by referring to approved current version of documents such as Standard Operating Procedure, Specifications, Standard Testing Procedure, General Test Procedure etc., Analytical data shall be recorded contemporaneously. Update all relevant logbooks concurrently.

The certificates received from the vendor/manufacturer shall be reviewed with respect to its specification/standards and completeness upon receipt. As a proof of verification of tests and results (wherever applicable), the reviewer shall put a stamp as per annexure T07.

Reviewer shall sign with date as a part of review on first page and remaining pages shall be stamped as reviewed. Attach all relevant analytical raw data obtained from instrument such as analytical balance weight prints, chromatograms, spectrums, Polari meter, refractometer, particle size analyzer, potentiometer, tap density apparatus, dissolution tester etc., with sign and date to the record of analysis/after labeling the same.

The analytical raw data where it is obvious to invalid the data/results, in the event of incident (e.g. system suitability failures, instrument malfunctions/errors), OOS and OOT or data returned by the reviewer during the empower sign-off 2 process; Such data shall be stamped as invalid data.

The reason for invalidation along with sign and date shall be mentioned on first page and remaining pages shall be stamped as invalid data as per annexure T07.

The invalid data shall be attached along with the batch testing record. Protect all documents from spillage of chemicals/solutions during performing analysis.

1.13 Incident, OOS and OOT

For Quality control operation related to receipt, handling, storage, analysis, calibration, all non-conformance classified and addressed as OOS/OOT/Incidents shall be investigated, evaluated and documented according to defined procedure to identify errors like analyst error, instrument malfunctioning, method error, improper peak shape, system suitability failure etc. according to procedure.

1.14 Self-Inspection (Internal Audits)

In order to verify compliance with the Quality control system, regular internal audits/self-inspection should be conducted in accordance with approved procedure.

1.15 Validation/Verification

Validation/Verification of analytical methods shall be followed as per applicable SOP.

1.16 Measures in Good Laboratory Practice

Establish and Follow Procedures: Develop approved procedures and inventory.

Maintain Your Proficiency: Analysts must have the education, training and experience, acquired through formal education or on-the-job training, sufficient to perform assigned analytic duties.

Validate Methods: Method should be validated before usages.

Use Traceable Standard Reference Materials (SRM): Reference material uses include validating methods that help ensure accurate data from individual test runs, calibrating instruments and assessing analyst proficiency. In the United States, a NIST standard reference material is considered the "gold standard" for that material. NIST has more than a thousand different SRMs covering diverse technologies. The results of analyses backed by NIST-traceable SRMs are widely accepted as valid.

Run in Duplicate: The purpose of duplicate (sometimes triplicate) testing is to add to the confidence that the test run has produced good data for the test object. Replicate data that is in agreement is a good measure of method reproducibility but does not prove data accuracy (validity).

Keep Original Data: Document everything and Maintain Good Records. Whether data is first recorded in electronic/digital form, in a notebook or on the closest piece of scrap paper, keep it.

Assign Instruments and Equipment to Analysts: A good practice is to formally assign that analyst the responsibility for keeping the instrument operational and for alerting management to malfunctions. When an instrument is used by multiple staff members, assign these responsibilities to a primary user, who should schedule usage time for other staff members, provide training and mentoring to new users, ensure that any instrument control charts are current and ensure that calibration and maintenance occur on schedule.

Calibrate Instruments and Equipment: Instrument should be calibrated to its working range.

Use Control Charts: Control charts are excellent tools for several uses, including those already noted. A control chart enables a laboratory to track the results of a reference material and/or control sample at the end of each test run. It gives the laboratory a snapshot of test run quality and a picture of the quality of the laboratory's results for that particular test over time.

1.17 Technical Transfer of Testing Methods in Good Laboratory Practice

Prior to transferring a test method, the transferring site should verify that the test method(s) comply with those as described in the Marketing Authorization or the relevant technical dossier. The transfer of testing methods from one laboratory (transferring laboratory) to another laboratory (receiving laboratory) should be described in a detailed protocol. The transfer protocol should include, but not be limited to the following parameters:

- Identification of the testing to be performed and the relevant test method(s) undergoing transfer;
- Identification of the additional training requirements;
- Identification of standards and samples to be tested;
- Identification of any special transport and storage conditions of test items;
- The acceptance criteria which should be based upon the current validation study.
- Deviations from the protocol should be investigated prior to closure of the technical transfer process;
- The technical transfer report should document the comparative outcome of the process and should identify areas requiring further test method revalidation, if applicable

1.18 On-going Stability Programme in Good Laboratory Practice

The on-going stability programme is to monitor the product over its shelf life and to determine that the product remains, and can be expected to remain, within specifications under the labelled storage conditions. The on-going stability programme should be described in a written protocol and should include, but not be limited to the following parameters:

- Number of batch(es) per strength and different batch sizes, if applicable;
- Relevant physical, chemical, microbiological and biological test methods;
- Acceptance criteria;
- Reference to test methods;

- Description of the container closure system(s);
- Testing intervals (time points);
- Description of the conditions of storage (standardised ICH/VICH conditions for long term testing, consistent with the product labelling, should be used);
- Other applicable parameters specific to the medicinal product.

The number of batches and frequency of testing should provide a sufficient amount of data to allow for trend analysis. Unless otherwise justified, at least one batch per year of product manufactured in every strength and every primary packaging type, if relevant, should be included in the stability programme (unless none are produced during that year). For products where on-going stability monitoring would normally require testing using animals and no appropriate alternative, validated techniques are available, the frequency of testing may take account of a risk-benefit approach. The principle of bracketing and matrixing designs may be applied if scientifically justified in the protocol. In certain situations, additional batches should be included in the on-going stability programme. For example, an on-going stability study should be conducted after any significant change or significant deviation to the process or package. Any reworking, reprocessing or recovery operation should also be considered for inclusion.

1.19 Documentation in Good Laboratory Practice

- Specifications;
- Procedures describing sampling, testing, records (including test worksheets and/or laboratory notebooks), recording and verifying;
- Procedures for and records of the calibration/qualification of instruments and maintenance of equipment;
- A procedure for the investigation of Out of Specification and Out Of Trend results;
- Testing reports and/or certificates of analysis;
- Data from environmental (air, water and other utilities) monitoring, where required;
- Validation records of test methods, where applicable;
- Trend analysis for test results and environment controls;
- All raw data such as laboratory notebooks and/or records should be retained and readily available.

1.20 Sampling in Good Laboratory Practice

Sampling may be required for different purposes, such as prequalification; acceptance of consignments; batch release testing in-process control; special controls; inspection for customs clearance, deterioration or adulteration; or for obtaining a retention sample. Sampling shall consist following contents:

- Approved procedure sampling plans and methods must be written and defined;
- Samples must be representative of the population;
- Samples or sampling plans must be based on appropriate statistical criteria, and;
- Representative of batch;
- Sample should be Samples must be properly identified and handled;
- Equipment to be used the amount of the sample to be taken;
- Instructions for any required sub-division of the sample;
- The type and condition of the sample container to be used;
- Identification of containers sampled;
- Any special precautions to be observed, especially with regard to the sampling of sterile or noxious materials; storage conditions;
- Instructions for the cleaning and storage of sampling equipment.

A. Approaches for sampling plans will be discussed for:

- Incoming Packaging Components
- Incoming Raw Materials
- Labeling Materials
- Non-sterile Liquid Products
- Sterile Products
- Creams, Suspensions, and Emulsions
- Powder Blends
- Tablets, Capsules, and Other Solid Dosage Forms

1.21 Testing in Good Laboratory Practice

Testing methods should be validated. A laboratory that is using a testing method and which are defined in the marketing authorization or technical dossier. The results obtained should be recorded. The tests performed should be recorded and the records should include at least the following data:

- Name of the material or product and, where applicable, dosage form;
- Batch number and, where appropriate, the manufacturer and/or supplier;
- References to the relevant specifications and testing procedures;
- Test results, including observations and calculations, and reference to any certificates of analysis;
- Dates of testing;

- Initials of the persons who performed the testing;
- Initials of the persons who verified the testing and the calculations, where appropriate;
- A clear statement of approval or rejection (or other status decision) and the dated signature of the designated responsible person;
- Reference to the equipment used.

1.22 Annexure

Table 1 Format for Quality Control Laboratory closedown checklist

Quality Control Laboratory Closedown Checklist			
Date :			Page No.
S. No.	Check Points	√/ X / NA	Remarks
	All water taps are closed.		
	All the glassware kept in designated place.		
	Exhaust fan of Titration room is switched OFF.		
	All acids and reagents kept in designated place.		
	Exhaust fan of Fume Hood is switched OFF & panel door is closed.		
	All the chemicals /Reagents/ Standards kept in designated place.		
	Sink is clean and clear.		
	Glassware drying oven switched off.		
	Area is tidy and samples kept on designated place.		
	All SOPs/Working documents are kept on designated place.		
	All instruments/equipment's/computers are in standby mode/switched OFF.		
	All columns and other accessories are kept in designated place.		
	GC gas lines are turned OFF (If GC is not running)		
	All lights are switched OFF.		
	GC gas line connection from cylinders located at Gas Bank shall be off or dismantle. (If GC is not running).		
Checked by:		Verified By:	
Date:		Date	

Table 2 Format for Microbiology Laboratory closedown checklist

Microbiology Laboratory Closedown Checklist			
Date :			**Page No.**
S. No.	**Check Points**	**√ /X/NA**	**Remarks**
	Media placed at designated place.		
	Instruments/Equipment are switched OFF.		
	Computers are shut down.		
	SOP and all other documents placed at designated place.		
	Unloaded articles placed at designated place.		
	LAF working bench cleaned.		
	LAF motor, UV light, Visible light switched OFF.		
	Cultures placed at designated place.		
	Observed plates removed.		
	BOD incubators working properly.		
	Pass Boxes are empty, cleaned and their door closed from both side.		
	Clean and dried Glassware kept in their respected place.		
	Doors of Autoclave are closed.		
	Area clean and tidy.		
	Lights switched OFF.		
Checked by: Date:		Verified By: Date	

Table 3 Format for Instrument/Equipment status board

Logo
INSTRUMENT /EQUIPMENT STATUS BOARD
Instrument/Equipment Id. No. : _______________________
Product Name   : _______________________
Batch No. /A.R. No.   : _______________________
Test   : _______________________
Under Analysis/Awaited for Results
Analyst Name/Sign   : _______________________
Date   : _______________________
Date   : _______________________
XXXX-00

Table 4 Format for desiccant replacement status label

Logo
DESICCANT REPLACEMENT STATUS
Item Name   : ___________________________
Replaced on   : ___________________________
Next Due Date   : ___________________________
Sign / Date   : ___________________________
XXXX-00

Table 5 Format for water replacement status label

Logo
WATER REPLACEMENT STATUS
Instrument/Equipment ID : ___________________________
Replaced on   : ___________________________
Next Due Date   : ___________________________
Sign/Date   : ___________________________
XXXX-00

Table 6 Format for daily work allotment and status report

Daily Work allotment and status report					
					Page No.
Analyst Name/Employee Code:					
Group/Section:					
Work Allotted by :					Date:
Shift	Work Allotted for the day/Previous day Carry over (AR No. / Batch No/Other details)	Test	Activity started (Hr.)	Analyst Sign/Date	Status (Remarks, if any)
Details of Incident, OOT or OOS, if any:					
Reviewed By (Sign & Date):					

Table 7 Format for stamps

1. Stamp for weight print

A.R. No.	: ____________
Test.	: ____________
Signature	: ____________
Date	: ____________

2. Stamp for Invalid Data

INVALID DATA
Reason   : ______________________
Sign/Date : ______________________

3. Stamp for Reviewed

REVIEWED
Sign/Date : ______________________

Table 8 Format for analysis status board

Logo
ANALYSIS STATUS BOARD
Product Name   : ______________________
Batch No./A.R. No.   : ______________________
Test   : ______________________
Under Analysis/Awaited for Results
Analyst Name/Sign   : ______________________
Date   : ______________________
XXXX-00

Table 9 Change History format

Change History	
Department :	Page No.:
Document Type	SOP
Document No	TAP029-00
Title	Good Laboratory Practices

Table 9 *Contd...*

Document No. & Version No.	Effective Date	Change Control No. & brief details of changes
TAP029-00	12-12-2017	Change Control No. 20172915 New SOP prepared.

Document No. Version No.	Effective Date	Change Control No. & brief details of changes
TAP029-00		Change Control No. 20172915 New SOP prepared.

Overview of Good Microbiology Practices in Pharmaceutical Industry

Introduction

Good laboratory practices in a microbiology laboratory consist of activities that depend on several principles: aseptic technique, control of media, control of test strains, operation and control of equipment, diligent recording and evaluation of data, and training of the laboratory staff. Because of the inherent risk of variability in microbiology data, reliability and reproducibility are dependent on the use of accepted methods and adherence to good laboratory practices.

2.1   Premises, Layout and Zone

Microbiology laboratory generally divided into clean or aseptic areas and live culture areas. If complete separation of live and clean culture zones cannot be accomplished, then other barriers (such as protective clothing, sanitization and disinfection procedures, and biological safety cabinets designated for clean or aseptic operations only) and aseptic practices should be employed to reduce the likelihood of accidental contamination. Areas in which environmental or sterile product samples are handled and incubated should be maintained completely free of live cultures, if possible. Separate air supply, air handling unit should be available for laboratories and production areas. Sterility testing should always be performed in a dedicated area. Access to the microbiological laboratory should be restricted to authorized personnel.

Laboratory activities, such as sample preparation, media and equipment preparation and enumeration of microorganisms, should be segregated by space or at least time, so as to minimize risks of cross-contamination and false positives. Where non-dedicated arrays are used, risk management principles should be applied.

Operations should be carried out preferably in the following zones:

Zone	Installation Grade	Proposed
Sample receipt	Unclassified	Unclassified
Media preparation	Unclassified	Unclassified
Autoclave loading inside the sterility testing area	Grade B	ISO 5(turbulent) and <10 cfu/m^3
Sterility testing – UDAF	Grade A	ISO 5 (UDAF) &<1 cfu/m^3
Sterility testing –back d round to UDAF	Grade B	ISO 5 (UDAF) &<10 cfu/m^3
Sterility testing-Isolator	Grade A (NVP and microbiology only)	ISO 5(UDAF) and < 1cfu/m^3
Sterility Testing-Background to isolator	Unclassified	Unclassified
Incubator	Unclassified	Unclassified
Enumeration	*Unclassified	*Unclassified
Decontamination	Unclassified	Unclassified
cfu, colony-forming unit.		
*A Critical steps should be done under laminar flow.		

Laboratory Environmental Monitoring

Appropriate an environmental monitoring programme should be in place i.e use of air settlement plates and surface swabbing, temperature and pressure differentials. Alert and action limits should be defined. Trending of environmental monitoring results shall be carried out.

2.2   Laboratory Equipment

Each item of equipment, instrument or other device used for testing, verification and calibration should be uniquely identified and should have a documented programme for the maintenance, calibration and monitoring of its equipment.

Laboratory equipment used in the microbiology laboratory should not be used outside the microbiology area, unless there are specific precautions in place to prevent cross-contamination.

2.3   Personnel

Each person engaged in laboratory should have the education, training, and experience to do his or her job. They should have basic training in microbiology and relevant practical experience before being allowed to perform work covered by the scope of testing. Personnel should be made aware of:

(a) The appropriate entry and exit procedures including gowning;

(b) The intended use of a particular area;

(c) The restrictions imposed on working within such areas;

(d) The reasons for imposing such restrictions; and

(e) The appropriate containment levels.

2.4   Media and Preparation

A. Media Procurement and Storage

Media should be purchase from be approved and qualified vendor. Media should be clearly labeled with batch or lot numbers, preparation and expiration dates, and media identification. Each media should consist of a certificate of analysis describing expiration dating and recommended storage conditions, as well as the quality control organisms used in growth-promotion and selectivity testing of that media. Growth promotion should be done on all media on every batch by the user. Media should always be storage under validated controlled condition to ensure its quality through to the expiry date and also to minimize the loss of moisture, control the temperature, prevent microbial contamination, and provide mechanical protection to the prepared media.

B. Media Preparation

Media should be prepared as per manufacture formula, instruction for routine dehydrated media and ready-made media preparation. Cleaned container and tools should be used for preparation of the media to prevent foreign substance entering in the preparation. Equipment used in the preparation of media should be appropriate to allow for controlled heating, constant agitation, and mixing of the media. Darkening of media is generally indication of overheating of media. Sterilization of media should be performed within the parameters provided by the manufacturer or validated by the user. Remelting of an original container of solid media should be performed only once to avoid media whose quality is compromised by overheating or potential contamination. It is recommended that remelting be performed in a heated water bath or by using free-flowing steam. The pH of each batch of medium should be confirmed by a flat pH probe at room temperature (20°–25°) by aseptically withdrawing a sample for testing. The pH of media should be in a range of ±0.2 of the value indicated by the manufacturer, unless a wider range is acceptable by the validated method.

Prepared media should be checked by appropriate inspection of plates and tubes for the following:

(a) Cracked containers or lids

(b) Unequal filling of containers

(c) Dehydration resulting in cracks or dimpled surfaces on solid medium

(d) Hemolysis

(e) Excessive darkening or color change

(f) Crystal formation from possible freezing

(g) Excessive number of bubbles

(h) Microbial contamination

(i) Status of redox indicators (if appropriate)

(j) Lot number and expiration date checked and recorded

(k) Sterility of the media

(l) Cleanliness of plates (lid should not stick to dish)

Growth promotion test should be performed on each lot prepared media based on manufacturer recommendation micro-organism or may include representative environmental isolates (but these latter are not to be construed as compendial requirements). Expiration dates on media should have supporting growth-promotion testing to indicate that the performance of the media still meets acceptance criteria up to and including the expiration date. The length of shelf life of a batch of media will depend on the stability of the ingredients and formulation under specified conditions, as well as the type of container and closure. Disposal of used cultured media (as well as expired media) should follow local biological hazard safety procedures.

C. Media Incubation Time

Incubation times for microbiological tests of less than 3 days' duration should be expressed in hours: e.g., "Incubate at 30° to 35° for 18 to 72 hours". Tests longer than 72 hours' duration should be expressed in days: e.g., "Incubate at 30° to 35° for 3 to 5 days". For incubation times expressed in hours, incubate for the minimum specified time, and exercise good microbiological judgment when exceeding the incubation time.

2.5 Reference Culture

Cultures for use in compendial tests should be acquired from a national culture collection or a qualified secondary supplier. They can be acquired frozen, freeze-dried, on slants, or in ready-to-use forms. Confirmation of the purity of the culture and the identity of the culture should be performed before its use in quality control testing. Ready-to-use cultures should be subjected to incoming testing for purity and identity before use. The confirmation of identity for commonly used laboratory strains should ideally be done at the level of genus and species. All cultures must be no more than 5 passages removed from the original stock culture. The number of transfers of working control cultures should be tracked to prevent excessive sub culturing that increases the risk of phenotypic alteration or mutation. Working stocks shall not be subculture to replace reference stocks. Reference strains may only be used as working cultures. Test culture are:

   (a) Candida albicans (ATCC No. 10231)
   (b) Aspergillus brasiliensis (ATCC No. 16404) (formerly Aspergillus niger)
   (c) Escherichia coli (ATCC No. 8739)
   (d) Pseudomonas aeruginosa (ATCC No. 9027)
   (e) Staphylococcus aureus (ATCC No. 6538)

2.6 Sampling

Sampling should only be performed by trained personnel. It should be carried out aseptically using sterile equipment. Appropriate precautions should be taken to ensure that sample integrity is maintained through the use of sterile sealed containers for the collection of samples where appropriate. It may be necessary to monitor environmental conditions for instance air contamination and temperature at the sampling site. Time of sampling should be recorded. The laboratory should record all relevant information, e.g.

   (a) date and, where relevant, the time of receipt;
   (b) condition of the sample on receipt and, when necessary, temperature; and
   (c) characteristics of the sampling operation (sampling date, sampling conditions, etc.).

2.7 Laboratory Testing

All testing in laboratories used for critical testing procedures, such as sterility testing of final dosage forms, bulk product, seed cultures for biological production, or cell cultures used in biological production, should be performed under controlled conditions. Isolator technology is also appropriate for critical, sterile microbiological testing. Isolators have been shown to have lower levels of environmental contamination than manned clean rooms, and therefore, are generally less likely to produce false-positive results. Proper validation of isolators is critical both to ensure environmental integrity and to prevent the possibility of false-negative results as a result of chemical disinfection of materials brought into or used within isolators Subculturing, staining, microbial identification, or other investigational operations should be undertaken in the live culture section of the laboratory. If possible, any sample found to contain growing colonies should not be opened in the clean zone of the laboratory. Careful segregation of contaminated samples and materials will reduce false-positive results.

Staff engaged in sampling activities should not enter or work in the live culture handling section of a laboratory unless special precautions are taken, including wearing protective clothing and gloves and careful sanitizing of hands upon exiting.

All test performed should be document as and when performed.

Discrepancy and failure investigations related to manufacturing and testing: documented; evaluated; investigated in a timely manner; includes corrective action where appropriate.

2.8   Disposal of Contaminated Waste

The procedures for the disposal of contaminated materials should be designed to minimize the possibility of contaminating the test environment or materials. It is a matter of good laboratory management and should conform to National/International environmental or health and safety regulations.

2.9   Laboratory Data Management

All laboratory records should be archived and protected against catastrophic loss. A formal record retention and retrieval program should be in place.

- Microbiologist training and verification of proficiency
- Equipment validation, calibration, and maintenance
- Equipment performance during test (e.g., 24-hour/7-day chart recorders)
- Media preparation, sterility checks, and growth-promotion and selectivity capabilities
- Media inventory and control testing
- Critical aspects of test conducted as specified by a procedure
- Data and calculations verification
- Reports reviewed by qualified responsible manager
- Investigation of data deviations (when required)
- Test results should include the original plate counts, allowing a reviewer to recreate the calculations used to derive the final test results
- If charts or graphs are incorporated into laboratory notebooks, they should be secured with clear tape and should not be obstructing any data on the page. The chart or graph should be signed by the person adding the document, with the signature overlapping the chart and the notebook page.

- Lab notebooks should include page numbers, a table of contents for reference, and an intact timeline of use.
- the laboratory write-up should include Date, Material tested, Microbiologist's name, Procedure number, Document test results, Deviations (if any), Documented parameters (equipment used, microbial stock cultures used, media lots used) and Management/Second review signature.

Chapter - 3

Quality Assurance Management in GLP Environment

Introduction

Quality Assurance Programme means a defined system, including personnel, which is independent of study conduct and is designed to assure test facility management of compliance with these Principles of Good Laboratory Practice. Principles of Good Laboratory Practice as "a defined system, including personnel, which is independent of test conduct and is designed to assure test facility management of compliance with these Principles of Good Laboratory Practice. The responsibilities of the Quality Assurance management of a test facility include ensuring "that there is a Quality Assurance Programme with designated personnel and assure that the quality assurance responsibility is being performed in compliance with these Principles of Good Laboratory Practice" QA's mandated role is that of an independent witness to the whole process and its organisation. To be effective QA must have access to staff, documents and procedures at all levels of the organisation, and be supported by a committed top management.

3.1 Requirement of Quality Assurance Management in GLP

- The test facility should have a documented Quality Assurance Programme to assure that studies performed are in compliance with these Principles of Good Laboratory Practice.

- The Quality Assurance Programme should be carried out by an individual or by individuals designated by and directly responsible to management and who are familiar with the test procedures.

- This individual(s) should not be involved in the conduct of the test being assured.

3.2   Responsibilities of the Quality Assurance Personnel

The responsibilities of the Quality Assurance personnel include, but are not limited to the following functions. They should:

- Maintain copies of all approved test plans and Standard Operating Procedures in use in the test facility and have access to an up-to-date copy of the master schedule;
- Verify that the test plan contains the information required for compliance with these Principles of Good Laboratory Practice. This verification should be documented;
- Conduct inspections to determine if all studies are conducted in compliance with these Principles of Good Laboratory Practice. Inspections should also determine that test plans and Standard Operating Procedures have been made available to test personnel and are being followed.

3.3   Qualifications of QA Personnel

- QA personnel should have the training, expertise and experience necessary to fulfil their responsibilities. They must be familiar with the test procedures, standards and systems operated at or on behalf of the test facility.
- Individuals appointed to QA functions should have the ability to understand the basic concepts underlying the activities being monitored. They should also have a thorough understanding of the Principles of GLP.
- In case of lack of specialized knowledge, or the need for a second opinion, it is recommended that the QA operation ask for specialist support. Management should also ensure that there is a documented training programme encompassing all aspects of QA work. The training programme should, where possible, include on-the-job experience under the supervision of competent and trained staff. Attendance at in-house and external seminars and courses may also be relevant. For example, training in communication techniques and conflict handling is advisable. Training should be continuous and subject to periodic review.
- The training of QA personnel must be documented and their competence evaluated. These records should be kept up-to-date and be retained.

3.4   Quality Assurance Activity in GLP

- Maintain copy of master schedule sheet of all studies conducted. These are to be indexed by test article and must contain the test system, nature of study, date the study was initiated, current status of each study, identity of the sponsor, and name of the study director.
- Maintain copies of all protocols pertaining to the studies for which QAU is responsible.

- Inspect studies at adequate intervals to assure the integrity of the study and maintain written and correctly signed records of each periodic inspection. These records must show the date of the inspection, the study inspected, the phase or segment of the study inspected, the person performing the inspection, findings and problems, action recommended and taken to resolve existing problems, and any scheduled date for re inspection. Any problems discovered which are likely to affect study integrity are to be brought to the attention of the study director and management immediately.

- Periodically submit to management and the study director written status reports on each study, noting problems and corrective actions taken.

- Determine whether deviations from protocols and SOPs were made with proper authorization and documentation.

- Review the final study report to ensure that it accurately describes the methods and SOPs and that the reported results accurately reflect the raw data of the study.

- Prepare and sign a statement to be included with the final study report that specifies the dates of audits and dates of reports to management and to the study director.

- Audit the correctness of the statement, made by the study director, on the GLP compliance of the study.

- Audit laboratory equipment.

3.5 QA and Non-Regulatory Test

- Compliance with GLP is a regulatory requirement for the acceptance of certain studies. However, some test facilities conduct in the same area studies which are and which are not intended for submission to regulatory authorities. If the non-regulatory studies are not conducted in accordance with standards comparable to GLP, this will usually have a negative impact on the GLP compliance of regulatory studies. Lists of studies kept by QA should identify both regulatory and non-regulatory studies to allow proper assessment of work load, availability of facilities and possible interferences.

- QA should have access to an up-to-date copy of the master schedule to assist them in this task. It is not acceptable to claim GLP compliance for a non-GLP test after it has started. If a GLP-designated test is continued as a non-GLP test, this must be clearly documented.

3.6   The QA Statement

- The Principles of GLP require that a signed quality assurance on any documents, protocol, report, specification SOP to approve and state it meets the regulatory expectation.

3.7   Regulatory Expectations in GLP

- A robust system for the continuous improvements necessary to assure the safety, identity, strength, quality, and purity of drug products.
- CGMP in practices for 24hrs.
- Linking Product, Process and Patient
- Adopting QbD design
- Quality Culture
- Quality System Elements
- Statistical Process Controls
- Continuous improvements

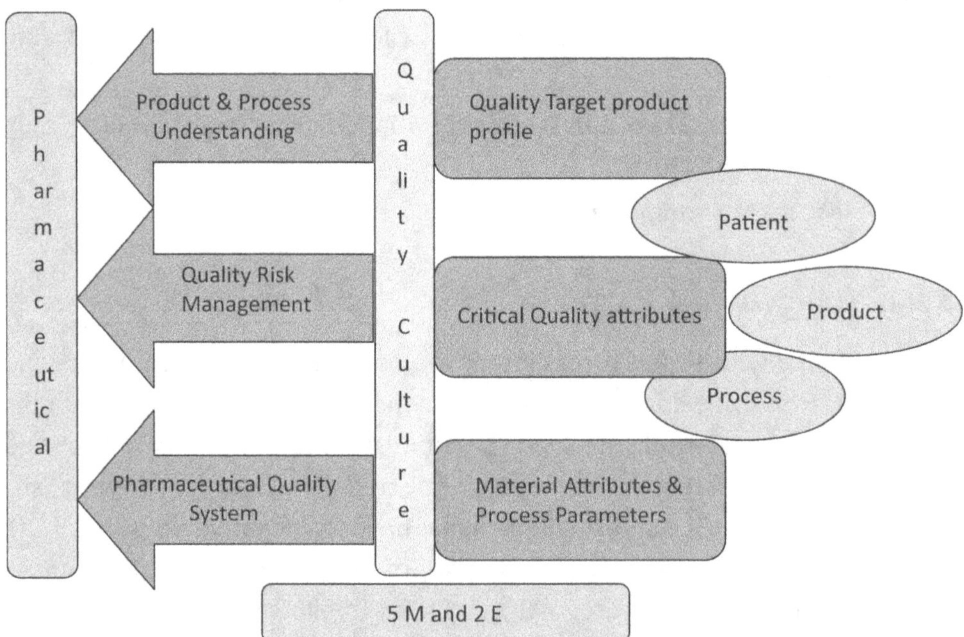

Chapter - 4

Laboratory Quality Manual

Introduction

The laboratory quality manual is a document that a facility writes to explain which portions of which regulations are applicable to the facility and which documents the policies, procedures, responsibilities, and documentation, that must be in place for the facility to comply with these regulatory responsibilities

A laboratory quality manual is required for implementing a quality management system. Such a system aims primarily at achieving customer satisfaction by meeting customer requirements through application of the system, continuous improvement of the system and prevention of the occurrence of nonconformities.

The organization should establish and maintain a laboratory quality manual that includes

- The quality policy
- The scope of the quality management system, including details of and justification for any exclusions
- The documented procedures established for the quality management system, or reference to them, and
- A description of the interaction between the processes of the quality management system.

4.1   Contents of the Laboratory Quality Manual

- A laboratory quality manual is a document that a facility writes to explain which regulations are applicable to the facility (i.e. which regulations will be followed by the facility.)

- The regulations that are to be followed are based on the process(es) that are performed at the facility.

- The facility's laboratory quality manual   should outline which regulations are going to be followed, how they are going to be followed, who is responsible for ensuring that the regulations are followed, and which of the companies approved procedures address the regulations to be followed. If all parts of a regulation are not going to be followed, a facility may want to include a brief explanation as to why a part of a regulation is not applicable to the facility and therefore will not be outlined in the laboratory quality manual.

- The laboratory quality manual   should be written in general terms with minimal specifics. The format of a laboratory quality manual   is usually different than the format used for the facility's other approved documents. The format should still include such things as a facility's name, version control, and approval signatures.

  (a) *Table of Contents:*   A list of the sections contained within the Quality Manuel and the page each section begins on.

  (b) *Introduction:* A brief description of the facility, the facility's purpose for writing a laboratory quality manual , and a brief description of the scope of the laboratory quality manual .

  (c) *Facility Background:* List the name of the facility, where the facility is located, what radiopharmaceutical products are produced at the facility, and how these will be distributed.

  (d) *Purpose:* Provide general statements explaining why and how the laboratory quality manual   will be used.

  (e) *Scope:* List the regulations that will be followed by the facility as well as any portions of the regulations that will not be followed. Provide a justification for why those portions of the regulations are not going to be followed.

  (f) *Quality Policies and Objectives:* Include a brief statement about the approach that the facility is taking in regards to quality, (i.e. a Quality Mission Statement) and list the quality objectives of the facility. The quality objectives should not be numerous (five to seven is a typical number) and should be briefly stated.

  (g) *Organization and Structure of Documentation:* Provide an explanation as to how the documentation structure at your facility is organized and managed in relation to the applicable regulations.

  (h) *Products:* This section should include a brief description of the products that are made and distributed by the facility including what the intended use of the products is to be. This description should be similar to a marketing type summary in that it does not list proprietary or explicit product information that would be detrimental to the facility if persons outside the facility read the description. This section should provide enough detail to justify the sections of the regulations that are and are not going to be followed.

(i) *References:* A list of all the different regulations that were sited within the laboratory quality manual.

(j) *Quality Policies for Specific Regulation Elements:* This section will encompass the majority of the manual. It will include each regulation that will be followed by the facility, a brief description of how the facility intends to follow the regulation, and a list of the approved documents (by document name and number) that the facility has in place which specifically address/discuss the facility's policies and objectives as stated in this section of the laboratory quality manual.

4.2 Topics that should be Covered in the Final Section on Quality Policies for Specific Regulation Elements

A. Quality Systems

The Quality System ensures that the product complies with its specified requirements and is manufactured in accordance with appropriate standards.

The Quality System is documented on four levels:

(a) Level 1 Quality Policy

(b) Level 2 laboratory quality manual

(c) Level 3 Standard Operating Procedures

(d) Level 4 Quality Records and Logbooks

B. Deviations and Events

- Event Recording
- Reviews and Investigations
- Remediation

C. Change Control

- Change Initiation and Approval
- Change Execution and Tracking
- Change Verification and Closure

D. Audit Management
- Planning
- Execution
- Review and Remediation

E. Document Management
- Storing and Classification
- Access and Viewing
- Creating New Documents
- Lifecycle Managements
- Printing

F. Training Management
- Employee Course Management
- Training Request and Approval
- Gap Management and Records

G. Equipment Management
- Inventory Management
- Preventative Maintenance
- Remedial Maintenance

H. General Requirements
- Security and Access Control
- Facility Management
- Corrective Action Preventive Action (CAPA)
- CAPA Initiation
- Investigations and Action Plan
- CAPA Closure

I. Validation
- Documentation Structure and Formats
- Change Control
- Planning and Scheduling
- Roles and Responsibilities
- Regulatory Requirements for Testing

J. A list of SOPs, Forms, Validation Documentation and other appendices

4.3   Format for Laboratory Quality Manual

<table>
<tr><td colspan="3" align="center">LABORATORY QUALITY MANUAL</td></tr>
<tr><td colspan="2">Function: Quality Laboratory Department</td><td>Effective Date:</td></tr>
<tr><td colspan="2">Document: Quality Manual</td><td>Version: 01</td></tr>
<tr><td colspan="2">Next Review Date:</td><td></td></tr>
<tr><td>Prepared by:
Sign and Date</td><td>Reviewed by:
Sign and Date</td><td>Approved by:
Sign and Date</td></tr>
<tr><td colspan="3" align="center">Contents</td></tr>
<tr><td colspan="2" align="center">SECTION AND TITLE</td><td align="center">Page No</td></tr>
<tr><td colspan="2">

1. Introduction
2. Organization
3. Manufacturing Facility
4. Quality System
5. Management Responsibility
6. Design and Change Control
7. Document Control
8. Purchasing and Vendor Assurance
9. Material and Product Identification and Traceability
10. Process Control Programmes
11. Inspection and Testing Programmes
12. Calibration Programmes
13. Inspection and Test Status
14. Control of Non-Conforming Product
15. Corrective and Preventive Action Systems
16. Handling and Storage of Materials and Products
17. Control of Quality Records
18. Internal Quality Audits
19. Training Programmes
20. Statistical Techniques
21. Computer Systems Management
22. Facility Control
23. Management of Equipment and Critical Process Utilities
24. Validation Programmes
25. Risk Assessment
26. Occupational Safety and Health Programmes

</td><td></td></tr>
</table>

Chapter - 5

Laboratory Quality Policy

Introduction

This Quality Policy has been prepared to be consistent with the guiding principles set forth in the pharmaceutical laboratory and Compliance Core Objective.

The quality requirements with which the pharmaceutical laboratory must comply are set forth in the following documents, separated into two levels based upon level of specificity:

- The Quality Policy Manual, which includes this Policy, the Quality Management Framework, and the individual Quality Policy Standards that state the required outcomes for each of the identified elements of a Quality System.
- The laboratory Quality Technical Standards that further define, in greater detail, the specific requirements for quality processes that are required to be uniform throughout GLP management.

These requirements documents:

(a) Provide a common foundation for quality systems across the laboratory.

(b) Form the requirements for quality systems that meet the expectations of the laboratory, with primary responsibility to assure safe, effective and quality products for our customers, and the operation of our businesses in compliance with regulations.

(c) Support leverage of knowledge and experience between laboratory area.

(d) This Policy shall be approved by the authorised personnel of the organization management.

5.1 Scope of GLP Quality Policy

Every GLP business unit, as defined below, must have procedures that comply with, and thereby implement, the requirements documents outlined in the above introduction.

The procedures shall contain instructions to perform actions or make decisions and shall be prescriptive and mandatory, and applicable to those individuals defined in the scope and executing the task. Such business units must also maintain records, as required, providing objective evidence that procedures have been executed.

This Policy is applicable to the research, design and development1, sourcing, production, storage, shipping, distribution, installation, service, marketing and post-market surveillance of any products marketed by, or bearing a company name, trade name, trademark belonging to any of GLP business units, whether or not the output is produced solely under its control or is partially or completely produced by a Supplier.

Throughout this Policy:

(a) Where the term 'product' is used, it also applies to labeling, raw materials, work in process, and outputs such as services, software, information and data, including that generated during research and development. Additionally, for products subject to installation and/or servicing, these processes are considered to be "products" and within the scope of this Policy.

(b) Where the term 'business unit' is used, it includes any organizational unit operating within the GLP business unit that is led by a senior executive, supported by a Management Board or Leadership Team, and responsible for quality related activities that trigger quality-related responsibilities.

(c) Where the term 'Quality' is used, it encompasses all aspects of Quality and Regulatory Compliance.

Each laboratory business unit is responsible for:

- The quality of, and meeting the applicable requirements for, the products that it develops, manufactures, distributes and or markets.
- Compliance with the elements of this Policy that are applicable to the activities that it undertakes.

5.2 GLP General Requirements

Each GLP business unit shall establish and maintain or operate under processes to perform applicable sections of the following:

- Establish and maintain or operate under a quality system that is applicable to the activities that it undertakes and meets the requirements of this Quality Policy, applicable regulatory requirements, and the requirements of any voluntary standards with which compliance is claimed or certification held.

- Provide sufficient visibility and triggers, within the quality system, to assure safety, efficacy, and quality of products throughout their full lifecycle.
- Provide the necessary resources for the implementation, maintenance, and improvement of the system. Define the roles and responsibilities necessary to ensure compliance with this Quality Policy in accordance with the GLP Quality Management Framework.
- Identify the processes needed for the quality system together with the criteria and methods for the operation and control of these processes.
- Maintain written agreements defining responsibilities.

5.3   Risk Management

Each GLP business unit shall establish and maintain or operate under processes to perform applicable sections of the following:

- Ensure that risks associated with products are identified throughout the product lifecycle and assessed based on scientific evidence.
- Ensure that measures are taken to minimize and control the risk through the quality system, and that remaining risk is understood by management and communicated effectively to customers, healthcare professionals and regulatory agencies wherever appropriate.
- Use established risk management principles that meet applicable regulatory expectations.

5.4   Documentation, Records and Data Controls

Each laboratory business unit shall establish and maintain or operate under processes to perform applicable sections of the following:

- A documentation system to assure the planning, operation and control of the quality system.
- Processes for the identification, completion, checking, storage, protection, retrieval, retention and destruction of records and data.
- Established standards for information asset management.

5.5   Management Responsibility

Management with Executive Responsibility, who direct a GLP business unit that has a quality function with regulatory responsibilities and who control the company's resources, shall, in accordance with regulatory requirements and GLP Quality Standards:

- Ensure that the elements of the quality system required to be performed at the company are established and maintained, that applicable quality requirements are met and that trends and changes in regulations or their interpretation are identified and addressed proactively.
- Establish the quality function as an independent function with the authority and expertise to fulfill applicable quality systems responsibilities, as required by regulation.

Specifically:

(a) The Quality Function shall be represented on the Company Management Board or Senior Leadership Team, and report directly to the next highest level of Quality leadership or the Chair of that Board.

(b) All Quality Functions that report into a GLP business unit shall also be accountable to, and have a reporting obligation to the GLP Chief Quality Officer.

(c) In all matters of quality and regulatory compliance escalation, the decision-making line is to the Quality & Compliance organization.

  - Empower the applicable Quality Functions to establish and administer quality-based decision-making processes for all matters of quality including, but not limited to, release of product and initiating field action.
  - Review the company's quality system at planned intervals to ensure its continuing suitability, adequacy and effectiveness.
  - Establish procedures to assure Significant Quality Issues can be reported by an employee to the Quality and Compliance organization for evaluation and escalation.
  - Set the expectations, tone and visibility for quality and compliance under their responsibility including, but not limited to, the effective implementation of this Quality Policy.
  - Perform Quality & Compliance (Q&C) Strategic/Business Planning, establish Q&C Goals & Objectives and conduct Quality Planning on an annual basis. These strategies, plans, goals and objectives will be communicated to the entire business unit in a manner such that employees are aware of the impact on their job.

5.6   Resource Management

Each business unit shall establish and maintain or operate under processes to perform applicable sections of the following:

Management Responsibilities	Management Resposibilities				
Resource Management	Organization personnel & Training	Facilities, Utilities and Equipment	Software and Computer Systems	Documentation and Data Management	Suppliers, Outsourced Operations
Product LifeCycle Management	Product design, Development, Transfer and Improvement · Process and method validation	Material Management Systems · Production & Process Controls	Laboratory Testing and Controls · Labelling and Packaging Systems	Product Disposition and Release · Storage Distributions Systems	Change Control · Incidents, Investigations and CAPA management · Process Performance and Product Quality Monitoring
Quality Systems Analysis and Improvement	Quality Risk Management		Audits and Regulatory Compliance		

- Identify resource requirements and provide resources, infrastructure and qualified personnel to establish and maintain the elements of the quality system within its responsibility. This includes management resources and those for the performance of work and verification activities.

- Establish and maintain or operate under processes to ensure that personnel have the competence (education and experience) and training to perform their responsibilities.

- Implement a due diligence process for any intended acquisition of a company or product into the GLP business unit that assesses the quality and compliance status of the potential acquisition, including any research, design and development, production, storage, shipping, distribution, marketing and post-market surveillance. The GLP business unit shall identify and make provisions for any additional resources necessary to mitigate any identified gaps.

- Within 60 days following acquisition, a post-acquisition assessment shall be carried out by the business unit with participation of GLP Compliance in order to complete a gap assessment to applicable regulatory requirements and GLP Policies and Standards (Enterprise, Sector, and Operating Company), and prepare a documented plan that includes provision of resources and timelines for identified actions.

- Obtain the agreement of the GLP Chief Quality Officer and applicable Sector Chief Quality Officer prior to the appointment of Quality & Regulatory Compliance Vice Presidents.

5.7   Product Realization

Each business unit shall establish and maintain or operate under processes to perform applicable sections of the following:

- Establish processes for research, design and development, appropriate for the stage of development and meeting customer and regulatory requirements.
- Ensure product and processes for clinical evaluation conforms to requirements.
- Ensure Controlled Conditions using appropriately designed and qualified facilities, utilities and equipment (computerized systems, calibration and maintenance), consideration of necessary microbiological and environmental controls, water quality and, where necessary, preservation of the product.
- Ensure that purchased materials conform to GLP Standards.
- Control label contents and labeling processes to ensure compliance with customer and regulatory requirements.
- Specific requirements and demonstrated evidence to assure the sterility of its sterile products in accordance with applicable GLP Standards
- Demonstrate traceability of products throughout the full product lifecycle.
- A system to control changes such as changes to materials, product, labeling, packaging, processes, systems, facilities, methods and equipment, including computerized systems, to ensure changes are assessed, recorded, reviewed, verified and or validated, and approved prior to implementation.
- Systems to ensure that transportation, storage, raw materials, packaging, production, laboratory and applicable support areas are clean, orderly, maintained and that contamination is controlled, to avoid mix-ups, damage, deterioration, contamination or other adverse effects to products.
- Systems to ensure that products are maintained in appropriate conditions during manufacture, storage and distribution.
- Systems to ensure that products meet safety, quality and efficacy or performance requirements throughout the stated shelf-life. Evidence to support the shelf-life shall be reviewed and approved prior to initial commercial release of product.
- Prevent diversion of product from its intended distribution channel and protect the distribution channel from the entry of counterfeited product.
- Systems to ensure that where product is designated for destruction, destruction is undertaken in a timely manner and records are maintained to confirm that destruction was effected.
- Systems to ensure that Product Returns or reverse logistics streams are handled appropriately.

- Systems to ensure that products selected for donation meet specification and requirements of the country where the donation is made. Products for donation have sufficient shelf life remaining at the time of donation, generally at least six months.
- Validation of products, processes, computerized systems and equipment prior to initial commercial release of the product.
- Where applicable, appropriate installation, servicing and repairs are handled in a controlled manner via an approved process or procedure.

5.8 Product Sourcing

Each business unit shall establish and maintain or operate under processes to perform applicable sections of the following:

- For products sourced outside business unit:
    (a) Be responsible for product quality and for meeting applicable regulatory requirements where operations include partially or completely outsourcing any of the following: research (including clinical research); design & development; production; processing; assembling; testing; packaging; labeling; sterilization; repackaging; or, distribution. This responsibility includes initial evaluation, authorization to proceed, and monitoring for on-going compliance. The extent of control shall be established based on the criticality of the outsourced activity and detailed in a written agreement.
    (b) Ensure that control of active pharmaceutical ingredients and/or finished goods GLP

- For products that GLP analysis for which an external company holds the marketing rights and/or regulatory approval to market:

    (a) Ensure that there is a quality unit with oversight responsibilities to see that all regulatory requirements for the market to be supplied are met.
    (b) Demonstrate and document compatibility with any associated GLP products.
    (c) Prepare a written agreement between all parties describing their response-bilities, including those for post-market surveillance, field action, and reporting, and ensure such agreements are appropriately incorporated into decision-making (e.g., authority for field action execution) for product.
    (d) Require approval from a GLP Quality function before distribution of any product on behalf of an external company.

5.9 Measurement, Analysis and Improvement

- Each Chief Quality Officer (Enterprise and Sector) shall ensure the establishment of:

    (a) An independent audit program.

(b) A process to ensure that significant corrective actions and preventive actions identified in one business unit are considered for implementation across the Sector or Enterprise, where applicable.

(c) Processes to provide oversight of field action activities.

(d) Quality Review processes that provide for consolidated review, at the sector and enterprise levels, of significant information from individual business unit management reviews.

- Each business unit shall establish and maintain or operate under processes to perform applicable sections of the following:

(a) An Internal Audit (self assessment/self inspection) program to monitor the effectiveness of the quality system.

(b) A comprehensive system for handling product complaints and reporting adverse events, as required by applicable regulation (e.g., Adverse Drug Experiences, Adverse Events, Medical Device Reportable Events or Medical Device Vigilance, medical occurrences). The system shall include the monitoring of trends and the establishment of triggers to allow escalation of Significant Quality Issues.

(c) Process for investigation and controlling nonconformities.

(d) Products failing to meet established specifications or made outside the established process are identified and segregated such that their use is prevented until a formal investigation is conducted to determine final disposition of the product and impact on marketed product. The final disposition is approved by the Quality Function and documented.

(e) Documented procedures for the conduct of field action activities.

(f) Processes for correction of non-conformities, correcting the cause of non-conformity (corrective action) and preventing occurrence of non-conformity (preventive action).

(g) Corrective actions and preventive actions will be appropriate to the potential impact of the non-conformity or potential non-conformity. Records of corrective actions and preventive action are maintained. Progress towards implementation, completion, and effectiveness of such actions shall be regularly reviewed and reported to management, to assure appropriate actions have been implemented in a timely manner.

(h) A periodic product review for pharmaceutical, medicinal, and applicable combination products. Records of such reviews are retained and results communicated to applicable marketing authorization holders, when required by applicable regulations.

- Program to proactively monitor regulatory agency requirements, shape external environment and monitor industry standards.

Chapter - 6

Laboratory Site Master Files

Introduction

Site Master file is self-explanatory notes to guide to the regulatory authority in planning and conducting GMP inspections for manufacturing site where all kind of manufacturing operations such as production, packaging and labelling, testing, relabeling and repackaging of all types of medicinal products are explained. A Site Master File should contain adequate information but, as far as possible not exceed 25-30 pages plus appendices. Simple plans outline drawings or schematic layouts are preferred instead of narratives. The Site Master File, including appendices, should be readable when printed on A4 paper sheets. The Site Master File should be a part of documentation belonging to the quality management system of the manufacturer and kept updated accordingly. The Site Master File should have an edition number, the date it becomes effective and the date by which it has to be reviewed. It should be subject to regular review to ensure that it is up to date and representative of current activities. Each Appendix can have an individual effective date, allowing for independent updating.

6.1 Content of Site Master File

A. General Information on the Manufacturer

I. Contact information on the manufacturer

- Name and official address of the manufacturer;
- Names and street addresses of the site, buildings and production units located on the site;
- Contact information of the manufacturer including 24 hrs telephone number of the contact personnel in the case of product defects or recalls.

- Identification number of the site as e.g. GPS details, or any other geographic location system, D-U-N-S (Data Universal Numbering System) Number (a unique identification number provided by Dun & Bradstreet) of the site1

II. Authorised pharmaceutical manufacturing activities of the site

- Copy of the valid manufacturing authorisation issued by the relevant Competent Authority in Appendix 1; or when applicable. If the Competent Authority does not issue manufacturing authorizations, this should be stated.
- Brief description of manufacture, import, export, distribution and other activities as authorized by the relevant Competent Authorities including foreign authorities with authorized dosage forms/activities, respectively; where not covered by the manufacturing authorization;
- Type of products currently manufactured on-site (list in Appendix 2) where not covered by Appendix 1 ;
- List of GMP inspections of the site within the last 5 years; including dates and name/country of the Competent Authority having performed the inspection. A copy of current GMP certificate (Appendix 3) should be included, if available.

III. Any other manufacturing activities carried out on the site

- Description of non-pharmaceutical activities on-site, if any.

B. Quality Management System of the Manufacturer

I. The quality management system of the manufacturer

- Brief description of the quality management systems run by the company and reference to the standards used;
- Responsibilities related to the maintaining of quality system including senior management;
- Information of activities for which the site is accredited and certified, including dates and contents of accreditations, names of accrediting bodies.

II. Release procedure of finished products

- Detailed description of qualification requirements (education and work experience) of the Authorised Person(s)/Qualified Person(s) responsible for batch certification and releasing procedures;
- General description of batch certification and releasing procedure;

- Role of Authorised Person/Qualified Person in quarantine and release of finished products and in assessment of compliance with the Marketing Authorisation;

- The arrangements between Authorised Persons/Qualified Persons when several Authorised Persons/Qualified Persons are involved;

- Statement on whether the control strategy employs Process Analytical Technology (PAT) and/or Real Time Release or Parametric Release;

III. Management of suppliers and contractors

- A brief summary of the establishment/knowledge of supply chain and the external audit program;

- Brief description of the qualification system of contractors, manufacturers of active pharmaceutical ingredients (API) and other critical materials suppliers;

- Measures taken to ensure that products manufactured are compliant with TSE (Transmitting animal spongiform encephalopathy) guidelines.

- Measures adopted where counterfeit/falsified products, bulk products (i.e. unpacked tablets), active pharmaceutical ingredients or excipients are suspected or identified.

- Use of outside scientific, analytical or other technical assistance in relation to manufacture and analysis;

- List of contract manufacturers and laboratories including the addresses and contact information and flow charts of supply-chains for outsourced manufacturing and Quality Control activities; e.g. sterilization of primary packaging material for aseptic processes, testing of starting raw-materials etc, should be presented in Appendix 4;

- Brief overview of the responsibility sharing between the contract giver and acceptor with respect to compliance with the Marketing Authorization.

IV. Quality Risk Management (QRM)

- Brief description of QRM methodologies used by the manufacturer;

- Scope and focus of QRM including brief description of any activities which are performed at corporate level, and those which are performed locally. Any application of the QRM system to assess continuity of supply should be mentioned;

V. Product Quality Reviews

- Brief description of methodologies used

C. Personnel

- Organisation chart showing the arrangements for quality management, production and quality control positions/titles in Appendix 5, including senior management and Qualified Person(s);
- Number of employees engaged in the quality management, production, quality control, storage and distribution respectively;

D. Premises and Equipment

(a) Premises

- Short description of plant; size of the site and list of buildings. If the production for different markets, i.e. for local, EU, USA etc takes place in different buildings on the site, the buildings should be listed with destined markets identified ;
- Simple plan or description of manufacturing areas with indication of scale (architectural or engineering drawings are not required);
- Lay outs and flow charts of the production areas (in Appendix 6) showing the room classification and pressure differentials between adjoining areas and indicating the production activities (i.e. compounding, filling, storage, packaging, etc.) in the rooms.;
- Lay-outs of warehouses and storage areas, with special areas for the storage and handling of highly toxic, hazardous and sensitising materials indicated, if applicable;
- Brief description of specific storage conditions if applicable, but not indicated on the lay-outs;

I. Brief description of heating, ventilation and air conditioning (HVAC) systems

- Principles for defining the air supply, temperature, humidity, pressure differentials and air change rates, policy of air recirculation (%);

II. Brief description of water systems

- Quality references of water produced
- Schematic drawings of the systems in Appendix 7

III. Brief description of other relevant utilities, such as steam, compressed air, nitrogen, etc.:

- Principles for defining the air supply, temperature, humidity, pressure differentials and air change rates, policy of air recirculation (%);

IV. **Brief description of water systems**

- Quality references of water produced
- Schematic drawings of the systems in Appendix 7

V. **Brief description of other relevant utilities, such as steam, compressed air, nitrogen, etc.**

b. **Equipment**

- Listing of major production and control laboratory equipment with critical pieces of equipment identified should be provided in Appendix 8.

I. **Cleaning and sanitation**

- Brief description of cleaning and sanitation methods of product contact surfaces (i.e. manual cleaning, automatic Clean-in-Place, etc).

II. **GMP critical computerised systems**

- Description of GMP critical computerised systems (excluding equipment specific Programmable Logic Controllers (PLCs)

E. **Documentation**

- Description of documentation system (i.e. electronic, manual);
- When documents and records are stored or archived off-site (including pharmacovigilance data, when applicable): List of types of documents/records; Name and address of storage site and an estimate of time required retrieving documents from the off-site archive.

F. **Production**

I. **Type of products**

(References to Appendix 1 or 2 can be made):

- Type of products manufactured including
  - (a) list of dosage forms of both human and veterinary products which are manufactured on the site
  - (b) list of dosage forms of investigational medicinal products (IMP), manufactured for any clinical trials on the site, and when different from the commercial manufacturing, information of production areas and personnel
- Toxic or hazardous substances handled (e.g. with high pharmacological activity and/or with sensitising properties);
- Product types manufactured in a dedicated facility or on a campaign basis, if applicable;

- Process Analytical Technology (PAT) applications, if applicable: general statement of the relevant technology, and associated computerized systems;

II. Process validation

- Brief description of general policy for process validation;
- Policy for reprocessing or reworking;

III. Material management and warehousing

- Arrangements for the handling of starting materials, packaging materials, bulk and finished products including sampling, quarantine, release and storage
- Arrangements for the handling of rejected materials and products.

G. Quality Control (QC)

- Description of the Quality Control activities carried out on the site in terms of physical, chemical, and microbiological and biological testing.

H. Distribution, Complaints, Product Defects and Recalls

I. Distribution (to the part under the responsibility of the manufacturer)

- Types (wholesale licence holders, manufacturing licence holders, etc) and locations (EU/EEA, USA, etc) of the companies to which the products are shipped from the site;
- Description of the system used to verify that each customer/recipient is legally entitled to receive medicinal products from the manufacturer
- Brief description of the system to ensure appropriate environmental conditions during transit, e.g. temperature monitoring/control;
- Arrangements for product distribution and methods by which product traceability is maintained;
- Measures taken to prevent manufacturers' products to fall in the illegal supply chain.

II. Complaints, product defects and recalls

- Brief description of the system for handling complains, product defects and recalls.

I. Self Inspections

- Short description of the self inspection system with focus on criteria used for selection of the areas to be covered during planned inspections, practical arrangements and follow-up activities.

6.2 Appendix

- Appendix 1 : Copy of valid manufacturing authorisation
- Appendix 2 : List of dosage forms manufactured including the INN-names or common name (as available) of active pharmaceutical ingredients (API) used
- Appendix 3 : Copy of valid GMP Certificate
- Appendix 4 : List of contract manufacturers and laboratories including the addresses and contact information, and flow-charts of the supply chains for these outsourced activities
- Appendix 5 : Organisational charts
- Appendix 6 : Layouts of production areas including material and personnel flows, general flow charts of manufacturing processes of each product type (dosage form)
- Appendix 7 : Schematic drawings of water systems
- Appendix 8 : List of major production and laboratory equipment.

6.3 Specimen Copy of Laboratory Site Master File

Site Name : XXXX			
Site master number and its Version number	**XXXX/01**		**Effective date:** **Page No.: XX of XX**
SITE MASTER FILE **Address :** **XXXX**			

	Section	**Content**	**Page No.**
	GENERAL INFORMATION		
	C.1.1	Brief Information of the site	XX
	C.1.2	Other manufacturing activities	XX
	C.1.3	Name and Address of the Site	XX
C.1	C.1.4	Short description of the Site	XX
	C.1.5	Total Employee strength	XX
	C.1.6	Description of the Quality Management system	XX
	C.1.7	Global Quality Policy	XX
	C.1.8	Responsibility of Quality Assurance function	XX
	PERSONNEL		
C.2	C.2.1	Organization Chart	XX
	C.2.2	Qualifications and Functional Responsibilities of Key Personnel	XX

Table *Contd...*

Site Name : XXXX			
Site master number and its Version number		**XXXX/01**	**Effective date:** **Page No.: XX of XX**
	C.2.3	Basic Training	XX
	C.2.4	Health requirements for personnel	XX
	C.2.5	Personnel hygiene Requirements including Clothing	XX
C.3	**PREMISES**		
	C.3.1	Premises	XX
	C.3.2	Nature of Construction and Finishes	XX
	C.3.3	Description of Ventilation Systems	XX
	C.3.4	Description of Water Systems	XX
	C.3.5	Maintenance and Servicing of Instruments & Equipment's	XX
	C.3.6	Major Process, Formulation and Analytical R&D Laboratory Equipment's	XX
	C.3.7	Qualification, Validation and Calibration	XX
	C.3.8	SAP for inventory control and Quality Notifications	XX
	C.3.9	Sanitation and Hygiene	XX
C.4	**DOCUMENTATION**		
	C.4.1	Arrangements for Preparation, Revision and Distribution of Documents	XX
	C.4.2	Documentation related to Quality system	XX
C.5	**DEVELOPMENT PROCESS**		
	C.5.1	Brief Description of development procedure	XX
C.6	**ANALYTICAL RESEARCH & DEVELOPMENT**		
	C.6.1	Description of Analytical Research & Development	XX
C.7	**CONTRACT MANUFACTURING**		XX
C.8	**COMPLAINT(S) HANDLING**		XX
ABBREVIATIONS			XX
APPROVALS			XX
ANNEXURE			
Annexure – I	Organization Chart		
Annexure – II	Key Personnel Details		
Annexure – III	Site Layout		
Annexure – IV	List of Equipment's & Instruments		
Annexure – V	SOP Index		
Annexure – VI	List of Approved laboratories for external testing		

Table *Contd…*

Site Name : XXXX		
Site master number and its Version number	**XXXX/01**	**Effective date:** **Page No.: XX of XX**

Annexure – VII	List of Approved SBPs for contract manufacturing
Whenever there is change in above listed annexures, same shall be updated every three months once. If there is no change same shall be continued.	

C.1   GENERAL INFORMATION:

C.1.1   Brief Information of the site:

XXXX is a global pharmaceutical company established in XXXX year. Since its inception the company has chosen the path of innovation and discovery. Active Pharmaceutical Ingredients (APIs), branded and generic formulations are exported and marketed in over 100 countries including US, Europe, Russia, Canada, Korea and China. XXXX follows the current regulatory requirements and cGMP procedures in its facilities so that the products manufactured are safe and efficacious.

This site is involved in the Process Development (APIs, Intermediates & NCEs), Formulation Development (NCEs & Generic drug products), Analytical Research & Development and Analytical Method Validations, Stability Management of pre-clinical & clinical molecules, Crystallization and Risk Assessment Lab is performing the investigation studies and developmental activities.

Developed methods/Process are transferred to manufacturing site for manu-facturing at Manufacturing units

This site is located at about XXXX km from XXXX Airport, Area Name XXXX

C.1.2   Other manufacturing activities:

No manufacturing activities other than those specified in C.1.1 are carried out at this site

C.1.3   Name and Address of the Site:

C.1.3.1 Name of Company with Postal Address:

XXXX

C.1.3.2 Contact Person, Telephone No. and e-mail address:

XXXX

C.1.3.3 Fax No. of Contact Person:

XXXX

C.1.3.4 24 hr contact Telephone No.:

XXXX

C.1.4   Short description of the Site:

This Unit was established in the year XXX. The site is located XXXX   meters off the XXXX

The site is well maintained with greenery. The surrounding area is dust free and Eco friendly. Inter block commuting roads are concrete ones and are well maintained.

North   :   Open Land

Table Contd...

Site Name : XXXX		
Site master number and its Version number	**XXXX/01**	**Effective date:** **Page No.: XX of XX**

South	:	NH4 National Highway
East	:	Open Land
West	:	Open Land

(Refer Site layout: Annexure – III)

C.1.5   **Total Employee strength:**

	Department	No. of Employees
C.1.5.1	Analytical Research & Development (Formulation)	XX
C.1.5.2	Analytical Research & Development (API)	XX
C.1.5.3	Finance	XX
C.1.5.4	Administration	XX
C.1.5.7	Engineering Services Department	XX
C.1.5.8	Formulation Research & Development	XX
C.1.5.9	Global Business Support & Analytics	XX
C.1.5.10	Knowledge Management	XX
C.1.5.11	Logistics	XX
C.1.5.12	Process Engineering	XX
C.1.5.13	Process Research & Development	XX
C.1.5.14	Program Management - API (India)	XX
C.1.5.15	Program Management - Formulations	XX
C.1.5.16	Purchase	XX
C.1.5.17	Quality Assurance	XX
C.1.5.18	Regulatory Affairs	XX
C.1.5.19	Safety Health & Environment	XX
C.1.5.20	Supply Chain Council	XX
C.1.5.21	Third Party Sourcing	XX
C.1.5.22	Warehouse	XX
	Total Employees	XX

C.1.6   **Description of the Quality Management system:**

Quality Management system is established, documented and maintained in accordance with the requirements of cGMP guideline ICH Q7. A quality system that

Table *Contd...*

Site Name : XXXX		
Site master number and its Version number	**XXXX/01**	**Effective date:** **Page No.: XX of XX**

includes Change control system, Incidents, Out of specifications (OOS), Document control, CAPA, Quality audits and continuous training at all levels are in place.

C.1.7   Quality Policy:

We are committed to establishing and maintaining quality standards that assure the safety and efficacy of all medicines marketed by us or by any of our partners. We are also committed to complying with all current national and international regulations, codes and standards applicable to our business.

We aim to achieve this through a well-designed quality management system that places balanced emphasis on:

1. Overarching responsibility of management
2. Appropriate management of all resources that we rely on in the manufacturing of our products
3. Management of quality through the entire product lifecycle Periodic reviews and continuous improvement

Information Security Management System (ISMS) Policy:

XXXX is committed to ensure Integrity, Confidentiality, Availability and Security of its Physical and Information Assets at all times for serving the needs of the customers and organization while meeting all legal, statutory and regulatory requirements.

C.1.8   Responsibility of Quality Assurance function:

- Documentation Control including archival & retrieval
- Review and approval of SOPs, Specification &MOA,of SBP , Development Report, Validation Protocol & Report, Qualification Protocol & Report, Optimization reports, Feasibility reports, PE packages, SHE packages, Campaign report etc.
- Compile and Maintaining Site Master File
- Imparting cGMP Training, scheduling and implementation.
- Review, evaluation and coordination with RA
- Change control Management
- Ensure OOS & Incidents are investigated and effective corrective and preventive actions are implemented.
- Internal audits/ self-inspections
- Co ordination with plant for complaints investigations.
- Ensuring stability data to support retest or expiry periods and storage condition of clinical drug product, APIs and / or intermediates, where appropriate.
- Providing timely response for DMF / Regulatory / Customer Deficiencies.
- Supporting for Vendor Qualification of clinical molecule.
- QA release of development API/samples to customer after reviewing relevant documents.

Table *Contd...*

Site Name : XXXX		
Site master number and its Version number	**XXXX/01**	**Effective date:** **Page No.: XX of XX**

- Quality overview of SBP for clinical product manufacturing.
- Technology Transfer

C.1.8.1 Elements of Quality Assurance System:

Quality Assurance Department operates independently and reports directly to Management of the company. All the activities are performed in accordance to written standard operating procedure and manuals. Development of all batches is performed as per the written instructions and documented in the Electronic Laboratory notebooks and LRBs (which ever applicable).

C.1.8.2 Audit Program/Self Inspection:

The Quality Assurance department along with cross functional team conducts self-inspections (internal audit) at scheduled intervals and covers all identified key departments of the unit. Self-inspections are carried out as per defined procedure and frequency. The purpose of carrying out audits is to have a check as to whether all departments are functioning in adherence and compliance to the established procedures. All aspects that are related to quality are audited department wise. An audit report is generated at the end of the audit and is distributed to the auditeedepartments. All actions proposed by Auditors to ensure for closure. The details pertaining to self-inspection are documented.

C.1.8.3 Other Standards:

ICH Q7 standard is adopted by the company to assess and audit the quality system within the company and similarly same standards are being used by the company to assess its suppliers/vendors.

C.1.8.4 Vendor Approval System:

All raw materials and packaging materials are procured from approved vendors. Vendors supplying the material are evaluated by a defined vendor approval system. The vendors are approved or rejected by Quality Assurance department based on Vendor Qualification SOP.

C.1.8.5 Product Stability Program:

Stability studies are designed for products developed at this facility. The stability studies are performed as per the approved protocol and analyzed as per the testing procedures at scheduled intervals. Stability studies of pre-clinical & clinical supplies are also carried out as per approved protocol.

C.1.8.6 Quality Review by Management:

Quality Review is done XXXX

C.1.8.7 Standard Operating Procedures (SOPs):

The SOP Index is attached as Annexure-V.

Table *Contd...*

Site Name : XXXX		
Site master number and its Version number	**XXXX/01**	**Effective date:** **Page No.: XX of XX**

C.2   **PERSONNEL:**

C.2.1   **Organization Chart:**

Organogram of functional departments such as Quality Assurance (QA), Analytical Research & Development, Process Research & Development, Engineering services department, Safety Health & Environment department Formulation Research & Developments are enclosed as Annexure – I.

C.2.2   **Qualifications and Functional Responsibilities of Key Personnel:**

Qualifications and functional responsibilities of key personnel attached in Annexure-II.

C.2.3   **Basic Training:**

Training is an integral part of the organizational building and development. Each employee undergoes training periodically depending upon the organizational need. The company has structured training program, which covers all employees. Training programs, Induction training and records are documented. Functional the training program will be coordinated by the department head or designee in assistance with the QA department. The training procedure is detailed in the SOP XXXX

C.2.4   **Health requirements for personnel:**

All the employees undergo medical examinations at the time of joining the company and also periodically i.e. once in a year. Health checkup of all employees is done by authorized medical practitioner. Emergency vehicle with a driver is always available in the facility to shift the casualty to nearby hospital immediately in case of accident/emergency. The company has provided Family Health Plan for all the employees.

C.2.5   **Personnel Hygiene Requirements including Clothing:**

Personnel hygiene is being followed as per procedures. The Company follows a code for the clothing of the various personnel of the company. Irrespective of the departments all the personnel shall wear safety equipment like safety goggles wherever necessary etc., when they enter into the process and/or analytical development laboratories. All visitors shall be provided with the necessary clothing and goggles when they visit various areas of process development. According to norms fixed for personnel hygiene, washroom facilities have been provided and are being maintained in good state of sanitation. General wash facilities are provided at work places and also in canteen.

- Aprons, safety goggles, helmets are provided, if needed safety-shoes are available. In addition, various kinds of respirators are provided as applicable.

- Pest control programs are in operation to control insects. Rodent control operations are also initiated as per the need in specific areas.

- First aid boxes with medicines are provided as applicable.

- Eyewash and body wash stations have been installed at various points.

<table>
<tr><td colspan="3">Site Name : XXXX</td></tr>
<tr><td>Site master number and its Version number</td><td>XXXX/01</td><td>Effective date:
Page No.: XX of XX</td></tr>
</table>

- Facilities are provided for employees to wash their hands or eyes or body whenever required with soaps/detergents.
- Emergency control room is located near security office to control / minimize the emergency.

C.3   PREMISES:

C.3.1   Premises:

Site is equipped with all utilities like Ventilation System, Nitrogen, Compressed Air, Backup Power etc.

C.3.2   Nature of Construction and Finishes:

The buildings are constructed with Steel columns. Development labs are covered with RCC structure. The roof of the buildings is made up of RCC slab and cement sheet on a steel structure having slope. The walls are of bricks with cement concrete plastered and painted with anti-peel-off paint. The flooring is made up of reinforced cement concrete. All surfaces are smooth, flushed to the walls and cleanable.

The buildings in the facility are of suitable size and facilitate easy cleaning. Buildings are constructed by providing free ventilation with reinforced concrete columns, mild steel columns, brick walls and corrugated cement sheet roofing. Emergency staircase has been constructed for departments wherever required.

C.3.3   Description of Ventilation Systems:

Fume hoods are constructed to allow free ventilation. Air supplied in the Fume hoods are filtered through appropriate filters. The volume of air delivered to the rooms is sufficient to maintain the general environmental conditions required to protect both the product and the personnel working in the area. Sufficient ventilation system is provided to achieve clean up, when rooms or areas during conditions such as spillage of powder, solvents etc.

C.3.4   Description of Water Systems:

Water is distributed from the sump through pumps is input to the water systems (Soft water and Treated water), Out let of treated water plant is fed to Milli Q system.

C.3.5   Maintenance and Servicing of Instruments & Equipment's:

Well-planned preventive maintenance procedure is maintained to avoid any incidental breakdown of the equipment's or instruments. Preventive maintenance procedure is designed in view with the manufacturer's manual and by considering the criticality of the machine parts.

Individual equipment's have a unique preventive maintenance procedure and scheduled frequency.

Preventive maintenance program is governed through written procedure. Preventive maintenance is carried out for critical equipment in process R&D, formulation R&D

Table Contd...

Site Name : XXXX		
Site master number and its Version number	**XXXX/01**	**Effective date:** **Page No.: XX of XX**

and Analytical R&D lab as per individual procedure. Preventive maintenance schedule is prepared at the start of the year by engineering department for equipment and for laboratory instrument by Analytical Research & Development. Preventive maintenance is carried out in-house as well as by contract agency. Contract agencies are qualified and carry out work in accordance to contract.

Checklist is used for carrying out preventive maintenance as per schedule. For each type of equipment separate check list is being prepared in accordance to manufacture's recommendation and/or based on the history of the equipment. Record for the preventive maintenance activity is maintained by engineering department. For laboratory instruments service/preventive maintenance record is kept by Analytical Research & Development.

C.3.6 Major Process, Formulation and Analytical R&D Laboratory Equipment's:

The process R&D equipment's include filters, nutsche filter, Vacuum tray dryers (VTD), Agitated nutsche Filter dryer (ANFD) Centrifuge, Round bottom flasks, Multimill etc. are used to carry out the various type's reactions, filtrations and drying operations.

The formulations equipment's include Rapid mixer granulator (RMG), Fluid bed Equipment (FBE), blender, Compression machine, capsule filling machine and in-process instruments etc.

The AR&D laboratory has adequate chemicals and glassware for routine analysis. Also equipped with sophisticated instruments such as HPLC, GC, FTIR, Digital polarimeter, UV spectrophotomer, Karl fisher, IR moisture balance, analytical balance, pH meter, IC, ICP-OES, GC-MS, LC-MS, DSC, Dissolution Test Apparatus, Disintegration Tester, Hot Air oven, Stability chambers, refrigerators etc..

(Refer List of Equipment's & Instruments: Annexure – IV)

C.3.7 Qualification, Validation and Calibration:

Qualification: New equipment is qualified by means of Installation, Operation and Performance Qualification as per the protocol and then put into use. Qualification is carried out by conducting the following activities:

- Design Qualification (DQ)
- Installation Qualification (IQ)
- Operational Qualification (OQ)
- Performance Qualification or Process validation (PQ)

Qualification includes the qualification of computerized systems, Equipment's/instruments facilities and buildings.

Qualification protocol includes:

- Test item and description of test
- Acceptance criteria

<table>
<tr><td colspan="3">Site Name : XXXX</td></tr>
<tr><td>Site master number and its Version number</td><td>XXXX/01</td><td>Effective date:
Page No.: XX of XX</td></tr>
</table>

- Reference to supporting documents
- Test results

All the qualification records are retained by Quality Assurance.

Calibration:

Equipment & Instruments being used are calibrated periodically as per the schedule. Annual calibration schedule is prepared which includes calibration of existing equipment's/instruments and master equipment's/instruments. The procedure for calibration includes; methodology for calibration, frequency and acceptance criteria. Calibration by external agency is carried out as per SOP. The calibration reports/certificates submitted by external agency are examined by concerned department.

The details of calibration are recorded in the equipment/instruments calibration record. After calibration, a calibration tag is attached to the equipment/instruments. When the equipment/instruments fails in calibration, this equipment/instrument is not used until it is repaired. In case of major changes made to equipment/instruments, the equipment/instruments is re-qualified.

C.3.8   **SAP – For Inventory control and quality Notifications:**

For all the raw materials, solvents and Intermediates Inventory Transaction, SAP is being used, which acts as tool for tracking mechanism and for Inventory Data Retrieval. Quality notifications like Change control, Incident and CAPA management are also being tracked through SAP.PM calibration are also traced through SAP.

C.3.9   **Sanitation and Hygiene:**

To avoid cross contamination during development activities the company follows procedures to maintain utmost cleanliness and sanitation of all the work areas. The premises & labs are cleaned effectively by trained persons as per Housekeeping procedure. The frequency of cleaning of various areas like floors, ceilings, doors, windows cupboards, glasses, wash-rooms etc is defined in procedure. Cleaning is done by sweeping, dry/wet mopping with lint free cloth or vacuum cleaners.

C.4   **DOCUMENTATION:**

C.4.1   **Arrangements for Preparation, Revision and Distribution of Documents:**

There are separate SOPs for the preparation of different types of documents with standard format. The documents (Product specifications, analytical methods, SOPs, Feasibility reports, Optimization reports, SHE documents etc.)

The documents are stored as per procedure mentioned in the SOP.

The documents are revised as per the procedure described in the SOP.

C.4.2   **Documentation related to Quality System :**

Apart from the above mentioned documents, other documents are Equipment Qualification, Validation, Training Records, Calibration Records, and Development

Site Name : XXXX		
Site master number and its Version number	**XXXX/01**	**Effective date:** **Page No.: XX of XX**

Reports etc. These records helps, to have picture of the past and present activities and serve as a future reference with respect to investigations, complaints and Incidents will be documented and retained in QA.

C.5   DEVELOPMENT PROCESS:

C.5.1   Brief description of development procedure:

Different types of development activities are carried out in site e.g. reaction, crystallization, filtration, drying, blending, formulation development etc. The facility is capable of carrying out each operation smoothly.

C.6   ANALYTICAL RESEARCH & DEVELOPMENT:

C.6.1   Description of Analytical Research & Development:

The Analytical Research & Development facility is well equipped with sophisticated instruments. The objective of Analytical Research & Development is to collect and analyze various materials such as raw materials, in process, intermediates and Finished Drug Substances as per the existing standard test procedures/MOAs.

Some of the prime activities are as follows:

- Analytical Method Development & validation.
- Approve or reject raw materials, Packing components and In-process materials based on the testing and analytical acceptance criteria and predetermined specification for development activities..
- Preparation or revision of specification, Analytical Methods and procedures, which directly or indirectly contribute to the identity, strength, quality and purity of the company's products.
- Investigation of stability OOS of Clinical molecules.
- Investigate complaints relating to analytical methods.
- Maintain a stability program, enter results in stability program formats, and make retest date/expiry date recommendations based on the results for clinical product
- Participate in the development, implementation, and maintenance of cGMP & Lab compliance.
- Calibration of instruments.

C.7   CONTRACT MANUFACTURING:

The company outsources contract manufacturing units for manufacturing of intermediates and/API. The Strategic Business Partners (SBPs) are qualified as per the approved SOP. The names of the Strategic Business Partners (SBPs) are given in Annexure-VII.

Table Contd...

<table>
<tr><td colspan="3">Site Name : XXXX</td></tr>
<tr><td>Site master number and its Version number</td><td>XXXX/01</td><td>Effective date:
Page No.: XX of XX</td></tr>
</table>

C.8   COMPLAINT(S) HANDLING:

QA is responsible for coordinating with manufacturing location QA in case of customer complaints, as described in SOP. All customer complaints are logged at plant-QA and investigated with cross functional team and implementation of CAPA is monitored whenever applicable.

ABBREVIATIONS:

AHU	Air Handling Unit
CAPA	Corrective and Preventive Action
HEPA	High Efficiency Particulate Area
HPAI	High potent Active Pharmaceutical Ingredient
HVAC	Heating, Ventilation and Air Conditioning
QA	Quality Assurance
UV	Ultra Violet

APPROVALS:

	Department	Name & Designation	Sign & Date
Compiled By	XX		
Checked By	XX		
Approved By	XX		

Chapter - 7

Validation Master Plan

Introduction

An analytical procedure is developed to test a defined characteristic of the drug substance or drug product against established acceptance criteria for that characteristic. The analytical method should be validated by research and development before being transferred to the quality control unit when appropriate. There should be specifications for both, materials and products. The tests to be performed should be described in the documentation on standard test methods

7.1   Types of Methods

A. Pharmacopoeialor Compendial methods

- USP/NF, EP, JP, BP, compendial methods are used for raw material and final product testing
- When pharmacopoeial methods are used, evidence should be available to prove that such methods are suitable for routine use in the laboratory (verifi cation).
- Pharmacopoeial methods used for determination of content or impurities in pharmaceutical products should also have been demonstrated to be specific with respect to the substance under consideration (no placebo interference).

B. Non-pharmacopoeial methods

- Non-pharmacopoeial methods should be appropriately validated.

7.2 Method Validation Requirements

- Validation should be performed in accordance with the validation protocol. The protocol should include procedures and acceptance criteria for all characteristics. The results should be documented in the validation report.
- Justification should be provided when non-pharmacopoeial methods are used if pharmacopoeial methods are available. Justification should include data such as comparisons with the pharmacopoeial or other methods.
- Standard test methods should be described in detail and should provide sufficient information to allow properly trained analysts to perform the analysis in a reliable manner. As a minimum, the description should include the chromatographic conditions (in the case of chromatographic tests), reagents needed, reference standards, the formulae for the calculation of results and system suitability tests.

7.3 Content of Analytical Procedure

The following is a list of essential information you should include for an analytical procedure.

A. Principle/Scope

A description of the basic principles of the analytical test/technology; target analyte(s) and sample(s) type.

B. Apparatus/Equipment

All required qualified equipment and components.

C. Operating Parameters

Qualified optimal settings and ranges (include allowed adjustments supported by compendial sources or development and/or validation studies) critical to the analysis (e.g., flow rate, components temperatures, run time, detector settings, gradient, head space sampler). A drawing with experimental configuration and integration parameters may be used, as applicable.

D. Reagents/Standards

The following should be listed where applicable:

- Description of reagent or standard
- Grade of chemical
- Source
- Standard/Reference Material
- Purity (for pure chemicals only)

- Potencies
- Storage conditions
- Directions for safe use
- Validated or documented shelf life

E.   Standards Control Solution Preparation and Sample Preparation

Procedures for the preparations for individual standard/sample tests with appropriate units of concentrations for working solutions (e.g., µg/ml or mg/ml) and information on stability of solutions and storage conditions.

A step-by-step description of the method (e.g., equilibration times, and scan/injection sequence with blanks, placeboes, samples, controls, sensitivity solution (for impurity method) and standards to maintain validity of the system suitability during the span of analysis) and allowable operating ranges and adjustments if applicable.

F.   System Suitability

Confirmatory test(s) procedures and parameters to ensure that the system (equipment, electronics, and analytical operations and controls to be analyzed) will function correctly as an integrated system at the time of use. The system suitability acceptance criteria applied to standards controls and samples, such as peak tailing, precision and resolution acceptance criteria, may be required as applicable.

G.   Calculations

The integration method and representative calculation formulas for data analysis (standards, controls, samples) for tests based on label claim and specification (e.g., assay, specified and unspecified impurities and relative response factors). This includes a description of any mathematical transformations or formulas used in data analysis, along with a scientific justification for any correction factors used.

H.   Data Reporting

A presentation of numeric data that is consistent with instrumental capabilities and acceptance criteria. The method should indicate what format to use to report results (e.g., percentage label claim, weight/weight, and weight/volume) with the specific number of significant figures needed

7.4 Validation Characteristics

- Specificity
- Linearity

- Accuracy
- Precision (repeatability, intermediate precision, and reproducibility)
- Range
- Quantisation limit
- Detection limit

7.5 Analytical Method Transfer

The transfer of method analytical (TAP) also referred to as method transfer, is the documented process that qualifies a laboratory (receiving unit) to use an analytical test procedure that originated in another laboratory (the transferring unit) thus ensuring that the receiving unit has the procedural knowledge and ability to perform the transferred analytical procedure as intended.

A. Types of transfer of Analytical procedure

TAP can be performed and demonstrated by several approaches.

(i) *Comparative testing*: A comparative testing performed on homogeneous lots of the target material from standard production batches or sample intentionally prepared for test (e.g. by spiking relevant accurate amounts of known impurities into sample.

(ii) *Co-validation*: Co-validation between laboratories, the complete or partial validation of the analytical procedures by the receiving unit, and the transfer waiver, which is appropriately justified omission of the transfer process.

The tests that will be transferred, the extent of the transfer activities, and implementation strategy should be based on a risk analysis that considers the previous experience and knowledge of the receiving unit, the complexity and specification

B. Elements for transfer of Analytical procedure

- Approved transfer protocol.
- Transferring unit should provide training to receiving unit.
- Transferring unit is responsible for providing analytical procedure.
- Receiving unit is responsible for the Qualified staff, facility and equipment.
- Both the transferring and receiving unit should compare and discuss data as well as any deviations from protocol. This discussion address any necessary correction or updates to the final report and the analytical procedure as necessary to reproduce the procedure.

C. Analytical Transfer Report

When TAP is successfully completed, the receiving unit should prepare a transfer report that describes the results obtained in relation to the acceptance criteria, along with conclusion that confirm that receiving unit is now qualified to run the procedure. Any deviation should be thoroughly documented and justified. If the acceptance criteria are met, the TAP is successful and the receiving unit is qualified to run the procedure. Otherwise the procedure cannot to considered transferred unit effective remedial steps are adopted in order to meet the acceptance criteria. An investigation may provide guidance about the nature and extent of the remedial steps, which may vary from further training and clarification to more complex approaches, depending on the particular procedure.

7.6   Analytical Revalidation

The degree of revalidation depends on the nature of the change. Revalidation shall be performed in following cases

- Change is made in analytical procedure
- Change is made in manufacturing procedure or formulation
- Change is made in piece of equipment
- changes in the synthesis of the drug substance
- changes in the composition of the finished product
- The degree of revalidation required depends on the nature of the changes. Certain other changes may require validation as well.

Type of analytical Procedure : Characteristics	Identification	Testing for impurities: quantitat.	limit	Assay - dissolution (measurement only) - content/potency
Accuracy	-	+	-	-
Precision				
Repeatability	-	+	-	+
Interm. Precision		+(1)	-	+(1)
Specificity (2)	+	+	+	+
Detection Limit	-	-(3)	+	-
Quantitation Limit	-	+	-	-
Linearity	-	+	-	+
Range	-	+	-	+

Table *Contd...*

Type of analytical Procedure : Characteristics	Identification	Testing for impurities: quantitat. limit	Assay - dissolution (measurement only) - content/potency
-	signifies that this characteristic is not normally evaluated		
+	signifies that this characteristic is normally evaluated		
(1)	in cases where reproducibility (see glossary) has been performed, intermediate precision is not needed		
(2)	lack of specificity of one analytical procedure could be compensated by other supporting analytical procedure(s)		
(3)	may be needed in some cases		

7.7   Analytical Procedure Life Cycle Management

Life cycle approach should be followed to assure validated or verified method remains fit for its intended purpose. Trend analysis on method performance should be performed at regular intervals to evaluate the need to optimize the analytical procedure or to revalidate all or a part of the analytical procedure.

The life cycle management approach is consistent with the concept of quality by design (QbD) as described in International Council for Harmonisation (ICH) Q8-R2, Q9, Q10, and Q11. In order to provide a holistic approach to controlling an analytical procedure throughout its lifecycle, one can use a three-stage concept that is aligned with current process validation terminology:

The lifecycle approach for analytical procedure:

(a) Analytical target profile (ATP): Structure and Application throughout the Analytical Lifecycle and Analytical Control Strategy.

(b) Comparable to the Quality Target Product Profile (QTPP).
- **Stage 1:** Procedure Design and Development (Knowledge Gathering, Risk Assessment, Analytical Control Strategy, Knowledge Management, Preparing for Qualification)
- **Stage 2:** Procedure Performance Qualification
- **Stage 3:** Continued Procedure Performance Verification (Routine Monitoring, Changes to an Analytical Procedure)

7.8 Definition

(a) Analytical Procedure

The analytical procedure refers to the way of performing the analysis. It should describe in detail the steps necessary to perform each analytical test. This may include but is not limited to: the sample, the reference standard and the reagents preparations, use of the apparatus, generation of the calibration curve, use of the formulae for the calculation, etc.

(b) Specificity

Specificity is the ability to assess unequivocally the analyte in the presence of components which may be expected to be present. Typically these might include impurities, degradants, matrix, etc.

Lack of specificity of an individual analytical procedure may be compensated by other supporting analytical procedure(s).

This definition has the following implications:

- Identification: to ensure the identity of an analyte.
- Purity Tests: to ensure that all the analytical procedures performed allow an accurate statement of the content of impurities of an analyte, i.e. related substances test, heavy metals, residual solvents content, etc.
- Assay (content or potency): to provide an exact result which allows an accurate statement on the content or potency of the analyte in a sample.

(c) Accuracy

The accuracy of an analytical procedure expresses the closeness of agreement between the value which is accepted either as a conventional true value or an accepted reference value and the value found.

This is sometimes termed trueness.

(d) Precision

The precision of an analytical procedure expresses the closeness of agreement (degree of scatter) between a series of measurements obtained from multiple sampling of the same homogeneous sample under the prescribed conditions. Precision may be considered at three levels: repeatability, intermediate precision and reproducibility.

Precision should be investigated using homogeneous, authentic samples. However, if it is not possible to obtain a homogeneous sample it may be investigated using artificially prepared samples or a sample solution.

The precision of an analytical procedure is usually expressed as the variance, standard deviation or coefficient of variation of a series of measurements.

- ***Repeatability***: Repeatability expresses the precision under the same operating conditions over a short interval of time. Repeatability is also termed intra-assay precision.
- ***Intermediate precision***: Intermediate precision expresses within-laboratories variations: different days, different analysts, different equipment, etc.
- ***Reproducibility***: Reproducibility expresses the precision between laboratories (collaborative studies, usually applied to standardization of methodology).

(e) Detection Limit

The detection limit of an individual analytical procedure is the lowest amount of analyte in a sample which can be detected but not necessarily quantitated as an exact value.

(f) Quantitation Limit

The quantitation limit of an individual analytical procedure is the lowest amount of analyte in a sample which can be quantitatively determined with suitable precision and accuracy. The quantitation limit is a parameter of quantitative assays for low levels of compounds in sample matrices, and is used particularly for the determination of impurities and/or degradation products.

(g) Linearity

The linearity of an analytical procedure is its ability (within a given range) to obtain test results which are directly proportional to the concentration (amount) of analyte in the sample.

(h) Range

The range of an analytical procedure is the interval between the upper and lower concentration (amounts) of analyte in the sample (including these concentrations) for which it has been demonstrated that the analytical procedure has a suitable level of precision, accuracy and linearity.

(i) Robustness

The robustness of an analytical procedure is a measure of its capacity to remain unaffected by small, but deliberate variations in method parameters and provides an indication of its reliability during normal usage.

Chapter - 8

Laboratory Computer-Software Validation and Qualification

Introduction

Computers are widely used during development and manufacturing of drugs and medical devices. Proper functioning and performance of software and computer systems play a major role in obtaining consistency, reliability and accuracy of data. Therefore, computer system validation (CSV) must be part of any good development and manufacturing practice computer systems must be validated at the level appropriate for their use and application

There are two essential parts of computerized systems:

- Infrastructure
- Applications

Validation of computer systems is not a onetime event. It starts with the planning, specification, stages of programming, testing, commissioning, documentation, operation, monitoring and modifying ,product or project requirements and setting user requirement specifications and cover the vendor selection process, installation, initial operation, going use, and change control and system requirement.

For new systems validation starts when a user department has a need for a new computer system and thinks about how the system can solve an existing problem. For an existing system it starts when the system owner gets the task of bringing the system into a validated state. Validation ends when the system is retired and all-important quality data is successfully migrated to the new system. Important steps in between are validation planning, defining user requirements, functional specifications, design specifications, validation during development, vendor assessment for purchased systems, installation, initial and ongoing testing and change control. In other words, computer systems must be validated during the entire life of the system.

8.1   GAMP 5: Good Automated Manufacturing Practice

GAMP covers all documentation, expectation, aspects, good practices and complies with GMP for computerized and software system. GAMP covers the life cycle and risk assessment approach of GxP computerized system. Based on the complexity of computerized system GAMP 5 software are categorized. It provide initial assessment as to the validation requirements.

GAMP Software Category 1 – Infrastructure Software

- **(a) Category 1 – Infrastructure software**

    Including operating systems, Database Managers, etc. Example: Windows XP

- **(b) Category 2: Firmware**

    Removed from GAMP 4 has been removed.

- **(c) Category 3 – Non configurable software**

    Including, commercial off the shelf software (COTS), Laboratory Instruments/Software where we can view, read, recipe can be selected and print

- **(d) Category 4 – Configured software**

    Include view, reading, recipe can be selected, get system generated print and capable for store and process and configure the data based on requirement. E.g: LIMS, SCADA, DCS, CDS, etc.

- **(e) Category 5 – Bespoke software/Customer Application**

    Design Electronic batch record.

    Following explanations are provided for better clarity on each category with example.

GAMP	Qualification	Password Control	Audit Trail Review	Software Back up or upgrade or Malfunctioning
Category-1	Yes	Preferred	Preferred	Preferred
Category -2			Removed	
Category-3	Yes	Compulsory	Compulsory	Preferred/Compulsory
Category-4	Yes	Compulsory	Compulsory	Preferred/Compulsory
Category-5	Yes	Compulsory	Compulsory	Preferred/Compulsory

8.2   Computer and Software Validation Master Plans

All validation activities must be described in a validation master plan which must provide a framework for thorough and consistent validation of computer. Key element of validation are: Requirements, Design Control, Testing, Change Control, Traceability and Monitoring.

Computer Validation master plans must include:

- Introduction with a scope of the plan, e.g., sites, systems, processes, Responsibilities by function
- Related documents, e.g., risk management plans, Products/processes to be validated and/or qualified

Validation approach, e.g., system life cycle approach, Risk management approach with examples of risk categories and recommended validation tasks for different categories, Vendor management, Steps for Computer System Validation with examples on type and extent of testing, for example, for IQ, OQ and PQ, Handling existing computer systems, Validation of Macros and spreadsheet calculations, Qualification of network infrastructure, Configuration management and change control procedures and templates, Back-up and recovery, Disaster recovery, Access control/user management, Data integrity including: prevention of deletion, poor transcriptions and omission, Authorized/unauthorized changes to data and documents, Critical Alarms handling (Process Data base management system), Network system, Error handling and corrective actions, Requalification criteria, contingency planning and disaster recovery, Maintenance and support, System retirement, Training plans (e.g., system operation, compliance),Validation deliverables and other documentation and Change Control

8.3   Computer Qualification

A. System requirement specifications (SRS) or user requirement specifications (URS)

- The vendor's specification sheets can be used as guidelines during SRS or URS.
- During SRS or URS all user department must be involved
- The SRS or URS control documents must state the objective of a proposed computer system, the data entered, stored, the flow of data, how it interacts with other systems and procedures, the information to be produced, the limits of any variables and the operating programme and test programme
- User requirements must have a couple of key attributes.

B. Design Qualification and Specifications

- Design qualification (DQ) defines the functional and operational specifications of the instrument and details the conscious decisions in the selection of the supplier.
- DQ must ensure that computer systems have all the necessary functions and performance criteria that will enable them to be successfully implemented for the intended application and to meet business requirements.

Steps for design specification normally include:

- Description of the task the computer system is expected to perform
- Description of the intended use of the system
- Description of the intended environment
- Includes network environment)
- Preliminary selection of the system requirement specifications, functional specifications and vendor
- Vendor assessment
- Final selection of the system requirement specifications and functional specification
- Development and documentation of final system specifications

C. Installation Qualification

Installation qualification establishes that the computer system is received as designed and specified, that it is properly installed in the selected environment, and that this environment is suitable for the operation and use of the instrument. The list below includes steps as recommended before and during installation.

Before installation

- Obtain manufacturer's recommendations for installation site requirements.
- Check the site for the fulfillment of the manufacturer's recommendations (utilities such as electricity, water and gases and environmental conditions such as humidity, temperature, vibration level and dust).

During installation

- Compare computer hardware and software, as received, with purchase order (including software, accessories, spare parts)
- Check documentation for completeness (operating manuals, maintenance instructions, standard operating procedures for testing, safety and validation certificates)

- Check computer hardware and peripherals for any damage
- Install hardware (computer, peripherals, network devices, cables)
- Install software on computer following the manufacturer's recommendation
- Verify correct software installation, e.g., are all files accurately copies on the computer hard disk. Utilities to do this must be included in the software itself.
- Make back-up copy of software
- Configure network devices and peripherals, e.g. printers and equipment modules
- Identify and make a list with a description of all hardware, include drawings where appropriate, e.g., for networked data systems.
- Make a list with a description of all software installed on the computer
- Store configuration settings either electronically or on paper
- List equipment manuals and SOPs
- Prepare an installation report
- Both the suppliers representative and a representative of the user's form must sign off the IQ documents

D. Operational Qualification

Operational qualification (OQ) is the process of demonstrating that a computer system will function according to its functional specifications in the selected environment

- During e OQ testing is done, the link between USR and DQ must be check
- System must be tested as per vender manual or operating procedure
- The general aspect must be also check such as power supply, temperature, magnetic disturbance
- Training must be conducted to the user
- Proper functioning of back-up and recovery and security functions like access control to the computer system and to data must also be tested.
- Full OQ test must be performed before the system is used initially and at regular intervals

E. Performance Qualification

Performance Qualification (PQ) is the process of demonstrating that a system consistently performs according to a specification appropriate for its routine use i.e testing of the system with the entire application. PQ activities normally can include

- Complete system test to proof that the application works as intended. For example for a computerized analytical system this can mean running a well characterized sample through the system and compare the results with a result previously obtained.
- Regression testing: reprocessing of data files and compare the result with previous result
- Regular removal of temporary files
- Regular virus scan
- Auditing computer systems and Audit trails
- Security
- Back up
- Accuracy checks
- Data storage
- Printout
- Change and Configuration management

8.4   Software Validation

Computerized analytical instruments contain integrated chips with low-level software (Firmware). Such instruments will not function without properly opening firmware, and users generally cannot alter firmware design or function. Firmware is therefore considered a component of the instrument itself. Indeed, the qualification of hardware is not possible without operating it via its firmware. Thus, when the hardware (that is, the analytical instrument) is qualified at the user's site, the integrated firmware is also essentially qualified. No separate on-site qualification of the firmware is needed. Whenever possible, the firmware version should be recorded as part of the IQ activities. Any changes made to firmware versions should be tracked through change control of the instrument.

A. Instrument Control, Data Acquisition, and Processing Software

Software for instrument control, data acquisition, and pro-sing for many of today's computerized instruments is loaded on a computer connected to the instrument. Operation of the instrument is then controlled via the software, the software is needed for data acquisition and post acquisition calculations. Thus, both hardware and software, their functions inextricably intertwined, are critical to providing analytical results.

The manufacturer should perform DQ, validate this software, and provide users with a summary of validation. At the user site, holistic qualification, which involves the en-tire instrument and software system, is more efficient than modular validation of the software alone. Thus, the user qualifies the instrument control, data acquisition, and pro-cessing software by qualifying the instrument according to the analytical Instrument qualification (AIQ) process.

B. Stand-Alone Software

An authoritative guide for validating stand-alone software, such as LIMS is available. The validation process is administers by the software developer, who also specifies the development model appropriate for the software. Validation takes place in a series in a series of activities planned and executed through various stage of the development cycle.

C. Software Validation Change Control

Changes to instruments, including software, become inevitable as manufacturers add new features and correct known defect. However, implementing all such changes may not always benefits users, Users should therefore adopt changes they deem useful or necessary and should also assess the effects of changes to determine what, if any, requalification is required. The change control process enables them to do this.

Change control may follow the DQ/IQ/OQ/PQ classification process. For DQ, evaluate the changed parameters, and determine whether need for the change warrants implementing it. If implementation of the change is needed, install the changes to the system during IQ. Evaluate which of the existing OQ and PQ tests need revision, deletion, or addition as a result of the installed change. Where the change calls for additions, deletions, or revisions to the OQ or PQ tests, follow the procedure outlined below.

D. Operational Qualification

Revise OQ tests as necessitated by the change. Perform the relevant tests affected by the change. This ensures the instrument's effective operation after the change is installed.

E. Performance Qualification

Revise PQ tests as necessitated by the change. Perform the PQ testing after installation of the change if similar testing is not already performed during OQ. In the future, perform the revised PQ testing.

For changes to firmware and to software for instrument control, data acquisition, and processing, change control is performed through DQ/IQ/OQ/PQ of the affected instrument. Change control for stand-alone software requires user-site testing of changed functionality

8.5   Validation Protocol/Report

- The validation protocol must be numbered, signed and dated , and must contain as a minimum the following information's:
- Objectives, Scope of coverage of the validation study ,
- Validation team, their qualifications and responsibilities
- Justification for validation'
- Risk Assessment,
- Type of validation: Prospective, concurrent, Retrospective, revalidation,
- A list of all equipment to be used with calibration, qualification, requalification and preventive maintenance details.
- Outcome of IQ, OQ for critical parameter,
- Critical parameters and their respective tolerance;
- Description of the processing steps : Challenges
- of sampling and sampling plans;
- Statistical tools to be used in the analysis of data,
- Forms and chart to be used for documenting results,
- Non-conference (Out of Specification /Out of Trend/Deviation)
- Change control
- Conclusion
- Summary
- Approval of study

8.6   Validation of Hardware and Software

Following aspect must be subjected to validation.

Hard ware	Soft ware
Types : Input device Output device Signal converter Central processing unit (CPU) Distribution System Peripheral devices	Level : Machine language Assembly language High-level language Application language
Key aspect : Location	Software identification: Language

Hard ware	Soft ware
Environment	Name
Distance	Function
Input device	Input
Signal conversion	Output
Operating system	Fix set point
Command overrides	Variable set point
Maintenance	Edits
	Input manipulation
	Programme overrides
Validation : Function Limits Worst case Reproducibility / Consistency Documentation Revalidation	Keys Aspects : Software Development Software Security
	Validation : Function Worst Case Repeats Documentation Revalidation

8.7   Computer System and Software Life Cycle Management

A systematic approach to computerized system validation, which begins with initial risk assessment and continues throughout the life of the system, must be defined to ensure quality is built into the computerized systems. Computer system and software validation takes place within the environment of an established Computer system and software life cycle. The s life cycle model that is selected should cover the software from its birth to its retirement. Risk management is applied throughout the life cycle it identify risks and to remove or mitigate them to an acceptable level.

   (a)   Quality Planning
   (b)   System Requirements Definition
   (c)   Detailed Software Requirements Specification
   (d)   Software Design Specification

(e)   Construction or Coding

(f)   Testing

(g)   Installation

(h)   Operation and Support

(i)   Maintenance

(j)   Retirement

8.8   Security Management

Security management is the process that ensure the confidentiality, integrity and availability of an organizations regulated systems, records and process. Physical and logical security controls must be applied to computerise in order to prevent unauthorized use and to ensure that data is adequately and securely protected against intentional or accidental loss, unauthorized change, damage or removal. The control may include use of key locks, pass card, biometric and username-password combinations. Security roles and responsibilities must be clearly defined and change made to authorization should be recorded.

In the event when the computerised system is used for batch release, the system should only allow authorization persons to certify the release of batch. There should be a clear identity of person releasing the batch.

The incidents should be formally documented, the cause's investigated and corrective action proposed, implemented and closed out. A regular review of incidents should be undertaken to determine is any trends or threats can be identified.

Organization should have security policy and should cover physical security, system access security including granting and revoking access, third party access, electronic messaging systems, shared network resources, internet resources, internet access, use of mobile computing resources, connectivity to external computer systems, anti-virus policies and intrusion detection

8.9   Electronic Records and Electronic Signature

Electronic records and signature are regards as equivalent to paper records and hand-written signature. System that generate, store or process electronic records or use electronic signature must be validated.

A.   Electronic Records

It must be possible for electronic records to be printed in a readable format. For release records, it should be possible to generate and print out the changes made to the original data.

B.   Electronic Signatures

Electronic signatures must be unique to one individual and there must be procedure to ensure the owners of the electronic signatures are aware of their responsibilities for their actions, i.e. users must be aware that electronic signatures have the same impact as hand-written signatures.

Electronic signature must be permanently linked to their respective electronic record to ensure that the signatures cannot be edited, duplicated or removed in anyway.

Electronic signature must clearly indicates

- The displayed/printed name of the signer
- The date and time when the signature was executed
- The reason for signing (such as review, approval, responsibility or authorship) associated

8.10 Handling of Audit Trial

An audit trial is a log generated by the computerised system that allows operator entries and actions that create, modify or delete GMP relevant electronic records to traced back to the original electronic records. A decision on whether to implement an audit trail must be based on documents risk assessment.

An audit trail is expected to have the following features:

(a) Access to audit trail data should be limited to print and or read only

(b) It must be protected against modification or deletion

(c) It should have the ability to capture a full history of all GMP related transitions, including the identity of the person making the change, the date and time of change and the reason for change.

(d) The data and time of the audit trial must be synchronised with a trusted date and time service e.g. main server

(e) It should be readable and readily available

(f) Confirmed during testing

(g) Reportable

(h) Permanently linked to record – confirmed during testing.

(i) Audit trail configuration logs must be reviewed regularly based on the criticality of the system.

Audit trail Assessments

The audit trail assessment is the first and the most critical steps to implement audit trail reviews. An inventory of all impacted systems need to be created to identify whether each individual system provide audit trails that are adequate and that can be used for performing the reviews. System level risk assessments need to be performed to identify whether the system is high, medium or low risk. The system risk needs to be used to prioritize the audit trail assessment and implementation of periodic reviews. Once the audit trail assessment are performed, system risk identified and all corrective actions are closed the audit trail reviews can be implemented. Prior to implementation the impact to resource need to be well understood based on the expected volume of work. Once this impact is understood hiring and reassigning of resource need to be completed prior to formal implementation.

Each functional area that have GxP computer systems need to perform the audit trail assessment to determine the following:

➢ Who has access to view the audit trails?

➢ Can the audit trail be printed from the application?

➢ Can the reviewer select a data range?

➢ Can the reviewer select a specific activity of interest during the audit trail review?

➢ Will it feasible to include the audit trail with the data results?

➢ Will it be feasible for QC systems to include the audit trail with the assay results?

➢ Are user's action time and date stamped?

➢ Does the audit trail records creation, modification and deletion of records?

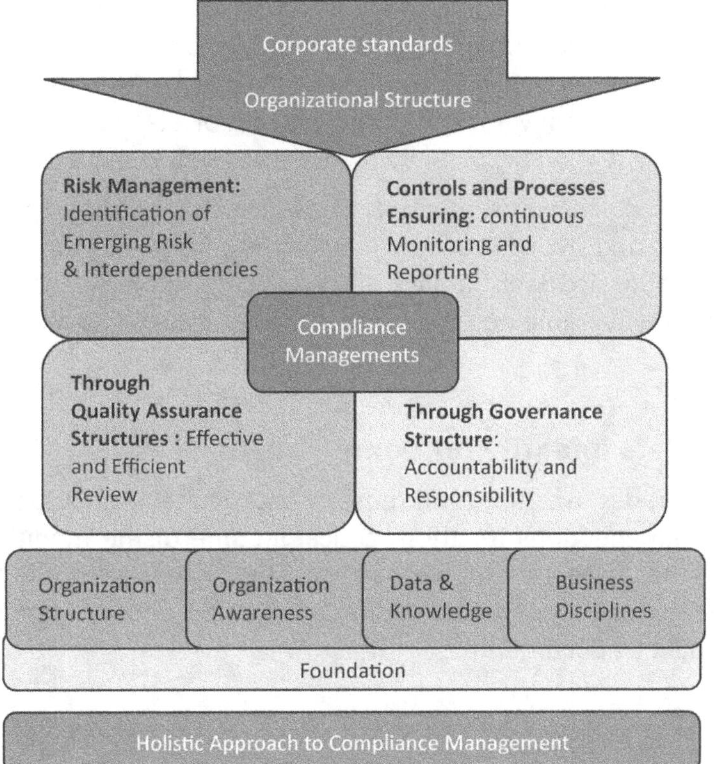

8.11 System Periodic Review

Computer system should be periodically reviewed in order to ensure and evaluate the system remain in a validated state i.e its adequacy to perform its designated task(s), compliance with current standards and industry trends and assurance the testing and associated documentation still supports the systems current use.

The results of the assessment should be formalised in a periodic review report stating, where applicable, all changes to system, deviation, incident, maintenance, security issue and overall validation status.

Based on the periodic review assessment results, revalidation may be required due to major changes outside the control of a regulated company that could have an effect on the process or product quality or due to the system being out of control. The extent of revalidation will depend on the significance of chances and it should be the absolute last resort and should be avoided by change management, configuration management, periodic review and evaluation

8.12 Data Back-up, Recovery, and Archiving

A backup means a copy of one or more electronic files created as an alternative in case the original data or system are lost or become unusable (for example, in the event of a system crash or corruption of a disk). It is important to note that backup differs from archival in that back-up copies of electronic records are typically only temporarily stored for the purposes of disaster recovery and may be periodically over-written. Back-up copies should not be relied upon as an archival mechanism. Backup of electronic records is to ensure disaster recovery.

There should formal procedure in place to ensure that routine back-ups of all GMP related data are made and stored in secure location at a frequency based on a risk analysis to prevent intentional or accident damage. There must also be periodic check in place to ensure backups are recoverable and data integrity and accuracy is maintained during the recovery process.

8.13 Electronic Data Integrity for Long Term in GLP

Electronic media degrades overtime, and reuse of backup media should in accordance with manufacturer recommendations for the practical lifetime of the media. The backup and restore procedure should be verified periodically. The frequency should be based on risk. Backup and restore procedure and documentation should be check during periodic review. The conclusion should be documented.

8.14 System Retirements Management

When a computerized system requires retiring, a retirement plan and retirement summary report must be develop to ensure all data is appropriately transitioned without loss of information. It is the process owner's responsibility to manage the replacement or withdrawal of a software system from use. Retirement formally closes the use of the system and ensures any records and data are available.

Following steps are involved in the system retirement:

A.   System Retirement plan

    (a)   Identification of system with its asset number, serial number, software name, location of the system, hardware and software components to be retired.

    (b)   Raised the change control and defined the method for deactivation, removal, or re-assignment of the system, preservation of data, records, etc.

    (c)   Approved the retirement plan.

B.   Data Archival and Disposal

Data must be either archived or migrated to another system for future regulatory requirement. Procedure must be in place and periodic checks must be performed to assure that data integrity and accuracy is maintain and recoverable throughout the retention period. Access to data should be ensured throughout the retention time. Once the archived data reaches the end of the retention period, the data may be destroyed or the retention period may be extended.

C.   Data Migration

Data migration shall be performed in case replacing of software or an application, upgrading of the system or retirement of system. The approved procedure should be available for data migration, storage and retrieval. There might be the need for data migration as part of the project activities. Data migration must include test to assure acceptable accuracy and completeness of data migration. A final report is recommended to summarize the status. Additional functional testing could also be required and will be specified in the validation plan. The data migration plan should describe the entire migration process, including as a minimum:

- Migration projection purpose and scope
- System description(s)
- Roles and responsibilities
- Required deliverables
- Risk management strategy
- Configuration management strategy, including the source, staging and target environments

- Software tool overview and strategy for ensuring compliance and fitness for intended use
- Migration steps and technical activities
- Data mapping and modelling activities
- Transformation rules
- Data verification strategy and acceptance criteria standard
- Cut over plan
- Rollback strategy

The data migration plan should be approved by system owner and Site Master expert (SME). Each time the plan is executed, data migration report should be created. Data should be verified each time it is moved or its state is transformed.

D.   Decommissioning

Formal decommissioning of the computerised system typically occur when the system is retired from use. However, in some circumstance, a computerised system may remain in a read-only state from a period of time to permit data accessing this, case decommissioning will occur when the requirement to access the data has expired.

E.   Retirement Summary

Software retirement summary shall be prepared on completion and acceptance of retirement activity. Summary shall include change control details, steps followed during the retirement and any non-conformance against plan. The retirement summary shall be approved.

8.15 Business Continuity Plan (BCP) and Disaster Recovery Plan (DRP)

A Business continuity Plan describes how to continue business after a disturbance. A Disaster Recovery Plan deals with recovering Information Technology (IT) assets after a disastrous disruption.

A. Business Continuity Plan includes

- Plans, measures and arrangements to ensure the continuous delivery of critical services and products, which permits the organization to recover its facility, data and assets.
- Identification of necessary resources to support business continuity, including personnel, information, equipment, financial allocations, legal counsel, infrastructure protection and accommodations
- Organization risk from potential disasters that include: Natural disasters such as tornadoes, floods, blizzards, earthquakes, fire, accidents, Interrupt, Power

and energy disruptions, Failure in Communications, transportation, safety and service sector, Environmental disasters such as pollution and hazardous materials spills and Cyber attacks and hacker activity.

Creating and maintaining a BCP helps ensure that an institution has the resources and information needed to deal with these emergencies

Steps involved in Business continuity plan:

(a) ***BCP Governance***: BCP committee, roles and responsibilities, Insurance, approval of plans, monitoring, provide strategic direction and communicate essential messages, resolve conflicting interests and priorities and approval of results.

(b) ***Business Impact Analysis (BIA)***: Identify the mandate and critical aspects of an organization, Prioritize critical services or products, Mitigating threats and risks, Analyze current recovery capabilities, Create continuity plans and Alternate facilities.

(c) ***Plans, measures, and arrangements for business continuity:*** Approved plans, committee, measures and effective check.

(d) ***Readiness procedures:*** Training, Set goals, artificial aspects and assumptions, Exercise Narrative, Communications, Testing and Post-Exercise Evaluation.

(e) ***Quality assurance techniques:*** Internal review, maintenance and auditing.

B. Disaster Recovery Plan (DRP)

Disaster Recovery (DR)is the process used to recover the access to software, data, and/or hardware that are needed to resume the performance of normal, critical business functions after the event of either a natural disaster or a disaster caused by humans. It is on-going process of planning, developing, testing and implementing Disaster Recovery management procedures and processes to ensure the efficient and effective resumption of vital business functions in the event of an unscheduled interruption.

Disaster Recovery plans focus on bridging the gap where data, software, or hardware have been damaged or lost, one cannot forget the vital element of manpower that composes much of any organization.

Steps involved in Disaster Recovery Plan

(a) ***Risk assessment:*** Identify the mandate and critical aspects of an organization, Prioritize critical services or products, Mitigating threats and risks, Analyze current recovery capabilities, Create continuity plans and Alternate facilities

(b) ***Data back-up and recovery procedures:*** Responsibility, Periodic backup, archival and retrieval Procedure.

(c) ***Implementation procedures:*** Effective implementation sand periodic assessment for implementation,

(d) **Plan Maintenance :** Recovery test for data management

(e) ***Disaster management plan:*** Procedure for Legal, Communications, Administration, Marketing and Sales, Human Resources and Technology management

(f) ***Post test review:*** Plan, schedule for effectiveness procedure implementation, Time line (Before and after task), logging of problem during the task and corrective and preventive action.

8.16 Approach to Managing Regulatory Risks and Achieving in GPL Compliance

Effective compliance management integrates risk management, controls and processes, and assurance and governance structures using tools and data. This should be backed by a strong organizational culture, an enterprise-wide awareness program, and business discipline. This holistic approach to the management of compliance will lead to greater awareness of regulation and help with the implementation of a focused plan to mitigate non-compliance and continuous process and not a one-time project

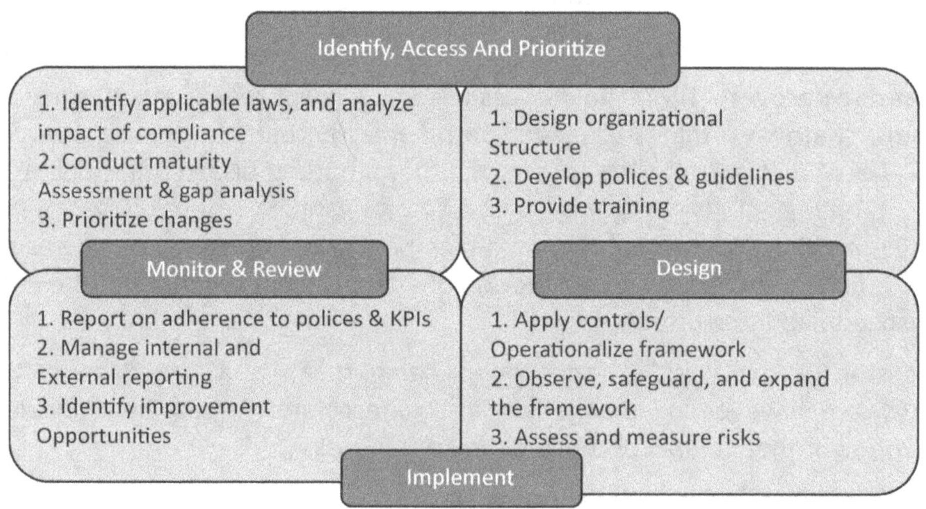

Chapter - 9

Control of Spreadsheets in GLP

Introduction

Spreadsheets have become commonly used within a wide range of applications within the pharmaceutical and biotechnology industries. Excel is an ideal tool for rapid application development and prototyping. Biggest advantage of spreadsheets is that anything can be configured, customized and a huge library of formulas is included to make the development process even easier for anyone. Control of spread sheet in GMP and GLP environment is very important throughout its life cycle.

9.1 Classification of Spreadsheet

Spreadsheet are classified based on the end-user(s) via a 'prototyping' methodology and GMP/GLP calculation.

(a) The single use, throw-away spreadsheet

(b) The calculator spreadsheet

(c) The Data spreadsheet

9.2 Control of the Spreadsheet

Spreadsheet which are used to make calculations that are used to support product release and quality based decisions then the spreadsheet must be controlled and verified to ensure that the spreadsheet is secure and that the calculations are accurate across the range. Spreadsheet cell should be protected where no variable, reading needs to be entered or where formula has been created for calculation of results.

Color coding or highlights should be done for locked cells. Lock all cells of excel calculation sheet, expect those needed by to user to input data. Master spread sheet calculation sheet are read only, with password protection and only authorized users can alter the excel calculations sheet. Spread sheet shall be designed in such a way that data outside acceptable condition are rejected.

Protect the worksheet and content of the locked cells by password. All excel should be version control and with unique chronological number. The spreadsheets should be maintained within a document management system or other secure location to ensure that the files can be access but cannot be modified or overwritten. Changes to the spreadsheets should be managed by Change Control.

The spreadsheet design standard should be more detail. Following approach cane consider.

(a) formatting

(b) calculations

(c) spreadsheet storage and distribution

(d) application of electronic records; electronic signatures

(e) version control

9.3   Validation of the Spreadsheet

FDA 21 CFR § 211.68(b) states: Input to and output from the computer or related system of formulas or other records or data shall be checked for accuracy. The degree and frequency of input/output verification shall be based on the complexity and reliability of the computer or related system.

A formal validation plan should where possible cover all spreadsheets in the company/department. This can be achieved by cross-referencing the spreadsheet inventory as a separate 'living' document. The production of separate validation plan for each spreadsheet should be avoided – this is simply duplicated work and additional documentation.

(a) Lock all cells of a spreadsheet, except those needed by the user to input data.

(b) Make spreadsheets read-only, with password protection, so that only authorized users can alter the spreadsheet.

(c) Design the spreadsheet so that data outside acceptable conditions is rejected (for example, reject non-numerical inputs).

(d) A qualification document is also produced once the spreadsheet design has been locked down

(e) Manually verify spreadsheet calculations by entering data at extreme values, as well as at expected values, to assess the ruggedness of the spreadsheet.

(f) Test the spreadsheet by entering nonsensical data (for example alphabetical inputs, <CTRL> sequences, etc.).

(g) Keep a permanent record of all cell formulas when the spreadsheet has been developed. Document all changes made to the spreadsheet and control using a system of version numbers with documentation.

(h) Periodically re-validate spreadsheets. This should include verification of cell formulas and a manual reverification of spreadsheet calculations.

(i) Validation of cell should be done by using both excel calculation sheet and a validated calculator

(j) Acceptable criteria for validation should be the results obtained using calculator should not differ from results reported using excel calculation sheet and the protected cells should not allow to enter any data.

A method of periodic evaluation is required. This can be run concurrent with normal use of the spreadsheet that at predefined intervals the calculations are independently verified.

9.4   Live use of Spreadsheet

Standard operating procedure should be available to incorporated qualified spread sheet qualification in day to day use:

(a) System administration including system

(b) Access/Authorisation

(c) Error logging/resolution

(d) Routine testing/Periodic Reviews/Re-qualification.

(e) Change Control/Configuration Management.

(f) Disaster Recovery/Business Continuity

(g) Records Retention

(h) Decommissioning/Data Migration procedures

9.5   Spreadsheet Inventory Management

Periodical inventory should be reviewed with the goal of minimising the workload and simplifying the inventory.

An inventory review should be performed with the following questions in mind:

(a) Are there any new requirements for existing spreadsheets?

(b) Can any spreadsheets be combined to produce a single template?

(c) Can spreadsheets be 'generalized' to allow future flexibility?

9.6 Data Storage and Security

Retaining data with a spreadsheet and being compliant with the 21 CFR Part 11 is a big challenge. The original purpose of a spreadsheet is not to retain data. Nevertheless it must be made sure that a backup of the spreadsheet exists and the integrity of the backup must be checked during the validation.

The backup functionality must be tested during the validation. The best way to test this during the validation is to install the spreadsheet, create a backup, delete the spreadsheet and restore it from the backup. Ideally the spreadsheet and the data printed out from it is archived in a secure place/environment, protected from uncontrolled change, destruction and usage. Any changes to the spreadsheet must follow a controlled change process.

9.7 Case Study for Master Spreadsheet Validation

Master Spread Sheet Validation		
Protocol Number and Version No :	**Effective Date :**	**Page No.**
Excel validation Performed for : XXXX Spread Sheet Number and Version No. :		
Done By: Sign and date	**Checked By: Sign and date**	**Approved by: Sign and date**
1. **Purpose:** The purpose of this document is to provide documentary evidence that a validated of a spread sheet calculation meets the defined criteria when is calculated manually and electronically. To calculate the average value of trend analysis or XXXX.		
2. **Scope:** To calculate the average value of trend analysis.		
3. **Acceptance criteria :** • The results obtained using calculator should not differ from results reported using excel calculation sheet. • The protected cells should not allow to enter any data.		
4. **Key Consideration :** (a) All tests shall be executed in accordance with the test instruction (b) The test specification with test results shall be reviewed and approved as a part of successful completion and approval of Master Spread Sheet.		
5. **Checklist for Master Spread Sheet :** Master Spreads Sheet No.: Test : Product : Specification no. : Formula :		

5.1 Table of Check point for Master Spread Sheet						
Sr. No	Test Instruction	Expected results	Actual Results	Pass/ Fail	Raw data	Tested By (Sign Date
1.	Check the excel sheet for entry of correct formulas and calculation, accepts those needed by the user to input data. Results of Electronic and Manual should be compare.	Correct formula to be entered in fields, other than input data user				
2.	Check the password protection for locked/calculated cells	Password protected cells are not editable.				
3.	Check Password protection for work sheet	Protect work sheet with password				
4.	Verify the spread sheet are read-only with password protection, such that only authorized users can alter the spread sheet	Alter option shall available only to authorized user and read only access shall available to users				
5.	Check the list of authorized users for altering the spread sheet.	Excel sheet allow only authorized user to alter				
6.	Manually verify spared sheet calculations by entering data at extreme values, as well as at expected values, to assess the ruggedness of the spread sheet.	Excel sheet allow the verification.				
7.	Design the spreadsheet, so that data outside acceptable conditions is rejected (i.e. Reject non-numerical inputs)	Data outside acceptable conditions is Rejected				
8.	Test the spreadsheet by entering nonsensical data (For example: Alphabetical inputs, <CTRL > sequence, etc)	Excel sheet does not allow to enter non sensical data.				
9.	Prove the compatibility of different version/ formats of excels	Verify different version/formats of excels.				

6. **Summary of Report :**

7. **Notification :**

8. **Change Control:**

9. **Over All Status (Pass/Fail):**

Reviewed By : Sign and Date :	Checked By : Sign and Date :	Approved By : Sign and Date :

Chapter - 10

Analytical Instrument and Equipment Qualification

Introduction

Analytical instrument and equipment in GLP should be well – designed, qualified, calibrated, maintained and regularly checked to ensure compliance with predetermined specifications and its suitability during on-going use. GLP regulations are intended to assure that the data submitted to the regulators are of high quality, integrity, and reliability. In order to fulfil this requirement, documentation and associated records are used to show that controlled conditions are maintained and procedures are followed. Each GLP laboratory should have a comprehensive, up-to-date list of all equipment used in the facility. This list should include both active and inactive equipment. The list of equipment should be version controlled and approved by quality assurance unit. The equipment list should include Asset number (or unique identifier) or serial number, Basic description, Model number, Location Department owner, Notation as to whether maintenance and calibration or standardization is required, Frequency of maintenance or calibration, Responsible site person for the equipment, Notation as to whether the equipment is maintained or calibrated by internal or external personnel and Equipment status (i.e., active or inactive).

If equipment is on loan from another department, facility, or customer, the equipment should be added to the equipment list. If the equipment is loaned to another department, facility, or customer, the equipment list should be revised to accurately reflect its temporary location.

10.1   GLP Qualification and Validation

The term Validation is used for manufacturing process, analytical procedure and software procedures and the term qualification is used for instruments. The phrase "Analytical Instrument qualification" (AIQ) is used for the process of ensuring that an instrument is suitable for its intended application in GLP environment.

Validation is a key principle and defined as a documented program that provides a high degree of assurance that a specific process, method, or system will consistently produce a result meeting predetermined acceptance criteria. All equipment used to create, revise, store, retrieve, or distribute GLP data must be qualified. This includes computerized equipment. Typically, computerized equipment is validated as a "system" that includes both the hardware as well as the software.

10.2   Components of Data Quality in GLP

Equipment qualification and validation of computerized systems cover the entire life of a product. It starts when somebody has a need for a specific product, and ends when the equipment is retired. Because of the length of time and complexity, the process has been broken down into shorter phases, so-called life cycle phases. Several life cycle models have been described for qualification and validation. The 4Q model, which is widely, used in pharmaceutical laboratories.

A. Analytical Instrument Qualification (AIQ)

AIQ is the collection of documented evidence that an instrument performs suitably for its intended purpose. Use of a qualified instrument in analyses contributes to confidence in the validity of generated data.

B. Analytical Method Validation (AMV)

Analytical method validation is the collection of documented evidence that an analytical procedure is suitable for its intended use. Use of a validated procedure with qualified analytical instruments provides confidence that the procedure will generate test data of acceptable quality.

C. System Suitability Test (SST)

System suitability tests verify that the system will performed in accordance with the criteria set forth in the procedure. These tests are performed along with the sample analyses to ensure that the systems performance is acceptable.

D. Quality Control Check Samples (QCCS)

Many analysts carry out their tests on instruments standards. Some analyses also require the inclusion of quality control check samples to provide an in-process or ongoing assurance of tests suitable performance. In this manner AIQ and analytical method validation contribute to the quality of analysis before analyst conduct the test. System suitability test and quality control checks help ensure the quality of analytical results immediately before or during sample analysis.

10.3   Analytical Instrument and Equipment Design

It is the equipment qualification that documents the ability of the equipment to meet its intended Equipment used in the generation of measurement, or assessment of data and equipment used for facility environmental control, shall be of appropriate design and adequate capacity to function according to the protocol and shall be suitably located for operation, inspection, cleaning, and maintenance. After establishing the user requirements, the functional requirements should be identified. Using the established functional requirements, the equipment qualification can then be performed use and capacity to function as required by the protocol.

10.4   Analytical Instrument and Equipment Qualification Phase

New systems should pass through all stages of qualification (V model of Qualification and flow Chart) including the preparation of user requirement specifications (URS), design qualification (DQ), installation qualification (IQ), operational qualification (OQ) and performance qualification (PQ) .Instrument qualification is not a signal continuous process, but instead results from several discrete activities. Some AIQ activities cover more than one qualification phase and analysts potentially could perform them during more than one of the phase as per below Timing, Applicability and Activity for each Phase of Analytical Instrument table.

Timing, Applicability and Activity for each Phase of Analytical Instrument.					
Design Qualification	Installation Qualification		Operation Qualification		Performance Qualification
Timing and Applicability					
Prior to purchase of a new model of instrument	At installation of each instrument (new, old, or existing unqualified)		After installation or major repair of each instrument		Periodically at specific interval for each instrument
Activities					
Assurance of manufacturers DQ	Description	↔	Fixed parameters		Preventive maintenances and change control

Table Contd...

Timing, Applicability and Activity for each Phase of Analytical Instrument.						
Design Qualification	**Installation Qualification**		**Operation Qualification**		**Performance Qualification**	
Assurance of adequate support availability from manufacture	Instrument delivery				Establish practices to address operation, calibration, maintenance and change control	
Instruments fitness for use in laboratory	Utilities/Facility	↔	Environment			
	Assembly and installation					
	Network and data storage	↔	Secure data storage, back up and archive			
	Installation verification	↔	Instrument function tests	↔	Performance check	

10.5   Steps involved in Qualification

The entire qualification process is broken down into four qualification phases: design qualification (DQ), installation qualification (IQ), operational qualification and performance qualification (PQ). The whole process for a specific project is outlined in the qualification plan and the results are summarized in a qualification report Qualification should be done in accordance with predetermined and approved qualification protocols. The results of the qualification should be recorded and reflected in qualification reports. Equipment should qualified and brought into routine use only once there is documented evidence that it is fit for its intended purpose. The relevant documentation associated with qualification including standard operating procedures (SOPs), specifications and acceptance criteria, certificates and manuals should be maintained and be traceable. Systems, utilities and equipment should be maintained in a qualified state and undergo periodic requalification where appropriate, as well as requalification after change when needed. Processes should be validated on qualified equipment as per **V**-Model for the Analytical Instrument.

V - Model for Direct Impact System

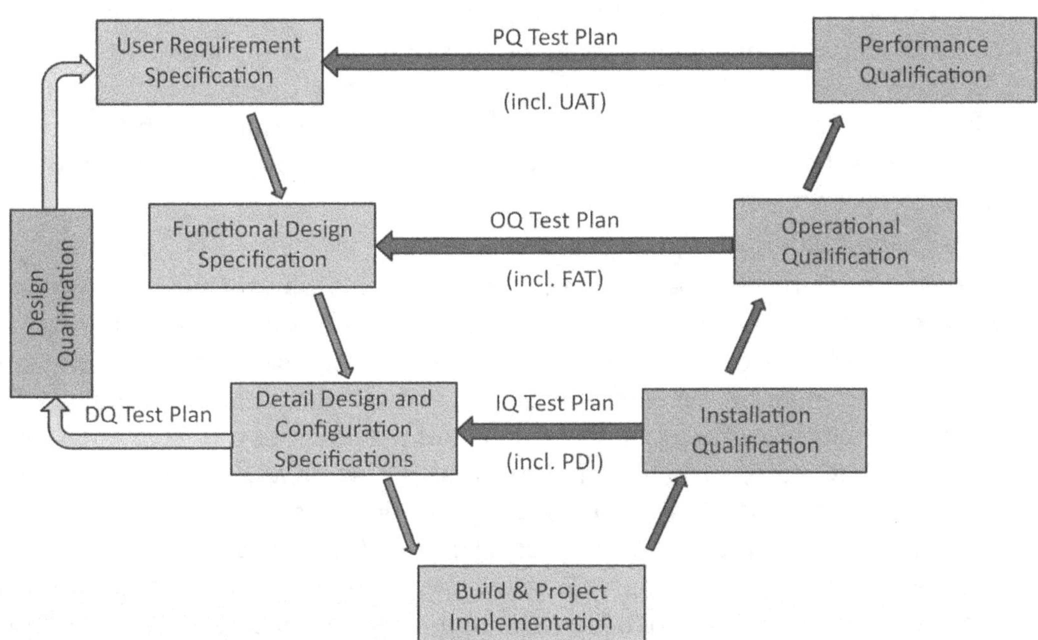

A. User Requirement Specification(URS)

URS should be used to verify, at a later stage, that the purchased and supplied system, utility or equipment is in accordance with the manufacturer's needs as specified.

B. Factory Acceptance Test (FAT) and Site Acceptance Test(SAT)

Where a system, utility or equipment is assembled, or partially assembled at a site other than that of the purchaser or end-user, testing and verification should be done to ensure that the checks and tests during assembly or partial assemble, should be recorded. System, utility or equipment is appropriate and ready for dispatch.

The acceptability of the assembly and overall status of the system, utility or equipment should be described in a conclusion of the report for the factory acceptance test (FAT), prior to shipment. When the system, utility or equipment is received at the end user, tests should be performed to verify the acceptability of the system, utility or equipment.

The results of the tests should be recorded and the outcome of the acceptability of the system, utility or equipment recorded in the conclusion of the report for the site acceptance test (SAT).

C.   Design Qualification(DQ)

In the DQ phase the user writes requirement specifications for the equipment. This includes all functions the instrument should have and the performance specifications the equipment should meet as required for the intended application. .Design qualification (DQ) is the documented collection of activities that define the functional and operational specifications of the instrument and criteria for selection of the vendor, based on the intended purpose of the instrument. Once DQ is completed, IQ may commence.

D.   Installation Qualification(IQ)

During the IQ phase the shipment is compared with the purchase order for completeness, and the vendor's installation instructions are executed. Installation qualification (IQ) is the documented collection of activities necessary to establish that an instrument is delivered as designed and specified, and is properly installed in the selected environment, and that this environment is suit able for the instrument. IQ applies to an instrument that is new or was pre-owned, or to any instrument that exists on site but has not been previously qualified. Relevant parts of IQ would also apply to a qualified instrument that has been transported to another location or is being reinstalled for other reasons, such as prolonged storage. The activities and documentation typically associated with IQ are as follows

- Description
- Instrument Delivery
- Utilities/Facility/Environment
- Assembly and Installation
- Network and Data Storage
- Installation Verification

IQ should include identification and verification of all system elements, parts, services, controls, gauges and other components. Measuring, control and indicating devices being installed should be calibrated. The calibration should be traceable to relevant national or international standards. Certificates should be available. The execution of the protocol should be recorded in the IQ report. The report should include at least the title, objective, site, details of the supplier and manufacturer, system or equipment name and unique identification number, model and serial number, date of installation, components and their identification numbers or codes, actual results of tests and measurements, spare parts list, relevant procedures followed for tests and certificates as applicable. All deviations and non-conformances from URS, DQ and acceptance criteria specified, observed during installation, should be recorded and investigated. The outcome of the IQ should be recorded in the conclusion of the report, before OQ is started.

E.   Operation Qualification (OQ)

At the end of the installation process, the IQ protocol is completed by recording the vendor, model, serial number and other relevant information. After a successful IQ, the instrument is ready for OQ testing Operational qualification (OQ) is the documented collection of activities necessary to demonstrate that an instrument will function according to its operational specification in the selected environment. Testing activities in the OQ in phase may consist of these test parameters.

- Fixed Parameters
- Secure Data Storage, Backup, and Archiving
- Instrument Function Tests

F.   Performance Qualification (PQ)

Performance qualification (PQ) is the documented collection of activities necessary to demonstrate that an instrument consistently performs according to the specifications compile defined by the user, and is appropriate for the intended use. The PQ may include the following parameter

- Performance Check
- Test for specified application
- Preventive maintenance
- On-going performance tests
- Calibration

10.6   Requalification Frequency

Requalification frequency depends on the instrument itself, the recommendations from equipment manufacturers, laboratory experience and the extent of use. Typically, a complete spectrometry system OQ performance test should be done every 6 to12 months.

10.7   Instrument and Equipment Operating Ranges

The system should be operate in qualified and calibration range. The relationships between design conditions, operating range and qualified acceptance criteria are given in below

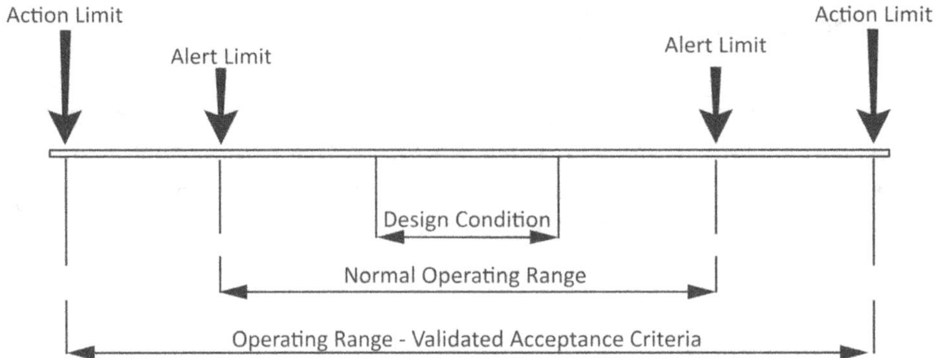

10.8   Categories of Analytical Instrument

On basis of need and user requirement specification analytical instrument are categories into three groups: Group A, Group B and Group C. The exact grouping of the instrument must be determined by users for their specific requirement. Depending on individual user requirements, the same instrument may appropriately fall into one group for one user and another group for another group user. Therefore, a careful selection of groups by users is highly encouraged.

I.   **Group A Instrument categorization:**

Group A includes standard equipment with no measurement capability or usual requirement for calibration, where the manufacturer's specification of basis functionality is accepted as user requirements. Conformance of Group A equipment with user requirement may be verified and documented through visual observation of its operation. Examples of equipment into this group are nitrogen evaporators, magnetic stirrer, vortex mixture and centrifuge.

II.   **Group B Instrument categorization:**

Group B includes standard equipment and instruments providing measured values as well as equipment controlling physical parameter (such temperature, pressure or flow) that needs calibration, where the user requirements are typically the same as the manufacturer's specification of functionality and operation limits. Conformance of Groups B instrument or equipment to user requirements is determined according to the standard operating procedure for the instrument or equipment and documented during Installation Qualification and Operation qualification. Example of instrument into this group are balance, variable pipets, refract meters, titrators and viscometers. Example of the equipment in this group are muffle furnace, ovens, refrigerator- freezers, water baths pumps and dilutors.

III.   **Group C Instrument categorization:**

Group C includes instrument and computerized analytical systems, where user requirements for functionality, operational and performance limits are specific for analytical application. Conformance of Group C instruments to user requirements is determined by specific function tests and performance test .Installing these instruments can be a complicated undertaking and may require the assistance of specialists. A full qualification process, as outlined in this documents, should apply to these instrument.

Example of Group C instruments in this group include the following.		
Atomic Absorption Spectrometers	Differential Scanning Calorimeters	X-ray Fluorescence Spectrometers
Electron Microscopes	Flame Absorption Spectrometers	High-Pressure Liquid Chromatographs
Mass Spectrometers	Thermal Gravimetric Analyzers	Micro plate readers
Dissolution Apparatus	X-ray Powder Diffract meters	Densitometer
Diode-Array Detectors	Elemental Analyzers	gas chromatographs
IR spectrometers	Near-IR spectrometers	Raman spectrometers
UV/Vis Spectrometers	Conductively coupled plasma-emission spectrometers	

10.9   Analytical Instrument and Equipment Standard Operating Procedure

The written standard operating procedure required ...shall set forth in sufficient detail the methods, materials, and schedules to be used in the routine inspection, cleaning, maintenance, testing, calibration, and or standardization of equipment and shall specify, when appropriate, remedial action to be taken in the event of failure or malfunction of equipment. The written standard operating procedure shall designate the person responsible for the performance of each operation.

Analytical instrument and equipment procedures should include the frequency for maintenance and calibration in-house and outside service organization. Procedure should include frequency of calibration and maintenance, Equipment and standards required, Acceptance criteria, Remedial action when if equipment is found out of calibration, Routine inspection, cleaning, maintenance, testing, calibration, and standardization and Health and safety precautions.

10.10 Calibration Management

Equipment shall be adequately inspected, cleaned, and maintained. Equipment used for the generation, measurement, or assessment of data shall be adequately tested, calibrated, and or standardized.

GLP laboratory must establish a maintenance and calibration program for equipment and instrument. The frequency of maintenance and calibration for each piece of equipment and maintenance must be specified in written procedures. The frequency of maintenance and calibration should be based on the equipment manufacturer's recommendation, the extent and frequency of equipment use, type of usage, potential impact from the environment or use, and the overall risk or potential impact of equipment malfunction.

Calibrated equipment must be tagged with a "date calibrated" and "next calibration due date", so that the calibration status is immediately recognizable along with the name of the individual or organization responsible for the calibration. If the equipment is out of service or out of calibration, it must also be tagged to clearly alert laboratory operators.

10.11 Planned Preventive Maintenance Program

Analytical instruments should be well-maintained to ensure proper on-going performance. Procedures should be in place for regular preventive maintenance of hardware to detect and fix problems before they can have a negative impact on analytical data. The procedure should describe:

- When maintenance should be done.
- How it should be done.
- What should be requalified after maintenance is done.
- For example, a PQ test should always be performed after instrument maintenance.
- How to document maintenance activities.

Planned maintenance activities should follow a documented instrument maintenance plan. Some vendor soffer maintenance contracts with services for preventive maintenance at scheduled time intervals. A set of diagnostic procedures is performed and critical parts are replaced to ensure on-going reliable system uptime.

Unplanned activities that are necessary in addition to the planned activities should be formally requested by the user of the instrument or by the person who is responsible for the instrument. The reason for the requested maintenance should be entered, as well as priority. All maintenance activities should be documented in the instrument's log book.

10.12 Usage Log of Analytical Instrument and Equipment

Usage log should be maintained for all Analytical instrument and equipment. All laboratory activities conducted on analytical instrument and equipment such as testing, operation, maintenance, calibration, shall be recorded on real time. There should be an equipment record assembled for each piece of major equipment.

10.13 Handling Defective and Nonqualified Instruments and Equipment

Defective and nonqualified instruments should be either removed from the laboratory area or when this is inconvenient, for example for large or permanently-installed instrument systems, clearly labelled as being defective or not qualified. Procedures should be available for most common problems. Such procedures should also include information if and what type of requalification is required after the repair.

10.14 Analytical Instrument and Equipment Audits

Audits should be conducted periodically to ensure : Equipment is included in the maintenance and calibration program, Non-functioning equipment is tagged as "Out of Service", Equipment is within its calibration interval, Equipment is labelled with it last calibration date and date due for calibration, Calibration records are readily available, Calibration records demonstrate that the equipment meets its specifications , SOPs are available for key equipment , SOPs or manuals are readily available in the area where the equipment is used, Laboratory personnel are trained on the equipment ,Equipment has been qualified prior to use and following any modification or major repair ,Operating ranges and limits of accuracy are readily available to users.

10.15 Definition

(i) *Design qualification:* Documented evidence that the premises, supporting systems, utilities, equipment and processes have been designed in accordance with the requirements of good manufacturing practices.

(ii) *Factory acceptance test:* A test conducted at the vendor's premises to verify that the system, equipment or utility, as assembled or partially assembled, meets expected specifications.

(iii) *Installation qualification:* The performance of tests to ensure that the installations used in a manufacturing process are appropriately selected and correctly installed and operate in accordance with established specifications.

(iv) *Operational Qualification:* Documented verification that the system or subsystem performs as intended over all anticipated operating ranges.

(v) *Performance Qualification:* Documented verification that the equipment or system operates consistently and gives reproducibility within defined specifications and parameters for prolonged periods. (In the context of systems, the term "process validation" may also be used.)

(vi) *Site acceptance test:* A test conducted at the site of use to verify that the system, equipment or utility, as assembled or partially assembled meets expected specifications.

(vii) ***System****:* A regulated pattern of interacting activities and techniques that are united to form an organized whole.

(viii) ***User Requirement Specifications****:* An authorized document that defines the requirements for use of the system, equipment or utility in its intended production environment.

(ix) ***Utility****:* A system consisting of one or more components to form a structure designed to collectively operate, function or perform and provide a service such as electricity, water, ventilation or other.

Chapter - 11

Electronic and Paper-Based Data Management in GLP

Introduction

In recent years there has been a significant increase in the number and types of data integrity issues that have been cited in regulatory inspections. Regulatory focus on the integrity of electronic and paper-based data has increased sharply. Data integrity issues have been increasing to become one of the most important GMP and GLP issues. This increase has led to red flags and triggering a more intensive investigation. Data integrity and Data governance system is fundamental in a pharmaceutical quality system which ensures that medicines are of the required quality. Systems should be designed in a way that encourages compliance with the Principles of data integrity.

Spectrometric instruments for elemental impurity analysis are connected to computer systems with specific application software that controls instrument parameters, acquires signal and spectral data, converts the original digital data into meaningful test results (such as concentrations), and finally prints results and stores and archives instrument and method parameters, original digital data and processed data for the required retention period. The objective of all these documents is to make sure that electronic record and signatures are as trustworthy and as reliable as paper records and handwritten signatures, and that the use of computer systems does not adversely impact the product quality and quality assurance when compared to manual systems.

"Confidence in the quality and the data generated integrity is a fundamental of Data reliability".

Before starting discussion, terms need to be understand such as:

- **Data:** Facts, figures and statistics collected together for reference or analysis (**ALCOA**)
- **Data Governance:** The sum total of arrangements to ensure that data, irrespective of the format in which it is generated, is recorded, processed, retained and used to ensure a complete, consistent and accurate record throughout the data lifecycle.
- **Data Integrity:** The extent to which all data are complete, consistent and accurate throughout the Data lifecycle.
- **Data Lifecycle:** All phases in the life of the data (including raw data) from initial generation and recording through processing (including transformation or migration), use, data retention, archive/retrieval and destruction

11.1 ALCOA

Data to be **A**ttributable, **L**egible, **C**ontemporaneous, **O**riginal and **A**ccurate.

A	Attributable	Who acquired the data or performed an action and when?
L		
C	Legible	Can you read the data?
O	Contemporaneous	Documented at the time of the activity
A	Original/Reliable	Written printout or observation or a certified copy thereof
	Accurate	No errors or editing without documented amendments

11.2 Data Integrity GLP Regulatory Expectations

Data integrity applies to all pharmaceutical quality management system and apply equally to manual (paper) and electronic data. Data integrity applies to all an organization with respect to people, systems and facilities to ensure data is complete consistent and accurate in all its forms, i.e. paper and electronic. Periodic data verification should done and documented for its correctness, accurate legibility and exiting controls are working fine. Regulators expect industry should balance data risk with other quality and compliance priorities. Risk to data integrity should communicate the senior management and action should be taken on the prioritization .Short term and long term action plan with measure shall be traced and review periodically for its effectiveness and completion. Procedure and system controls should be available to review and control the data integrity. Periodic personnel training plays an important role in data integrity.

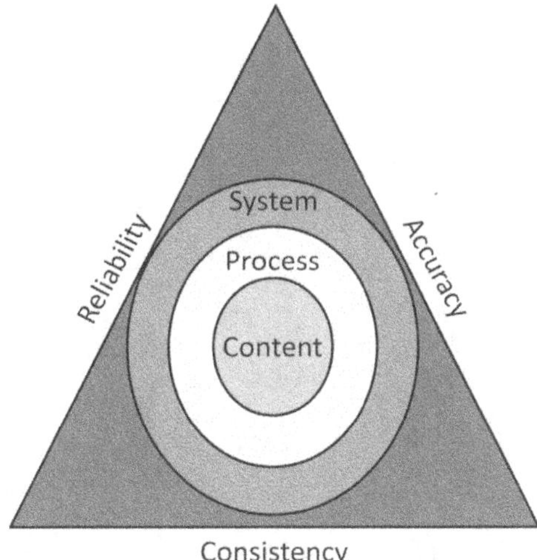

11.3 When does Electronic Data become a cGMP and GLP Record?

When generated to satisfy a CGMP and GLP requirement, all data become a CGMP record which cannot be modified and static or dynamic nature of the original records. Static is used to indicate a fixed-data document such as a paper record or an electronic image, and dynamic means that the record format allows interaction between the user and the record content.

Data integrity requirements applicable to:

- API and FP manufacturers, including contract manufacturing
- Testing units, including contract laboratories
- Outsourced GMP activities such as equipment qualification and calibration

11.4 Confidentiality, Integrity and Availability Module (CIA)

CIA triangle or **AIC** triangle (availability, integrity and Confidentiality) is a model designed to guide policies for information security within an Organization. The elements of the triangle are considered the three most crucial components of data security: **confidentiality** - set of rules that limits access to information, **integrity** - assurance that the information is trustworthy and accurate, and **Availability**-A guarantee of reliable access to the information by authorized people.

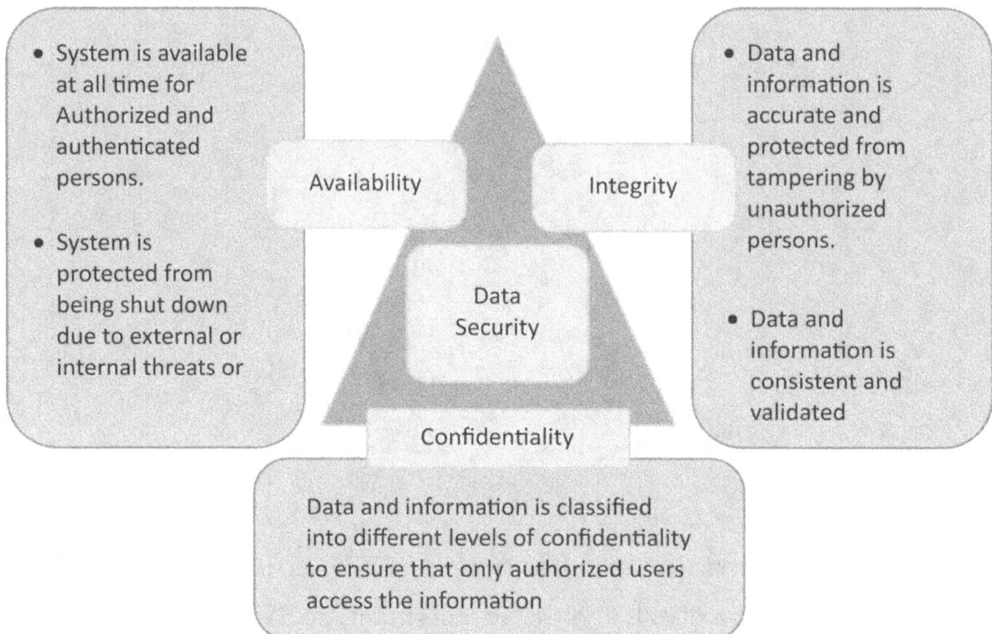

11.5 GLP Data Governance System

The data governance system should be integral to the pharmaceutical quality system and shall be controlled organizational (procedural) and technically (electronically).The data information may be qualitative or quantitative or numerical values or Text or images or drawings or audio or video records or it can come from a variety of sources. The key is that the data must be responsibly managed and secure. Data management encompasses several different tasks such as **6 D concept module system:** Data selection, Data analysis, Data handling, Data reporting, Data Issuance and **Data ownership.**

Two other important aspects of data governance system include the validated state of a process or a computerized system (ensuring accuracy of generated or recorded data), and the management of critical authorizations (protection of data to avoid integrity breaches during operation).

Risk management approach can be applied based on Data criticality and its risk.

11.6 Designing GLP Systems to Assure Data Quality and Integrity

Systems should be designed in a way that encourages compliance with the **Principles of data integrity**. Examples include:

- Access to clocks for recording timed events
- Accessibility of batch records at locations where activities take place
- Control over blank paper templates for data recording ,

- User access rights which prevent (or audit trail) data
- Automated data capture or printers attached to equipment such as balances
- Access to sampling points
- Access to raw data for staff performing data checking activities.

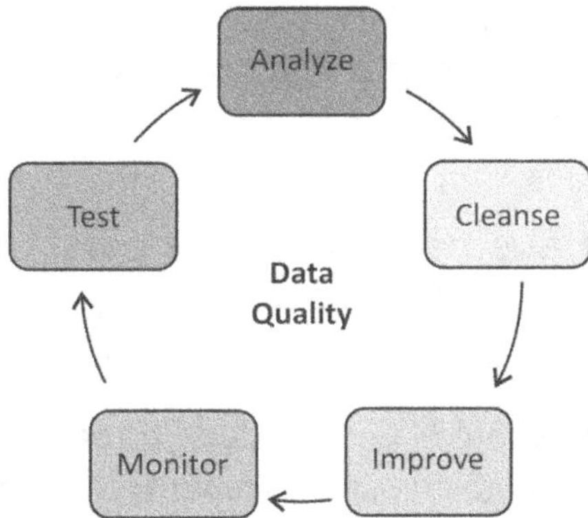

11.7 GLP Triggers of Data Integrity Loss

Triggers of data integrity loss involve attacks on data integrity which can be intentional and non-intentional, during the data life cycle. Triggers of data integrity loss may due to Changes to access permissions and privileges, Inability to track the use of privileged passwords, passwords shared, End-user error that impacts production data, Vulnerable code-in applications, Weak or immature change control and accreditation processes, Mis configuration of security devices and software, Incorrectly or incompletely applied patches, Unauthorized devices and Unauthorized applications connected to the corporate network, Inadequate or not applied segregation of duties, Processing and/or deriving data Deleting, removing and destroying data, replicating and distributing data, Archiving and recalling data, Backing up and restoring data

11.8 Data Integrity Metrics in GLP

Data integrity matrix framework is based on three complementary approaches.

- First: looks at specific known and uses attacks to analyze their probability,
- Second: addresses this limitation by applying cryptographic "provable security" results
- Third approach building information system components by establishing, implementing, measuring, and maintaining and review data integrity across an organization

The message for data integrity is clear. It's not a new concept; it's about getting back to the roots of training all staff on the importance of data integrity in cGMP documentation and honesty. It is critical to ensure employees understand the accountability and traceability requirements for retention of raw data and the consequences of data manipulation. Training operators and analysts to document the performance of a task by recording what happened at the time it occurs, including information about the person who performed it along with clearly documenting and investigating deviations, is vital for patient safety and product efficacy. This is achievable by providing the training and creating a company culture that promotes and rewards ethical behaviour as a core value from top to bottom of an organization.

11.9 Electronic Data Management in GLP as per USFDA Part 11 and EU GMP Annex 11

A.   Electronic Data Management Risk assessment

Risk management should be applied throughout the life cycle of the computerized system, taking into account patient safety, data integrity and product quality. As part of a risk management system, decisions on the extent of validation and data integrity controls should be based on a justified and documented risk assessment of the computerized system.

Key steps to follow in Electronic Data Management Risk assessment:

(a)   Determine the risk level of the computerized system: high, medium, or low. Criteria are: impact of the system on data integrity, (medicinal) product quality and patient safety.

(b)   Apply the appropriate type and extent of compliance activities according to the defined category, for example for the extent of validation and the frequency of revalidation, whether electronic audit trail should be implemented or not and the frequency of backup.

B.   Computer System validation

Computer systems used to generate, maintain and archive electronic records should be validated to ensure accuracy and reliability of the records.

Key steps to follow in Computer System validation:

(a)   Follow the life cycle model for validation.

(b)   Apply the concept of risk-based validation.

(c)   Documentation that should be generated and available during inspections.

C.   GLP Data Accuracy

Systems exchanging data electronically with other systems should include appropriate built-in checks for the correct and secure entry and processing of data, in order to minimize the risks. For critical data entered manually, there should be a check on the accuracy of the data. This check may be done by a second operator or by validated electronic means.

Key steps to follow in GLP Data Accuracy:

    (a) Validate the accuracy of data transfer between computer systems. This can be done during operational or performance qualification.

    (b) Use software functionality, if available, to check the plausibility and accuracy of manual data entries.

    (c) If there is no software functionality to validate manual data entries, for high risk data verify the accuracy through a second person.

D.   System limited access to authorized users and authority checks

Procedures and technical controls should be in place to limit access to systems and data to authorized individuals. Suitable methods of preventing unauthorized entry to the system include the use of keys, pass cards, personal codes with passwords, biometrics, and restricted access to computer equipment and data storage areas. The system should conduct authority checks to ensure that only authorized individuals can use the system, electronically sign a record, alter a record or perform the operation. Systems should be designed to record the unique identity of operators entering, changing, confirming or deleting data including date and time.

Key steps to follow in System limited access to authorized users and authority checks:

    (a) Develop policies for generation, distribution, use and maintenance of passwords.

    (b) Develop procedures for limited access to the system to individuals, for example through user ID and password.

    (c) Each employee should have his/her unique user ID and should select his/her own unique password.

    (d) Develop procedures for limiting access to the system operational functions and data.

    (e) Make sure that the installed software can be used to implement your company's password policies and procedure.

    (f) Configure the system to implement company's password policy.

    (g) Configure the system to implement your company's procedure for limited access to systems and data.

    (h) Verify correct implementation of password policy and procedures for limiting access to the system and data.

    (i) Establish a list of authorized users.

E.   GLP raw data and copies of records

Many documents exist in hybrid forms, that is, some elements as electronic and others as paper-based. Regulated users should define which data are to be used as raw data. At least, all data on which quality decisions are based should be defined as raw data. When copies of records are made, the copies should be accurate and complete or should provide the content and meaning of the original record.

Key steps to follow in GLP raw data and copies of records

For each application, define what raw data are. For example, original electronic records, intermediate processed data or computer printouts.

(a)   For spectrometric systems, define original electronic records as raw data.

(b)   Keep electronic raw data for review and copying by the agency after printing results.

(c)   Contact the agency if there are any questions regarding the ability of the agency to perform this review and copying of electronic records.

F.   GLP Protection of records

Records should be protected to enable their accurate and ready retrieval throughout the required records retention period. Data should be secured by both physical and electronic means against loss or damage. Stored data should be checked for accessibility, readability and accuracy. Access to data should be ensured throughout the retention period. Regular backups of all relevant data should be done. Integrity and accuracy of backup data and the ability to restore the data should be checked during validation, and monitored periodically.

Key steps to follow in GLP Protection of records:

(a)   Follow your company's procedure for retention of electronic records.

(b)   Check the required retention period for your records according to requirements as specified in relevant GxP regulations.

(c)   Migrate the electronic records when the systems are upgraded or replaced by new ones. Ask the vendor for validated file conversion routines.

(d)   Regularly check availability and integrity of the electronic records during the entire archiving period.

(e)   Make a backup of electronic records.

(f)   Validate backup and restore procedures.

G.   Computer-generated time-stamped audit trails

Secure, computer-generated, time-stamped audit trails should be used to independently record the date and time of operator entries and actions that create, modify or delete critical electronic records. Record changes should not obscure previously recorded information, that is, the modified record should be saved as a new version of the record. The audit trail documentation should be retained for a period at least as long as that required for the subject electronic records, and should be available for agency review and copying. The audit trail documentation should be regularly reviewed. For records supporting batch release, it should be possible to generate printouts indicating if any of the data has been changed since the original entry.

Key steps to follow in Computer-generated time-stamped audit trails:

(a) Include electronic audit trail in the user requirement specification for computerized spectrometric systems.

(b) Specify the requirements for the audit trail to include: what was changed, who made the change, when the change was made by date and time and as an option, the reason for the change.

(c) Make sure that audit trail documentation is available in human-readable form.

(d) Make sure that changed records do not replace or obscure original records.

(e) Regularly review electronic audit trail documentation.

(f) Verify correct functioning of the electronic audit trail.

(g) Retain the audit trail documentation for as long as the required retention period for subject records.

(h) Use software that can recognize changed records based on printouts. If such software is not available, a manual procedure.

H.   Electronic signatures

Part 11 and Annex 11 allow the signing of records electronically. Part 11 has more specific requirements for the execution of electronic signatures. Information associated with the electronic signature should include:

(a)   The printed name of the signer.

(b)   The date and time when the signature was executed.

(c)   The meaning of the signature. For example, review, approval, responsibility and ownership.

(d)   Handwritten and electronic signatures should be permanently linked to their retrospective electronic records.

(e)   Decide and document the decision whether to sign records electronically or through handwriting.

(f)   If electronic signatures are to be used, include the required software functions in the system's user requirement specifications document.

(g) If electronic signatures are to be used, send a letter to the FDA that your company will be using electronic signatures (only applies to US FDA regulated industries).

(h) Train personnel on the meaning and accountability of electronic signatures.

(i) Configure the system for use with electronic signatures.

(j) Validate correct functioning of electronic signatures.

I. Periodic evaluation of GLP Data

Computerized systems should be periodically evaluated to confirm that they remain in valid **Key steps to follow in Periodic evaluation of GLP Data**

Develop and implement procedures to periodically review computerized systems.

(a) Set the time frequency to twice per year.

(b) Include in the evaluation the current range or functionality, deviation records, incidents, problems, upgrade history, reliability, security and validation reports, especially the results of regular performance qualifications.

Laboratory Training and Qualification Management Program

Introduction

Training & Development is essential to remove performance deficiencies by Firstly, they improve the skills of the people for specific job requirement and secondly they add to the job satisfaction. Systematic and effective training management helps to ensure that the all personnel in the company are trained on their duties or responsibilities according to approved standard operating procedures. When employees know/trained how to perform their duties or responsibilities, the results are higher quality products and can reduce the risk of nonconformance. So effective employee training is essential part for their organization's success.

"Training is an improving tool to understand and implement the approved procedure and reduce the non-conformance "

Training is a continuous learning process that involves the acquisition of knowledge, sharpening of skills, concepts and rules. Under a quality system, continued training is critical to ensure that the employees remain proficient in their operational functions and in their understanding of CGMP regulations.

12.1 Training Program Management

Initial steps of the training program is identification of training which requires brain storming is to produce multiple or multitude ideas for identification of training needs.

Training program should address the regulatory requirements, policies, processes, procedures, and written instructions related to operational activities, the product/service, the quality system, the desired work culture (e.g., team building, communication, change, behavior), employees' specific job functions, safety, Health and Occupation ,clothing and hygiene and the related CGMP regulatory requirements.

Personnel working in areas where contamination is a hazard, e.g. clean areas or areas where highly active, toxic, infectious or sensitizing materials are handled, should be given specific training.

Visitors or untrained personnel should, preferably, not be taken into the production and quality control areas. If this is unavoidable, they should be given information in advance, particularly about personal hygiene and the prescribed protective clothing. They should be closely supervised.

12.2 Star Bugs of Training Program Management

Training is an important element of cGMP and GLP which build up the integral part of the laboratory management. It is important or essential that people with right competency are brought together to complete the assigned task. Training programs should be prioritization according to need.

The most important aspect in training progrman is whether the trainees learned the required skill sets through the training for which it was intended. If trainees have failed to learn the required skill sets and knowledge, then it can be inferred that the training system has failed rather than the trainees. Thus the modification in the training system is required.

Three important element of training are accurate identification of training needs; accurate selection of participants and appropriate course content.

Star bugs involved in training management program

    A. Identification and Evaluation of training needs
    B. Identification of Trainer.
    C. Training Program Material ,Assets, Methods and Tools
    D. Training Program Management Plan:
        (a) Induction Training
        (b) On-Job Training
        (c) On-Going Training
        (d) Continuous process
        (e) Retraining or Refresher Training
        (f) Technical Training (Skill Based, Leadership, Software, Personnel Development etc.)
    E. Evaluation of effectiveness of training
    F. Documentation of training and/or re-training

12.3 Identification and Evaluation of Training Needs

The first step is to determining its training needs to understand who needs what training, at an individual or as well as group level. Sending employees to unnecessary external trainings or conducting useless internal trainings not only wastes monetary resources but also wastes time for work processes. Training programs should arranged in such a way that employees have the opportunity to apply it on the job, whether through new work methods or adapting new organizational/departmental strategies. Evaluation of training helps to know who needs what training, so in the future, where to invest more can be asset. Successful identification and Evaluation of training needs, helps the organization to compare productivity and profits before and after training, in their own analytic and strategic way.

12.4 Identification of Trainer

A trainer plays a critical roles in training management program, which act as a facilitator, as a role model, as a motivator, and as a leader. Trainers must have the necessary experience, knowledge of the subject or must be subject matter expert, understanding of the sector in which they are providing training , sincere enthusiastic, Has a sense of humor, Has an understanding of the audience, Is willing to involve the group Gives clear instructions, Ensures practicality of subject matter, Offers individual assistance and patient.

Trainer's main objective is to develop or modify observable skills, attitudes, and behaviors of employees on the job. Trainers should have a structured approach to their tasks, be able effectively to transfer knowledge, and to ensure that their trainees—the members of the training audience—acquire the needed skills to demonstrate competency on the job as well as compliance with regulations

A. **Trained the Trainer Programme**

Trained the Trainer program should be executed periodically to well-constructed and well-executed training program mainly in three areas: knowledge and information, attitudes, and skills and behaviors of the training audience.

Trainer should be certified by internal or external means and list of authorized trainer should be maintained.

B. **Training Consultants**

Consultants should have adequate education, training, and experience, or any combination thereof, to - advise on the subject for which they are retained. Records should be maintained stating the name, address, qualifications, and type of service provided by these consultants.

12.5 Training Program Material, Assets, Methods and Tools

Training program is the activity of combination of the trainer(s), trainee(s) with various skill sets and dispositions, training materials and assessment materials, training organization , facilities (allocated space, allotted time, utilities), and auxiliary materials (instruments and equipment, raw and in-process materials used in the training), etc. Training incorporates several delivery modalities, such as e-learning, mentoring, audio-video, Visual, Presentation, Self-learning, and classroom delivery

12.6 Training Program Management Plan

A. Induction Training

Training give to a new recruit, which is carried out at the time of starting of the employment to understand the organization systems, policy, code of ethics, Data integrity procedure and standards.

B. On-Job Training

Training given to an employee on the procedure or task in the actual work environment/Shop floor. On-Job training is important to gained skill from training and implement. in day-to-day performance

C. On-Going Training

Ongoing training is establish to trained employees on current regulatory requirement to ensure that at all personnel are trained to adequately perform their assigned responsibilities.

D. Continuous process

Training is a continuous process. Training programmes therefore, should not be used as a gimmick. Training needs must be continuously reviewed for experienced workers, Workloads may fluctuate. New workers may be recruited, improve the present product line or may introduce a new one, Job procedures may change or New technology may be introduced.

E. Retraining or Refresher Training

Retraining or Refresher training shall be carried out annually or periodically as per the Annual Training Plan.

F. Technical Training

Specific job related training, motivation training and personnel development shall be performed to increase the skill and confident of the people.

12.7 Evaluation of Effectiveness of Training

The process of examining a training program is called training evaluation. Training evaluation program is checks whether training has had the desired effect. Training evaluation ensures that whether candidates are able to implement their learning in their respective workplaces, or to the regular work routines. Training can be evaluated by two means : firstly by formalized i.e., evaluation by questionnaires (Quick assessment) and secondly by impact.i.e By analysis the personnel for their skill, long time effect.

Evaluating the effectiveness of training can be by four levels..

A.   **Level 1: Trainee Reaction Evaluation**

The trainees 'reaction to training are important to known personnel reaction, training or learning experience, enjoyment of the training, relevant to their work and good use of their time. Special consideration should be given ton trainee reaction as a important tool in training evaluation. A questionnaire are preferable method for the evaluation of the learning.

B.   **Level 2: Trainee Learning Evaluation**

This is the second area of measurement, achieving learning objectives is a type of post training evaluation of knowledge and skill gained through the training intervention and which will ultimately translate to improving job performance. A positive emotional reaction and increase practical skill and knowledge of functional concept are indication of successful training.

Learning can be described as the degree to which training has impacted on employee's work related attitude. It also connotes the level at which employee's skill is broadened and knowledge widened as consequences of training.

C.   **Level 3: Trainee Behaviour Evaluation**

The third measurement level is defined as behaviour. Behaviour outlines a relationship of learning (the previous measurement level) to the actualization of doing. Behaviour evaluation is the extent to which the trainees applied the learning and changed their behaviour, and this can be immediately and several months after.

D.   **Level 4: Training Results**

The fourth measurement level, results, is the expected outcomes of most educational training programmes such as reduced costs, reduced turnover, and improved performance of the trainee.

12.8 Documentation of Training Management

Proper documentation of all types of training is essential to ensure compliance. A record is a controlled document. Completion of training is not valid unless it is properly documented on the appropriate form. A training record should not be signed until all entries are complete. Answered to the question shall be control and should available in secure place where authorized person can be access.

12.9 Automated Training Management

Training Management program can be done automatically by providing the training task, job code, random exam question and optional verification. Training Management program shall be handled by authorized person and should be made of different hierarchy level such as Administrator, Operator, and Trainee. Training Management program shall be qualified for its intended used. Audit trail shall be verified for the changes. Following can be considered during the qualification of the leaning management.

A.   Automates Employee Training Tasks

Automates assignment, monitoring, and verification of training tasks. Provides trainees access to records that show past training, course due dates, and required future training for better employee training management.

B.   Random Exam Questions

Course exams can be configured so questions are randomly selected. A user who has taken an exam will not get the same questions in the same order during a second try. Random selection guarantees a trainee understands the task before proceeding to the next training level.

C.   Optional Training Verification

Streamlines basic courses that do not require verification by reducing the burden on supervisors who typically must act as verifiers and perform employee training management.

12.10 Analyst Trainee Training Qualification

After effective training the analyst shall undergo analyst qualification before on-job or before performing analysis. Analyst qualification is a valuable tool for analyst to increase scientific expertise, knowledge, and other skills which are used in the day to day in laboratory .Analyst Qualification help to reduce course the failure management system. Analyst Qualification can be performed on Quantitative Base or Qualitative Base. List of Quantitative (Non Complex) and Qualitative (Non Complex) Qualification test shall be maintained. Analyst Qualification shall be performed on the approved sample with known analytical values. Analysis qualification shall be performed at least on three different sample and result shall be compare with initial analysis. Difference between two results shall be within defined acceptance criteria. Qualification is based on difference between results by the qualified trainer and trainee analyst.

If he difference between trainer and trainee is not within the acceptance criteria ,then investigate and repeat the same analysis after specific time period(After week) by explaining hi steps or Do Don'ts etc. If the trainee re-attempted for the same analytical techniques fails maximum three time then trainee. Trainee shall goes to specific training

A.   Steps involved in the analyst qualification

Assign the Analyst Trainee Qualification number in "Analyst Trainee Qualification Record" such as ATQ/Year/001(Serial No).

Identify the samples of known analytical value

- Assign an analytical code number for analysis qualification.
- Do not disclose the initial analytical results.
- Analyst trainee shall independently performed the analysis as per approved procedure under the supervision of trainer.
- Supervisor or trainee shall verify the analyst trainee is performing the correct unit operation as per testing specification and record in the Analyst trainee Qualification Checklist.
- Number of tests shall be selected based on the job carried out by the particular analyst.
- Attach chromatograms and strip charts with analyst qualification report.
- Compare the results with the initial results.
- If the results are within the acceptance limit the analyst will be deemed as capable of performing the analysis.
- If the acceptance level then, train the analyst and requalify.

B.   Analysts Trainee Acceptance Criteria

Industrial Acceptance criteria normally used		
Sr. No	TEST	ACCEPTANCE CRITERIA
	Assay by HPLC (on Anhydrous/dried basis)	± 0.5 % of the initial value
	Assay by Non Aqueous Titration (on Anhydrous/dried basis	± 0.5 % of the initial value
	Infra Red Spectra /U.V Spectra	Shall be comparable
	Melting Range	± 2°C
	TOC	± 15 % of the initial value

Table *Contd...*

Industrial Acceptance criteria normally used		
Sr. No	**TEST**	**ACCEPTANCE CRITERIA**
	Residual solvent :GC	± 15 % of the initial value
	Particle Size by Malvern	± 15 % of the initial value
	Optical Rotation by Polari meter	± 2°
	Microbial limit test	Shall be comparable with initial results
	Pathogens	Shall be comparable with initial results
	Aerobic count	± 20 % of the initial value.

C.   Planning, reviewing and documenting qualification activities

All qualification activities have to be planned and reviewed periodically. The review and detection of the need of new and/or additional qualification can be performed during an appraisal interview, if it is done periodically. The documentation of the qualification activities has to be stored (preferably in the personnel file of the employee).

12.11 Reviewer Qualification

As per current regulatory excitation reviewer should be qualified in the respective category prior to review of the documents. Reviewer should be educated, trained and experience or any combination to know a deviation or error form the intended process or from the value expected from the testing. Review is a process of verification of the data provided and the examination for its accuracy and correctness.

A.   Steps followed in the Reviewer

Assign the Analyst Reviewer Trainee Qualification number in "Reviewer Trainee Qualification Record" such as RTQ/Year/001(Serial No).

Identify the Master Protocol with Known Error.

- Do not disclose the error and its categorization.
- Reviewer trainee shall independently perform the review as per approved procedure under the supervision of trainer.
- Supervisor or trainee shall verify the review trainee is performing the correct unit operation as per review procedure.
- Reviewer trainee shall identify the number of error and record in the observation sheet of protocol.

- Trainer shall compare the results with the initial results.
- If the results are within the acceptance limit the analyst will be deemed as capable of performing the review.
- If the acceptance level then, train the reviewer and requalify.

B.   Classification of error

- **Critical:** Error which leads to significant impact on quality of product.
- **Major:** Error which leads to impact on quality of product.
- **Minor:** Error which leads to no impact on quality of product.

C.   Acceptance criteria

- All critical and major errors should be identified 100%.
- Minor errors should not be less than 80%.
- If an individual fails to meets the acceptance criteria, training should be provided and qualified again. If any individual fails three times, individual shall be dropped for qualification.
- Requalification of reviewer shall undergo qualification within three year.

D.   Evaluation of reviewer qualification

- Randomly verified the few documents which are reviewed by reviewer periodically for its effectiveness review. During effectiveness check "Qualified error" fail to identify the critical and major errors during the routine review, the reviewer is recommended to prequalified.

12.12 Format used for Training Qualification Program Management

A. Format for Induction Training and Evaluation Records

Induction Training and Evaluation Records
Name of Employee : **Employee Number :** **Department :** **Date of Joining :** **Mail Id :**
Education Detail :

Details of Previous Employees :
  • **Organization Name:**
  • **Period**
  • **Job Profile**

Table Contd...

Induction Training and Evaluation Records				
Induction Training Program				
Sr.No.	Department	Trainer: Name, Employee ID. Sign and Data	Training Date /Time	Remark
	Polices, Ethics			
	Human Resource • Introduction • Personnel Hygiene • Medical Examination • Data Integrity Policy • Housekeeping			
	Safety • Introduction • Overview of Safety procedure and initiative			
	Quality Assurance • Introduction • Overview of Assurance Quality procedure and initiatives			
	• Laboratory Management • Introduction • Overview of Laboratory Management procedure and initiatives			
	Warehouse Management • Introduction • Overview of Warehouse Management procedure and initiatives			
	Production Management • Introduction • Overview of Production Management procedure and initiatives			
	Packaging and Labelling Management • Introduction • Overview of Packaging and Labelling Management procedure and initiatives			

Table *Contd...*

Induction Training and Evaluation Records			
Facility Management • Introduction • Overview of Facility Management procedure and initiatives			
Any other Department :			

Certification for Induction Training:
This is to Certified that Mr./Ms ______________________________ Has completed the Induction training program and understood. He/she is eligible to go on On-Job Training. • Name of Person : • Employee Code: • Sign/Date :

B. Format Used for On Job Training and Evaluation Records

ON JOB Training and Evaluation Records			
• **Name of Employee :** • **Employee Number :** • **Department :** • **Date of Joining :** • **Mail Id :**			

Sr.No	Contents of Training Program/Version No	Trainer Sign/Date	Remark

Certification for On-Job Training:
This is to Certified that Mr./Ms ______________________________ Has completed the Induction training program and understood. He/she is eligible to do On-Job work. • Name of Person : • Employee Code: • Sign/Date :

C. Format for Training Attendance Record

<table>
<tr><td colspan="6" align="center">Training Attendance Record</td></tr>
<tr><td colspan="2">Date of Training :</td><td colspan="2">Trainer :</td><td colspan="2">Training Time</td></tr>
<tr><td colspan="6">Training Topic :</td></tr>
<tr><td>Sr.No</td><td>Name</td><td>Employee Code</td><td>Department</td><td>Trainee (Sign/Date)</td><td>Remark</td></tr>
<tr><td></td><td></td><td></td><td></td><td></td><td></td></tr>
<tr><td></td><td></td><td></td><td></td><td></td><td></td></tr>
<tr><td></td><td></td><td></td><td></td><td></td><td></td></tr>
<tr><td></td><td></td><td></td><td></td><td></td><td></td></tr>
<tr><td colspan="6">Evaluation/Comment/Summary of Training Program :
Trainer Sign/Date</td></tr>
</table>

D. Format for Annual Training Calendar

<table>
<tr><td colspan="5" align="center">Annual Training Calendar for Year-2017</td></tr>
<tr><td>Sr. No</td><td>Topics of Training Program</td><td>Training Schedule Date</td><td>Training Completion Date</td><td>Remark</td></tr>
<tr><td></td><td></td><td></td><td></td><td></td></tr>
<tr><td></td><td></td><td></td><td></td><td></td></tr>
<tr><td></td><td></td><td></td><td></td><td></td></tr>
<tr><td></td><td></td><td></td><td></td><td></td></tr>
<tr><td></td><td></td><td></td><td></td><td></td></tr>
<tr><td colspan="2">Prepared By:
Date and Sign:</td><td>Checked By:
Date and Sign:</td><td colspan="2">Approved By:
Date and Sign:</td></tr>
<tr><td colspan="5">Summary of Annual Training Calendar :</td></tr>
<tr><td colspan="5">Closure of Annual Training Calendar Date :</td></tr>
<tr><td colspan="2">Prepared By:
Date and Sign:</td><td>Checked By:
Date and Sign:</td><td colspan="2">Approved By:
Date and Sign:</td></tr>
</table>

E. Format for Surprise Spot Check Record

		Surprise Spot Check Record					
Sr.No	**Date**	**Surprise Spot : Name, Employee code, Done By and Date**	**Employee :Name, Employee code and Activity Performed**	**Training Record Verification (Yes/No)**	**Observation**	**CAPA**	**Remark**

F. Format for list of Trainers Record

			List of Trainers Record			
Sr. No	**Trainer Name**	**Employee Code**	**Qualification**	**Department**	**Trainer Certification**	**Remark**
Prepared By: Date and Sign:			Checked By: Date and Sign:		Approved By: Date and Sign:	

12.13 Format Used for Analyst Trainee Qualification Program Management

A. Analyst Trainee Qualification Inward Record

					Analyst Trainee Qualification Inward Record					
Sr. No	**Date**	**ATQ. No**	**Analyst Name**	**Employee Code**	**Analyst Techniques**	**Test**	**Material**	**Batch No**	**Qualified/ Not Qualified**	**Evaluated by Sign /Date**
1.										
2.										
3.										
4.										

B. Analyst Trainee Qualification Record

Analyst Trainee Qualification Record						
Name of Analyst :			Employee Code :			
Sr.No	Date	Analytical Techniques	Analyst Trainee Sign/Date	Trainee Sign/Date	Qualified/Not Qualified	Remark
1.						
2.						
3.						
4.						

C. List of Qualified Analyst Record

List of Qualified Analyst Record				
Version No:				
Sr.no	Analyst name	Employee Code	Qualified on Test	Remark
1.				
2.				
3.				
Prepared By : Sign/Date :		Checked By : Sign/Date		Approved By : Sign/Date

D. List of Quantitative (Non Complex Test) Record

Quantitative (Non Complex Test) Records	
Version No:	Effective Date :
Sr. No	Test
1.	Description/Appearance
2.	Solubility
3.	Limit test
4.	Preparation of reagent/Volumetric solution
5.	pH of solution
6.	Colour and Clarity
7.	Sulphated ash

Table Contd...

Quantitative (Non Complex Test) Records	
Version No:	**Effective Date :**

Sr. No	Test
8.	Melting point
9.	Bulk/Tapped Density
10.	Tablet Hardness
11.	Thickness and Diameter
12.	Conductivity
13.	Distillation test
14.	Heavy Metal Test
15.	Disintegration time
16.	Friability
17.	Thin Layer chromatography

Prepared By : Sign/Date :	Checked By : Sign/Date	Approved By : Sign/Date

E. List Qualitative (Complex Test) Qualification Record

Quantitative (Complex Test) Records	
Version No:	**Effective Date :**

Sr. No	Test
1.	Chromatography-HPLC
2.	Chromatography-GC
3.	Water Content
4.	Potentiometric
5.	AAS
6.	FTIR
7.	UV
8.	XRD
9.	Specific Optical Rotation
10.	TGA or DSC
11.	Viscosity
12.	TOC
13.	Laszer Diffraction

Table *Contd...*

Quantitative (Complex Test) Records

Version No: **Effective Date :**

Sr. No	Test
14.	Delivery Dose
15.	Sterility
16.	MLT
17.	Microbial Assay
18.	BET

Prepared By : Sign/Date :	Checked By : Sign/Date	Approved By : Sign/Date

F. Analyst Qualification Record

Analyst Qualification Record

Analyst Name : Employee Code :	Analysis Techniques (Test):	Material Name: Batch No:
ATQ.No	Date :	Specification No:

Trained on Procedure : Yes/No

Procedure Name and Number :

Test Preparation Details (Name and Lot no.: Chemical , Standard used, Instrument detail):

Calculation Details:

Results :

Tested By : Sign/Date	Checked By : Sign/Date

Comparison of Analyst Trainee Qualification Results with

Sr.no	Test	Analyst Trainee Results	Initial Results	Variance Difference	Acceptance Criteria

Non Conformance :

Conclusion: Analyst Qualified/Not Qualified

Remark :

Checked By :	Approved By :

Table *Contd...*

<table>
<tr><td colspan="4" align="center">Analyst Qualification Record</td></tr>
<tr><td colspan="2">Analyst Name :
Employee Code :</td><td>Analysis Techniques (Test):</td><td>Material Name:
Batch No:</td></tr>
<tr><td colspan="2" align="center">ATQ.No</td><td align="center">Date :</td><td align="center">Specification No:</td></tr>
<tr><td colspan="3">Sign/Date</td><td>Sign/Date</td></tr>
<tr><td colspan="4">Certification for Qualified Analyst</td></tr>
<tr><td colspan="4">

This is to Certified that Mr./Ms ________________________________ Has completed the Induction training program and understood. He/she is eligible to do ____________ analysis

- **Name of Person :**
- **Employee Code:**
- **Sign/Date :**
</td></tr>
</table>

G. Check List for Analyst Qualification

<table>
<tr><td colspan="6" align="center">Check List for Analyst Qualification</td></tr>
<tr><td colspan="3">Analyst Name :
Employee Code :</td><td colspan="3">Analysis Techniques (Test):
Date :</td></tr>
<tr><td>Sr. No</td><td colspan="2" align="center">Check Point</td><td>Yes</td><td>No</td><td>Remark</td></tr>
<tr><td>1.</td><td colspan="2">Is analyst used Approved procedure?</td><td></td><td></td><td></td></tr>
<tr><td></td><td colspan="2">Is analyst use calibrated instrument and equipment?</td><td></td><td></td><td></td></tr>
<tr><td>2.</td><td colspan="2">Is analyst use Qualified Analytical standard such as Reference, Chemical, Reagent and Volumetric Standards?</td><td></td><td></td><td></td></tr>
<tr><td>3.</td><td colspan="2">Is analyst used appropriate glassware as per procedure?</td><td></td><td></td><td></td></tr>
<tr><td>4.</td><td colspan="2">Is analysis took safety precaution?</td><td></td><td></td><td></td></tr>
<tr><td>5.</td><td colspan="2">Is analyst store the solution as per storage condition?</td><td></td><td></td><td></td></tr>
<tr><td>6.</td><td colspan="2">Is analyst followed safety precaution?</td><td></td><td></td><td></td></tr>
<tr><td>7.</td><td colspan="2">Is analysis followed real time documentation?</td><td></td><td></td><td></td></tr>
<tr><td>8.</td><td colspan="2">Is analyst followed data integrity procedure</td><td></td><td></td><td></td></tr>
<tr><td>9.</td><td colspan="2">Is analyst followed the disposal procedure?</td><td></td><td></td><td></td></tr>
<tr><td>10.</td><td colspan="2">Specific remark on the Analytical Techniques?</td><td></td><td></td><td></td></tr>
<tr><td colspan="6">

Remark :
The independent analysis performed by analyst trainee is Satisfactory/Non- Satisfactory.
Conclusion :
Trainer
Sign/Date
</td></tr>
</table>

12.14 Format Used for Reviewer Qualification Program Management

A. Analyst Trainee Qualification Inward Record

Reviewer Trainee Qualification Inward Record										
Sr. No	Date	RTQ.No	Reviewer Name	Employee Code	Documents	Test	Material	Batch No	Qualified/Not Qualified	Evaluated by Sign /Date
1.										
2.										
3.										
4.										
5.										

B. Reviewer Trainee Qualification Record

Reviewer Trainee Qualification Record						
Name of Analyst :			Employee Code :			
Sr. No	Date	Reviewer Techniques	Analyst Trainee Sign/Date	Trainee Sign/Date	Qualified/Not Qualified	Remark
1.						
2.						
3.						
4.						

C. List of Qualified Reviewer Record

List of Qualified Reviewer Record				
Version No:				
Sr.no	Reviewer name	Employee Code	Qualified on Test	Remark
1.				
2.				
3.				
4.				
Prepared By : Sign/Date :		Checked By : Sign/Date		Approved By : Sign/Date

D. Reviewer Qualification Record

<table>
<tr><td colspan="6" align="center">Reviewer Qualification Record</td></tr>
<tr><td colspan="2">Reviewer Name :
Employee Code :</td><td colspan="2">Analysis Techniques :</td><td colspan="2">Material Name:
Batch No:</td></tr>
<tr><td colspan="2">RTQ. No</td><td colspan="2">Date :</td><td colspan="2">Specification No:</td></tr>
<tr><td colspan="6">Trained on Procedure : Yes/No
Procedure Name and Number :</td></tr>
<tr><td colspan="6">Review Trainee Observation :
Results :</td></tr>
<tr><td colspan="3">Reviewed By :
Sign/Date</td><td colspan="3">Checked By :
Sign/Date</td></tr>
<tr><td colspan="6">Comparison of Reviewer Trainee Qualification Results with</td></tr>
<tr><td>Sr. No</td><td>Test</td><td>Reviewer Trainee Results</td><td>Initial Results</td><td>Variance Difference</td><td>Acceptance Criteria</td></tr>
<tr><td></td><td></td><td></td><td></td><td></td><td></td></tr>
<tr><td></td><td></td><td></td><td></td><td></td><td></td></tr>
<tr><td colspan="6">Non Conformance :
Conclusion: Analyst Qualified/Not Qualified
Remark :</td></tr>
<tr><td colspan="3">Checked By :
Sign/Date</td><td colspan="3">Approved By :
Sign/Date</td></tr>
<tr><td colspan="6" align="center">Certification for Qualified Reviewer</td></tr>
<tr><td colspan="6">This is to Certified that Mr./Ms _______________________________ Has completed the Induction training program and understood. He/she is eligible to do ___________ reviewer.
• Name of Person :
• Employee Code:
• Sign/Date :</td></tr>
</table>

Chapter - 13

Reviewer Qualification Management

Introduction

The pharmaceutical quality control laboratory serves one of the most important functions in pharmaceutical production and control. A significant portion of the CGMP regulations (21 CFR 211) pertain to the quality control laboratory and product testing. Similar concepts apply to bulk drugs. Reviews and evaluations depend on accurate and authentic data that truly represents the product.

A reviewer qualification is based on the past experience and competencies of the employee.

13.1 Documentation Requirements for Reviewer Qualification

The documents required for qualification process shall be prepared by user department as dummy documents having critical, major and minor errors. Depending on the criticality and level of review, the dummy documents required for qualification shall be prepared as per below categories.

Category	Type of Documents	Detail description
I	Master Documents	The scope of this category covers review of documents such as Master Batch manufacturing record, Master batch packing record, Master equipment cleaning record, Specification and standard testing procedures.
II	ExecutedDocuments	The scope of this category covers review of documents such as executed batch records (manufacturing and packaging), executed equipment cleaning records, Analytical data like executed record of analysis, chromatograms and any relevant test reports.

Table *Contd...*

Category	Type of Documents	Detail description
III	Validation and Qualification Documents	The scope of this category covers review of documents such as Process validation protocols and reports, Water system, HVAC and other utilities qualification protocols and reports, Equipment qualification documents, Analytical method validation protocols and reports, cleaning validation protocols and reports.

The errors in dummy documents shall be documented in "Master Observation Sheet". This document can be used as reference during evaluation and approval of documents reviewed by person under reviewer qualification. The dummy documents along with master observation sheet must be reviewed and accepted by Quality Assurance

The dummy documents and master observation sheet shall be archived under custody of Quality Assurance. The copy of dummy document shall be issued by Quality Assurance on request of user department. During issuance of this document QA shall stamp first page of document along with sign date and remaining pages as "Document for reviewer qualification".

13.2 Qualification Procedure

Department head shall identify the individual and nominate for intended category of qualification.

Below mentioned points must be considered during identification of personnel for reviewer qualification process:

- **Academic Qualification:** Personnel shall have minimum education up to 12th class/diploma in science/pharmacy/technology or related subjects or suitable academic qualification.

- **Experience:** A minimum of 2 years work experience in relevant area/department in pharmaceutical organisation/industry regulated by the GMP principles. Relevant area/department work experience in current organisation shall be preferred.

- **Training:** Personnel shall have necessary training on the procedures and on job training in relevant area wherever applicable.

- Department head/designee shall mention details in a protocol for intended. Each individual can be nominated for more than one category based on requirement during qualification program. Department head/designee or Qualification Assurance shall provide dummy documents which is intended for reviewer qualification to evaluate reviewer's capabilities.

- Based on the type of document provided for review, maximum allotted time for review shall be included in the protocol. Reviewer under qualification shall complete the review of chosen or allocated documents in specified time and summarise all the observations and submit to department head/designee for evaluation.

13.3 Evaluation and Approval

Department head or designee or other qualified reviewer shall evaluate the documents reviewed by the individual. The process/procedure flow shall be discussed as part of evaluation and identified shortfall in the reviewed document shall be explained to the individual. After assessment, recommendation shall be documented and submitted to Quality Assurance. Quality Assurance head/designee shall evaluate the adequacy of the review and recommendation. Based on satisfactory evaluation, shall approve the qualification report and provide the certification.

13.4 Acceptance Criteria for Qualification

- Identification of all critical and major gaps.
- Score minimum 80% in identification of minor gaps.
- Review shall be completed within the timeline defined in the protocol.
- In case reviewer fail in Critical/Major errors identification as per acceptance criteria during initial qualification then person shall be retrained in relevant work area by department head or designee and shall undergo qualification procedure.
- In case reviewer fail only in minor errors identification during initial qualification as per acceptance criteria then need for retraining shall be evaluated by department head and person shall be subjected to qualification procedure.
- If the reviewer is being qualified for more than one category, based on the assessment in each category, reviewer qualification can limit to the category meeting acceptance criteria.

User department shall prepare the list of qualified reviewer and schedule for requalification in department for various categories. This document shall be version controlled and change control is not required for any revision to the list. (Addition/deletion shall be updated as and when required).

The master copy of this list shall be archived under custody of Quality Assurance. The documents pertaining to the individual qualification along with the attachments shall be retained with training file of an individual.

13.5 Requalification

Reviewer shall undergo requalification within three years. Requalification shall be done as per procedure followed for initial qualification. Different document shall be provided (other than used for initial qualification for intended category) for requalification activity. Requalification may be initiated earlier than the schedule on need basis. The concerned department head/Quality Assurance Head can recommend reviewer to undergo requalification or removal from approved list of reviewer for the specific category in cases but not limited to following,

- Qualified reviewer found failing to identify critical or major errors during routine review which have significant impact on product quality/manufacturing consistency/decision making ability.
- Outcome of investigation where root cause identified as reviewer error.

In case of requalification or removal from approved list of qualified reviewer, concerned department head in consultation with quality assurance head shall evaluate the need for revisiting the documents already reviewed by individual.

The person failed twice in reviewer qualification during initial qualification shall be dropped from qualification program. Such person can be selected for requalification based on HOD recommendation in consultation with QA and post ensuring necessary retraining in relevant work area. The protocol for reviewer qualification in case of re-selected person shall be approved by Head QA and appropriate justification for selection shall be documented in protocol under remarks section.

13.6 Loss of Qualification

If the qualification is lost, a new qualification plan and process like the approach for qualifying members of the approved team who change or enlarge duties has to be established and the qualification has to be successfully passed through. Under special circumstances the qualification process can be shortened, if justified.

13.7 Planning, Reviewing and Documenting Qualification Activities

All qualification activities have to be planned and reviewed periodically. The review and detection of the need of new and/or additional qualification can be performed during an appraisal interview, if it is done periodically. The documentation of the qualification activities has to be stored (preferably in the personnel file of the employee.

13.8 Reviewer Qualification Management

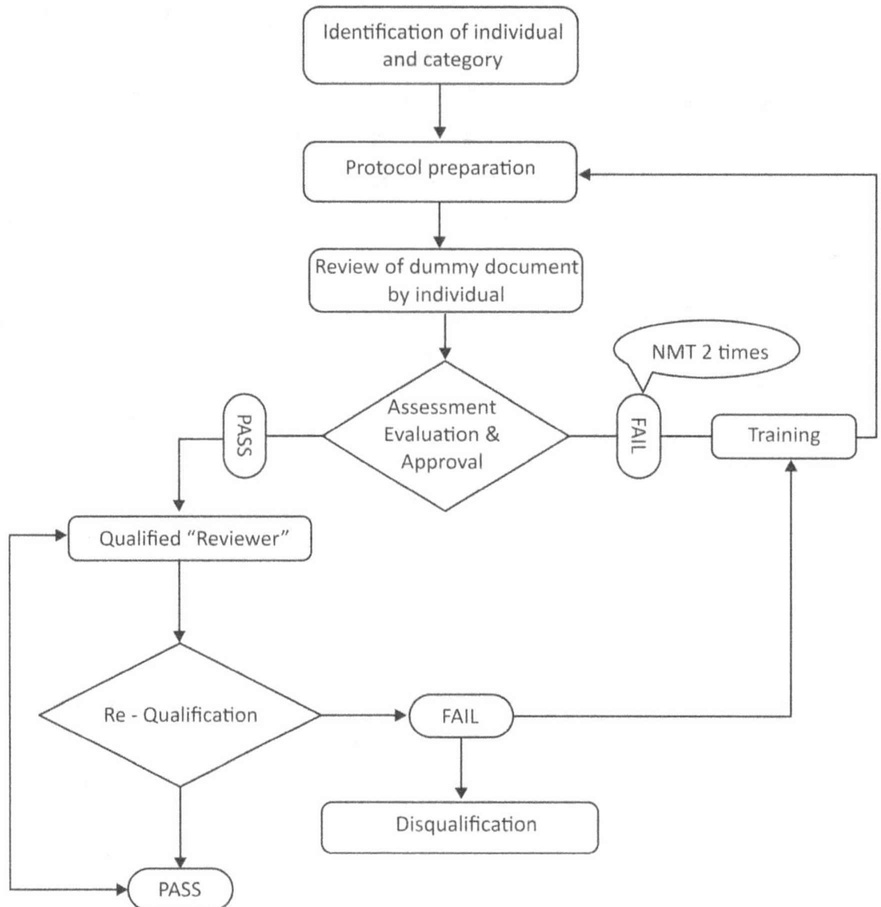

13.9 Definition

- **Reviewer:** A qualified person capable of correlating the data provided, evaluates the documentation practices and ensures data recorded is consistent & in accordance with the instructions or laid down procedures. The word 'reviewer' is applied to persons, who ensures completion of the record and signs for the complete record.

- **Qualification:** Qualification is a process of proving that any process or equipment or personnel works correctly and consistently and produces expected results.

- **Critical Errors:** Errors which leads to significant impact on product quality, manufacturing consistency, decision making ability.

- **Major Errors:** Errors which may have impact on product quality, manufacturing consistency, decision making ability.

- **Minor Errors:** Errors which doesn't have impact on product quality, manufacturing consistency, decision making ability.

Chapter - 14

Label Management in GLP Environment

Introductions

Label management in GLP should be accurate, integrated and controlled. Label are used to know the information of particular element. Label are any display of written, printed or graphical matter on the immediate container or packed or affixed to any articles. Label are pasted on the articles to provide information like product name, batch number expiry date, etc which gives the identification of the subject.

14.1 Master Specimen Label Copy

Master Specimen label should be prepared by User department with "Title of Label", colur combination, dimension and relevant required information. Sign and date should be part of the label. Each label should be control by the unique footer number with version control at the base of label body. Quality Assurance should approve the label. Master Specimen label list and album should be prepared for the each label used in laboratory. Master Specimen label album should have one copy of the Master specimen label .Any change in the Master specimen label or addition of label or deletion of label should be routed through change control.

14.2 Procurement, Receipt, Storage, Issuance, Reconciliation and Destruction of Label

Procurement of label should be done from the qualified vendor. Master specimen label should be share to the vendor while indenting the label. Each label should have unique serial number.

On the receipt of label from the vendor label should be check against the Master specimen copy for the accuracy and correctness. After approval, label should be stored under lock and key .In case of printing label, the printed label should be checked for its integrity, print quality or any misprint due to mechanical error. In case, reprinting is required, the reason of re-printing should be recorded before re-printing of label.

Issuance of the label should be recorded into the issuance log. Each every label should be reconciled. Destruction of label should be done by diagonal strike out mark with sign and date and send to the instruction.

14.3 Format Used for Label Management

    A. Master Specimen Label Record
    B. Master Specimen Label Album

A. Master Specimen Label Record

Master Specimen Label Record				
Version No.				**Page No.**
Sr. No	**Title of Label**	**Dimension**	**Control No**	**Effective Date**
	Received Label			
	Sampled Label			
	Raw Material Sample Label			
	In-Process Sample Label			
	Bulk Finished Product Label			
	Finished Product Label			
	Under Label			
	Approved Label			
	Rejected Label			
	Reserve Sample Label "Not to be Sale"			
	Sample Status Label			
	Reagent /Chemical Status Label			
	Working Standard Label			
	Reference Standard Label			
	Volumetric Solution Number			
	Reagent/Indicator Solution Label			
	Calibration Status Label			
	Preventive Maintenance Label			
	Breakdown Status label			
Prepared By : **Sign/Date:**	**Checked By :** **Sign/Date:**		**Approved By :** **Sign/Date:**	

B. Master Specimen Label Album

Master Specimen Label Album			
Sr. No	**Title of Label**	**Version No.**	**Page No.**
	Received Label		
	Sampled Label		
	Raw Material Sample Label		
	In-Process Sample Label		
	Bulk Finished Product Label		
	Finished Product Label		
	Under Label		
	Approved Label		
	Rejected Label		
	Reserve Sample Label "Not to be Sale"		
	Sample Status Label		
	Reagent /Chemical Status Label		
	Working Standard Label		
	Reference Standard Label		
	Volumetric Solution Number		
	Reagent/Indicator Solution Label		
	Calibration Status Label		
	Preventive Maintenance Label		
	Breakdown Status label		
Prepared By : **Sign/Date:**	**Checked By :** **Sign/Date:**	**Approved By :** **Sign/Date:**	

Master Specimen Label Album
1. Received Label: Affixed after chemical, Reagent are received.

Received Label
Received on :
Sign &Date :
Opened On:
Sign &Date :
Expiry :
XX/00

Table *Contd...*

Master Specimen Label Album

2. **Sample Label:** Affixed after sampling is performed container/Bags/Consignment.

Sample Label
Sample By :   **Sign &Date :**
XX/00

3. **Raw Material Sample Label:** Label affixed to the Polythene bag before Sampling for Raw Material. Where GRN: Good Receipt Note, where raw material received and inventory is maintained in ware house department and A.R no. : Analytical Reference Number used testing the material.

Raw Material Sample Label	
GRN No.:	**A.R No. :**
Item Name :	**Sample Qty:**
Item Code :	**Sample By :**
Batch No :	**Sign &Date :**
No. of Container	
XX/00	

4. **In-process Sample Label:** Label affixed to the Polythene bag before Sampling of in-process batch .Where A.R no. : Analytical Reference Number used testing the material.

In-process Sample Label:	
Item Name :	**A.R No. :**
Batch No :	**Sample Qty:**
Stage :	**Sample By :**
No. of Container	**Sign &Date :**
XX/00	

Table *Contd...*

Master Specimen Label Album

5. **Bulk Finished Product label:** Label affixed to the Polythene bag before Sampling of bulk batch .Where A.R no. : Analytical Reference Number used testing the material.

Bulk Finished Product Label	
Item Name :	A.R No. :
Batch No :	Sample Qty:
Stage :	Sample By :
No. of Container	Sign &Date :
XX/00	

6. **Finished Product label:** Label affixed to the Polythene bag before Sampling of Finished Product. Where A.R no. : Analytical Reference Number used testing the material.

Bulk Finished Product Label	
Item Name :	A.R No. :
Batch No :	Sample Qty:
No. of Container /Packs:	Sample By :
	Sign &Date :
XX/00	

7. **Under Test Label:** Label affixed to Container/Consignment/Bags after Sampling. Where A.R no. : Analytical Reference Number used testing the material.

UNDER TEST LABEL	
GRN No.:	A.R No. :
Item Name :	Sample Qty:
Item Code :	Sample By :
Batch No :	Sign &Date :
No. of Container	
XX/00	

Table Contd...

Master Specimen Label Album

8. **Approved Sample Label:** Label affixed to Container/Consignment/Bags after completion of analysis and material is passed as per approved Specification. Sampling .Where A.R no. : Analytical Reference Number used testing the material.

Approved Sample Label	
GRN No.:	A.R No. :
Item Name :	Sample By :
Item Code :	Sign &Date :
Batch No :	
No. of Container	
XX/00	

9. **Rejected Sample Label:** Label affixed to Container/Consignment/Bags after completion of analysis and material is failed to meet approved Specification. Sampling .Where A.R no. : Analytical Reference Number used testing the material.

Rejected Sample Label	
GRN No.:	A.R No. :
Item Name :	Sign &Date :
Item Code :	
Batch No :	
No. of Container	
XX/00	

10. **Reserve Sample Label "Not To Be Sold"** : Affixed to reserve Sample

Reserve Sample Label "Not To Be Sold"
Sign &Date :
XX/00

Table *Contd...*

Master Specimen Label Album

11. Sample Status Label: Affixed to Glassware before test solution preparation.

Sample Status Label
Product:
Test:
A.R No.:
Sign &Date :
XX/00

12. Chemical Standard Label: Affixed to Chemical or Reagent Bottle/Container.

Chemical/Reagent Standard Label
Chemical Name :
Batch No.:
Chemical/Reagent No.:
Validity Date :
Sign &Date :
XX/00

13. Working Standard Label : Affixed to working standard Bottle/Container.

Working Standard Label	
Working Standard :	**Purity :**
WS No.:	**Storage Condition :**
Prepared on :	**Sign &Date :**
Validity Date :	
XX/00	

14. Reference Standard Label: Affixed to reference standard Bottle/Container.

Reference Standard Label	
Reference Standard :	**Purity :**
RS No.:	**Storage Condition :**
Prepared on :	**Sign &Date :**
Validity Date :	
XX/00	

Table Contd...

<table>
<tr><td colspan="2" align="center">Master Specimen Label Album</td></tr>
<tr><td colspan="2">

15. Volumetric Solution Label: Affixed to volumetric solution Bottle.

Volumetric Solution Label	
Name : Solution No. : Prepared on : Expired on:	Molarity/ Normality: Storage condition Sign &Date :
XX/00	

</td></tr>
<tr><td colspan="2">

16. Reagent/Indicator Solution Label: Affixed to Reagent/Indicator Bottle.

Reagent/Indicator Solution Label	
Name : Solution No. : Prepared on : Expired on:	Storage condition Sign &Date :
XX/00	

</td></tr>
<tr><td colspan="2">

17. Calibration Status Label: Affixed to Instrument or Equipment to know the calibration status.

Calibration Status Label	
Inst/Equt Name : Inst/Equt. : Calibration Done on : Calibration Due on:	Sign &Date :
XX/00	

</td></tr>
</table>

Table *Contd...*

<table>
<tr><td colspan="2" align="center">Master Specimen Label Album</td></tr>
<tr><td colspan="2">

18. Preventive Maintenance Status Label: Affixed to Instrument or Equipment to know the preventive maintenance status.

<table>
<tr><td colspan="2" align="center">Preventive Maintenance Status Label</td></tr>
<tr><td>Inst/Equt Name :
Inst/Equt. :
Preventive Maintenance Done on:
Preventive Maintenance Due on:</td><td>Sign &Date :</td></tr>
<tr><td colspan="2">XX/00</td></tr>
</table>

</td></tr>
<tr><td colspan="2">

19. Break Down Status Label: Kept on Instrument or Equipment to know the break down status.

<table>
<tr><td colspan="2" align="center">Preventive Maintenance Status Label</td></tr>
<tr><td>Inst/Equt Name :
Inst/Equt. :
Break down Done on :</td><td>Sign &Date :</td></tr>
<tr><td colspan="2">XX/00</td></tr>
</table>

</td></tr>
</table>

Chapter - 15

Glassware Management in GLP

Introduction

Glass is an inorganic material (mostly silicates) or mixture of materials which when heated up and then cooled, solidifies without crystallization and therefore it has no melting point as such. Glassware is found in abundance in laboratories and comes in all shapes and sizes. Glass is relatively inert, meaning it will not react with the chemicals or substances placed inside. There is a vast variety of different glass apparatuses in a laboratory, and they can be manufactured from various types of glass depending on the purpose. Glassware may be transparent or amber color (light resistance), but should be easily monitoring and heat-resistant.

In laboratory management overall management of laboratory glassware, the key processes include:

- Glassware Procurement
- Glassware Verification
- Glassware Use and Cleaning

15.1 Classification of Glass as per United State of Pharmacopeia

Glass containers are classified into Type I glass, Type II glass, Type III glass and Type IV glass based on their degree of chemical/hydrolytic resistance to water attack. The degree of attack is dependent on the degree of alkaline release under the influence of the attacking media.

A. Type I glass containers (Borosilicate glass/Neutral glass)

This is a type of glass container that contains 80% silica, 10% boric oxide, small amount of sodium oxide and aluminium oxide. It is chemically inert and possess high hydrolytic resistant due to the presence of boric oxide. It has the lowest coefficient of expansion and so has high thermal shock properties.

Uses of Type I glass containers

- Type I glass is suitable as packaging material for most preparations whether parenteral or non-parenteral.
- They can also be used to contain strong acids and alkalis

B. Type II glass containers (soda-lime-silica glass/treated soda-lime glass/De alkalized soda lime glass)

This is a modified type of Type III glass container with a high hydrolytic resistance resulting from suitable treatment of the inner surface of a type III glass with sulfur. This is done to remove leachable oxides and thus prevents blooming/weathering from bottles. Type II glass has lower melting point when compared to Type I glass and so easier to mould.

Uses of Type II glass containers

- They are suitable for most acidic and neutral aqueous preparations whether parenteral or non-parenteral.

C. Type III glass containers (Regular soda lime glass)

This is an untreated soda lime glass with average chemical resistance. It contains 75% silica, 15% sodium oxide, 10% calcium oxide, small amounts of aluminium oxide, magnesium oxide, and potassium oxide. Aluminium oxide impacts chemical durability while magnesium oxide reduces the temperature required during moulding.

Uses of Type III glass containers

- They are used as packaging material for parenteral products or powders for parenteral use ONLY WHERE there is suitable stability test data indicating that Type III glass is satisfactory.
- They used in packaging non-aqueous preparations and powders for parenteral use with the exception of freeze-dried preparations
- It is also used in packaging non-parenteral preparations

D. Type IV glass containers (Type NP glass/General-purpose soda lime glass)

This type of glass container has low hydrolytic resistance. This type of glass containers are not used for products that need to be autoclaved as it will increase erosion reaction rate of the glass container.

Uses of type IV glass containers

- It is used to store topical products and oral dosage forms

15.2 Types of Glassware used in Laboratory Management

There is a vast variety of different glass or plastic wares apparatuses used in a laboratory, and they can be manufactured from various types of glass depending on the purpose .The most important is glassware used in laboratory should be accurate for its intended use. Glassware used in laboratory management glassware can be divided into two groups

A. Non-volumetric glassware

- Beaker
- Flask

B. Volumetric Glassware

- Volumetric Flask
- Graduated Cylinder

15.3 Common Glassware used in the Laboratory Management

- **Bulb and graduated pipettes.** These are used to transport specific amounts of fluids from one place to another.
- **Burettes.** These are used to dispense exact quantities of liquid into another vessel.
- **Beakers.** Simple containers used to hold samples and reagents.
- **Volumetric flasks.** Similar to beakers, these are used to hold samples, but usually come in a conical or spherical shape with a tapering neck.
- **Condensers.** Specifically used to cool heated liquid or gas.
- **Retorts.** These are used for distillation purposes.
- **Funnels.** The tapered neck of a funnel allows easy pouring of a liquid into a narrow orifice.
- **Petri dishes.** Shallow dishes used to culture living cells.
- **Graduated Cylinders.** Similar to beakers, these cylindrical vessels have volumetric markings to allow for monitoring of volume.
- **Vials.** Small bottles used to store samples or reagents.
- **Slides.** Used to hold items under a microscope for inspection and study.
- **Stirring Rods.** Used to mix solvents and samples together.
- **Desiccators.** A container designed to absorb moisture from a substance.
- **Drying pistols.** Similar to a desiccator, the pistol is a more direct method of removing moisture from a sample.

15.4 Glassware Classification by Accuracy

Graduated and volumetric glassware are calibrated to deliver (pipettes, graduated cylinders) and contain (volumetric flask) a known amount of liquid. Depends on accuracy vvolumetric glassware, are classified on A and B. Class B has twice the error limit of class A

A. Class A Glassware

Class A glassware are the highest accuracy in Volumetric Glassware with permanently marked "A" and complies with volumetric tolerances defined in DIN EN ISO or ASTM E694. Class A glassware is always supplied with the corresponding Batch Certificate and certificate consist of information such as 'Mean Value' and the 'Standard Deviation'. All Class A volumetric glassware is actually glass; volumetric plasticware is not eligible for Class A status.

B. Class B Glassware

Volumetric glassware that is permanently marked "B" has volumetric tolerances twice those of Class A (with the exception of Graduated Cylinders).These tolerances comply with Class B tolerances as defined in DIN EN ISO.

15.5 Laboratory Glassware Procurement

The success and accuracy of the assays and tests depends utmost on the accuracy and cleanliness of the glassware used. Only Class A glassware needs to be procured and used for quantitative tests. The certificates have to be assessed for accuracy as described on the certificate provided by the manufacturer and needs to be retained as per departmental procedure.

15.6 Laboratory Glassware Verification

Verification of glassware shall be performed categorically based on the glassware brand name (eg. Borosil, J-sil, etc...) and volume (e.g. 5mL, 10mL etc... volumetric flasks, 1mL, 2mL etc... pipettes, etc...). A single representative sample from each category is selected at random and verified as a onetime exercise. For example, if a 100 mL volumetric flask of Borosil make is verified and meets the acceptance criteria, then any 100mL volumetric flask received from approved vendor shall be considered as verified.

A set of 25mL of volumetric flasks of one make is considered as one category, and a set of 25mLvolumetric flasks of second make is considered as a another category. Single glassware from each category needs to be verified at least once. The water used for verification testing needs to be freshly boiled and cooled so as to expel any dissolved gases which has impact on density of water.

15.7 Glassware Calibration

Calibration of glassware is performed to determine the accuracy of glassware.

A. Class A Glassware

Class A glassware not required calibration and can be used directly used for analysis based on the certificate.

B. Class B Glassware

Calibration of Class B glassware is required to know accuracy and tolerance limit as defined in DIN EN ISO .Class B glassware are calibrated by using purified water at $25°C$ and calculated the volume by taking correction factor 0.99602 gm(i.e 1 ml of purified water at $25\,°C = 0.99602g$).

General procedure followed for Class B Glassware calibration.

- Weigh empty dried glassware (W1).
- Filled dried glassware with purified water up to mark with the help of bulb and wipe dry the outside of the glassware.
- Weigh Filled dried glassware with purified water (W2).
- Subtract weight of dry empty glassware (W1) from with Filled dried glassware with purified water (W2).
- Divide the weight with purified water density correction factor 0.99602 gm at $25\,°C$ (D) to get the results (W2-W1/D)
- Tolerance limit should be defined in approved procedure. Record of calibration shall be maintained.

15.8 Glassware Use and Cleaning

A. Use of Glassware

A clean and intact glassware is a prerequisite for obtaining accurate results. The analyst has to ensure the following prior to use of glassware for analysis:

- Only Class A glassware needs to be used for quantitative estimations.
- The glassware needs to be inspected prior to use for following :
    - (a)   Absence of any extraneous matter.
    - (b)   Absence of any visible damage like chipping, deformity etc.
    - (c)   Presence of identification mark for volume, meniscus etc.
- Analyst needs to ensure proper handling and storage of glassware during course of analysis and after cleaning to eliminate the possibility of cross contaminations. (E.g. proper labelling of in use glassware, covering the glassware, keeping stoppers on clean surface etc.)

- For optical measurements special care is required with repeated rinsing cycles of appropriate solvents since small contaminants may have impact on results obtained.

B. Additional Rinsing Steps Prior to use of Glassware

- The purpose of this step is to provide additional guidelines for care to be taken during usage of glassware. This step is recommended to be followed prior to performing critical analytical techniques e.g. HPLC, GC, CE, AAS, UV measurements etc. This will assist in reducing the risk of any potential interference due to presence of unknown contaminants from the glassware, which can have impact on the final result generated. The following are the recommended steps: The previously cleaned glassware should be first rinsed 3 times with Water followed by 3 times with methanol and then again with 3 times water. The water (Milli Q or equivalent) and methanol (HPLC grade) should be used.

- All these rinsed glassware should be dried preferably by keeping the same in inverted positions.

C. Empty out and Rinsing of Glass Apparatus after Analysis

The following steps are recommended to be followed by the analyst:

- Discard the solutions (Reference, Specificity & Sample, etc) from respective glass containers to the appropriate waste containers.

- Rinse the respective glassware with a suitable diluent by analyst (Diluent is a solution which was used for preparation of respective solutions during course of analysis) when sample is prepared in non polar diluent or sample matrix is in the form of creams, ointments or difficult to clean products. Keep these rinsed glass wares in the trays dedicated for keeping the glassware for washing. If the sample is prepared in polar solvent or easy to clean products, analyst can directly keep the empty glassware after use in the trays dedicated in the trays dedicated for keeping the glassware for washing.

- After completion of analysis the solvent in glassware shall be discarded as appropriate procedure and where the high potent or cytotoxic solution are available, initially it should be neutralized as per procedure and then transfer to "To be Cleaned Glassware "glassware tray or place. Precaution shall be taken during glassware washing such as wear safety apparels such as cut resistant gloves, nose mask, and apron and safety goggles. Before glassware cleaning, glassware shall be rinse with water, check for crack or damage.

Cleaning solution shall be prepared with appropriate validated concentration detergent solution. Preparation of cleaning solution shall be recorded Cleaning solution should properly labelled with Name, date of preparation and validity of solution with signed and date.

D. Laboratory Glassware Cleaning

- Glassware cleaning shall be performed manually or using automated glassware washer (For the Use of Automated glassware washer refer the applicable departmental procedure).

- Suitable neutral liquid cleaning agents shall be used for cleaning. The cleaning agent should be unaffected by acid solutions and solutions containing salt of heavy metals. This cleaning agent should be adequately diluted as per manufacturer's recommendation before use. A suitable equivalent cleaning agent may be used.

- For cleaning, diluted cleaning agent solution should be freshly prepared every day.

- Suitable scratch proof brushes made of soft and fine bristles of inert material may be used to facilitate removal of difficult to clean placebo matter.

- Suitable sponges may be used for cleaning external surfaces and internal surfaces of big

- Volume apparatus like beakers, dissolution jars etc.

- The glassware needs to be air dried by placing it in inverted position to drain out the liquid where possible or alternately may be dried in oven at low temperature.

   (i) *Manual glassware cleaning:* Glassware to be cleaned shall be rinsed inside and outside with potable water. Use only soft sponges or plastic core brushes with non-abrasive bristles and if required scrub to clean the inner surface and Conner of the glassware .Clean the glassware toughly till foam of cleaning solution is removed. Finally rinsed glassware with purified water. After final cleaning glassware shall transfer to the "Glassware drying oven "where temperature is maintained at 60°C, until its dries. Unload the cleaned and dry glassware from Glassware drying oven and keep glassware in "Cleaned and Dry "label tray/area. Cleaned and dried glassware shall be stored in dust free storage area.

(ii) ***Automatic glassware cleaning:*** Automatic glassware cleaning machine are used for cleaning glassware. Number of cycle used to cleaning of glassware shall be validated with validation concentration of solution. The unit is designed with a laboratory glassware washer, which is integrated with a hot air drying system.

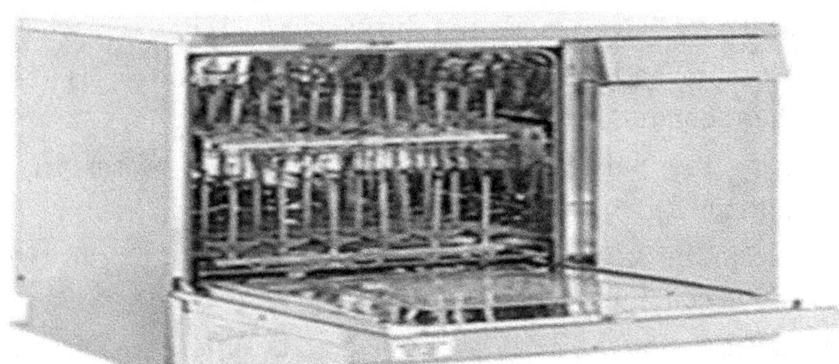

15.9 Glassware Care and Storage

Care should be taken to ensure that volumetric glassware is used and maintained in a way that does not cause damage to it nor alter its calibration. Extremes of temperature, including high temperature cleaning, and oven or hot air drying, which might lead to permanent changes in the capacity, shall be avoided. Certain solvents, strong acids/alkalis or surfactants may attack or alter the wetting characteristics of the glass, which in turn may affect draining properties. Any contamination that is not water-soluble should be removed with an appropriate solvent before the glassware is washed. Any guidance available from the supplier should be followed. Before use, the tips of pipettes and burettes shall be examined for mechanical damage and possible obstruction.

Storage of Glassware: The separate almirahs and drawers have been setup for systematic storage of the glassware. It helps in the smooth functioning.

15.10 Verification of Laboratory Glassware Procedure

A.   Verification of Pippetes:

A volume of water delivered by a pipette is obtained directly from the weight of the water and its density. The verification of the pipettes is checked in terms of actual volume delivered.

Procedure:

- Ensure that the pipette under verification check is absolutely clean and dry.
- If small beads of water adhere to the inner surface of the pipette after delivering water, the pipette is considered dirty and must be cleaned before proceeding further.
- Keep the pipette and water at room temperature at least for 1hr to attain thermal equilibrium.
- Determine the mass of the empty, stopper receiver with the help of an analytical balance and print the weight (W1) of the receiver.
- Measure and record the temperature of water (T) using calibrated thermometer.
- With the pipette, transfer full volume of water to the receiving container.
- Weigh the stopper receiver and its contents and print the weight (W2) of the container.
- Calculate the mass of water delivered by difference i.e. (W2 -W1).
- From the delivered mass of water, calculate the volume delivered by applying the density of water at measured temperature using following formula. Refer for density of water at different temperatures.

Formulas:

MV = (W2-W1) /D

E = MV – NV

Where,

E = Error, MV = Measured volume in mL, NV= Nominal volume in mL,
D = Density of Water at temperature (^{o}C)

Acceptance criteria for Pipette: The results must be within the range as given in the following Table

Sr. No.	Nominal capacity of pipette (X), ml	Limit of error (E), mL
1.	X ≤ 1.0	0.007
2.	1.0 < X ≤ 2.0	0.010
3.	2.0 < X ≤ 6.0	0.015
4.	6.0 < X ≤ 15.0	0.02
5.	15.0 < X ≤ 35.0	0.03
6.	35.0 < X ≤ 70.0	0.05
7.	X > 70.0	0.08

B.   Verification of Burettes

A volume of water delivered by a burette is obtained directly from the weight of the water and its density.

Procedure:

- Ensure that the burette under verification check is absolutely dry and clean.
- Keep the burette and water to reach room temperature for 1hr to attain thermal equilibrium.
- Measure and record the temperature of water using calibrated thermometer.
- Fill the burette up to the mark with the volume of water and make sure that the air bubbles are not trapped in the stopcock or tip. Wipe out any traces of liquid adhering to exterior to burette.
- Wait for 10min and recheck the volume. There should be no change in the meniscus.
- Take an absolutely dry container with the stopper and place it on the analytical balance and print the weight (W1) of the container.
- Slowly transfer the water of the specific volume from the burette to the dry container with the stopper and the weight is recorded.
- Determine the weight (W2) of container filled with water using analytical balance.
- Calculate the weight of dispensed volume from the burette by subtracting weight of container from the weight of the filled container.
- Calculate the actual volume of water in the volumetric flask by applying the density at
- Measured temperature using formula mentioned in piptter section and refer the density of water at different temperatures.

Acceptance criteria: The results must be within the range as given in the following Table

Sr. No.	Capacity of Burette, mL	Transferred volume, mL	Limit of error, mL
1.	10	3.0	± 0.01
		10.0	± 0.02
2.	20	6.0	± 0.01
		20.0	± 0.02
3.	25	7.5	± 0.01
		25.0	± 0.05
4.	50	15.0	± 0.02
		50.0	± 0.05

C. Verification of Volumetric Flasks

A volume of water contained in a volumetric flask, is obtained directly from the weight of the water and its density.

Procedure:

- Ensure that the interior as well as exterior of the volumetric flask under verification check is absolutely dry and clean.
- Keep the volumetric flask and water at room temperature at least for 1hr to attain thermal equilibrium.
- Measure and record the temperature of water using calibrated thermometer.
- Place the volumetric flask on the analytical balance with stopper and print the weight (W1) of the flask. Add the exact volume of the water till the verification mark of the volumetric flask.
- Wipe out any traces of liquid adhering to exterior to the volumetric flask and stopper it and print the weight (W2) of filled volumetric flask.
- Calculate the weight of water in the volumetric flask by subtracting weight of volumetric flask from the weight of the filled volumetric flask.
- Calculate the actual volume of water in the volumetric flask by applying the density at measured temperature using the formula mentioned in the section Pipette. Refer for density of water at different temperatures.

Sr. No	Nominal capacity of volumetric flask, mL	Limit of error, mL
1.	2.0-10.0	±0.025
2.	15.0-30.0	±0.04
3.	35.0-70.0	±0.06
4.	75.0-170.0	±0.10
5.	175.0-400.0	±0. 15
6.	450.0-650.0	±0.25
7.	750.0-1400.0	±0.40
8.	1500.0-2000.0	±0.60

D. Appendix for Density of Water at Different temperature

Temp. °C	0.0	0.1	0.2	0.3	0.4	0.5	0.6	0.7	0.8	0.9
	Density, gm/mL									
19	0.998405	0.998385	0.998365	0.998345	0.998325	0.998305	0.998285	0.998265	0.998244	0.998224
20	0.998203	0.998183	0.998162	0.998141	0.998120	0.998099	0.998078	0.998056	0.998035	0.998013
21	0.997992	0.997970	0.997948	0.997926	0.997904	0.997882	0.997860	0.997837	0.997815	0.997792
22	0.997770	0.997747	0.997724	0.997701	0.997678	0.997655	0.997632	0.997608	0.997585	0.997561
23	0.997538	0.997514	0.997490	0.997466	0.997442	0.997418	0.997394	0.997369	0.997345	0.997320
24	0.997296	0.997271	0.997246	0.997221	0.997196	0.997171	0.997146	0.997120	0.997095	0.997069
25	0.997044	0.997018	0.996992	0.996967	0.996941	0.996914	0.996888	0.996862	0.996836	0.996809
26	0.996783	0.996756	0.996729	0.996703	0.996676	0.996649	0.996621	0.996594	0.996567	0.996540
27	0.996512	0.996485	0.996457	0.996429	0.996401	0.996373	0.996345	0.996317	0.996289	0.996261
28	0.996232	0.996204	0.996175	0.996147	0.996118	0.996089	0.996060	0.996031	0.996002	0.995973
29	0.995944	0.995914	0.995885	0.995855	0.95826	0.995796	0.995766	0.995736	0.995706	0.995676

15.11 Definitions

- **Calibration:** The demonstration that the instrument/equipment produces results within specified limits when compared to those produced by a calibration standard or a standard which is traceable to a national or international standard, over an appropriate range of measurements.
- **Verification:** The activity that involves making a comparison, to confirm the continued accuracy/functionality of a process or instrument/equipment. (e.g., a check of an already calibrated piece of equipment or, the daily check of a balance)

15.12 Pictorial Diagram of Glassware used in Laboratory Management

Erlenmeyer flask (conical flask)

Graduated cylinder

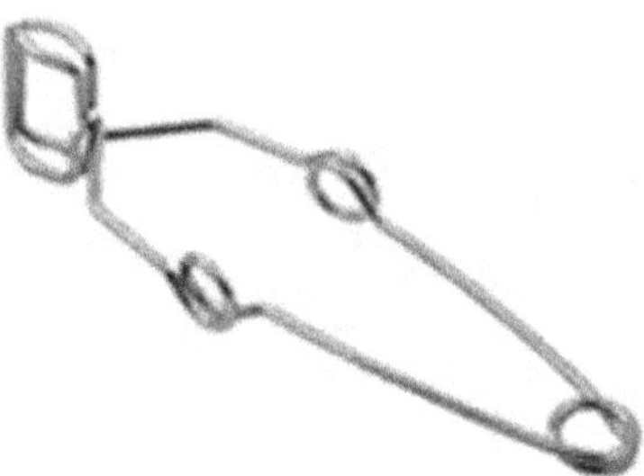

Test tube holder

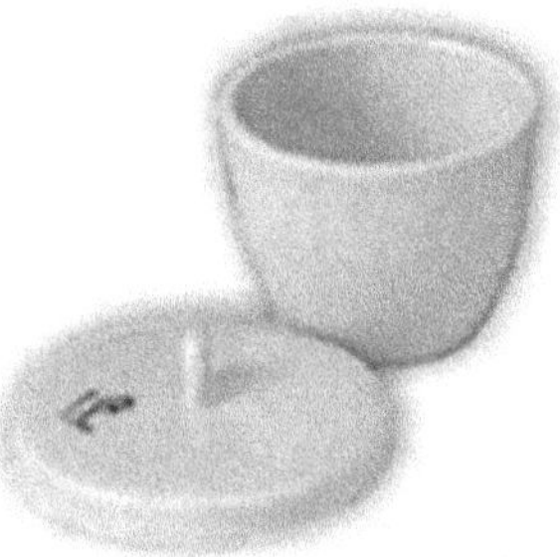

Crucible

Mortar and Pestle

Round bottom flask

Volumetric flask

Pipette

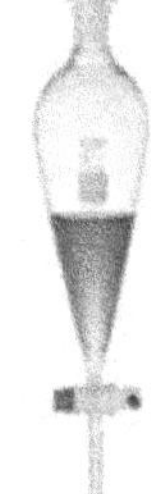

Separatory funnel

Desiccator

Funnel

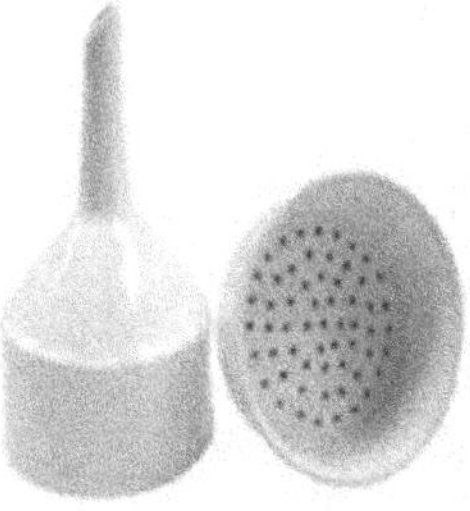

Buchner funnel

Watch glass

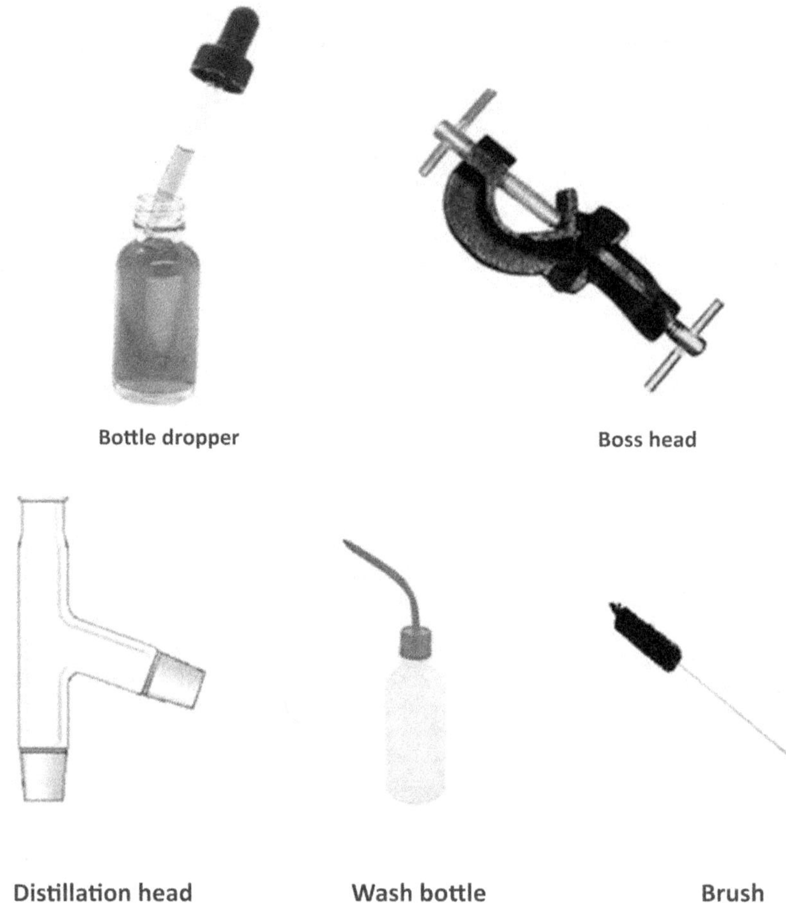

Bottle dropper

Boss head

Distillation head

Wash bottle

Brush

Condenser

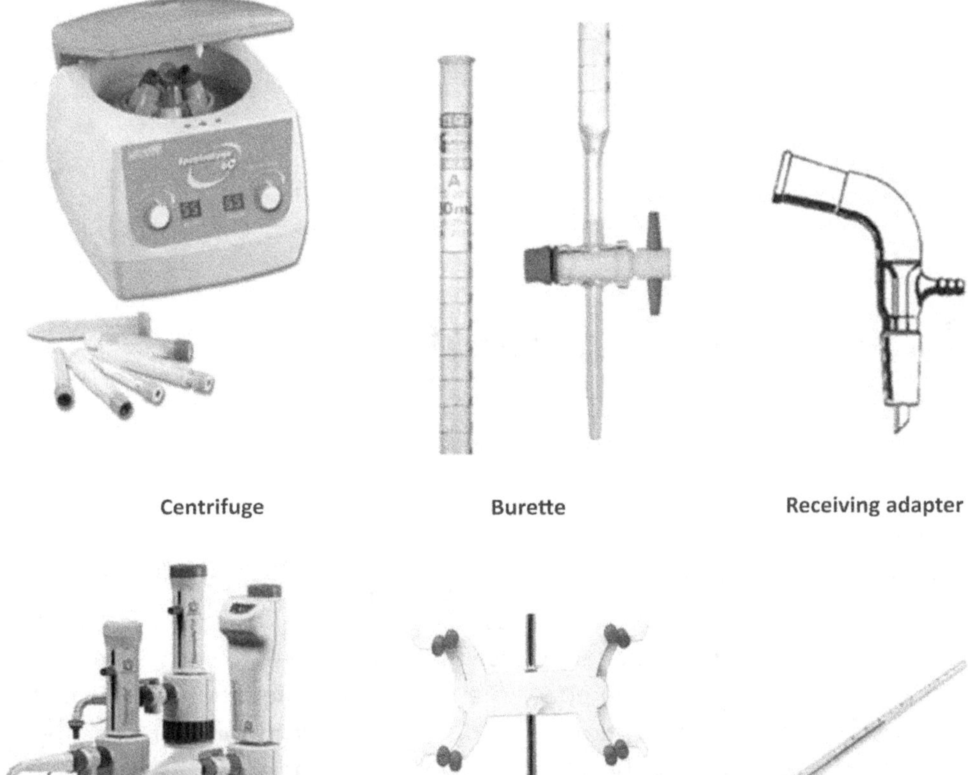

Centrifuge

Burette

Receiving adapter

Bottle top dispenser

Burette Clamp

Thermometer

Chapter - 16

GLP Column Management

Introduction

Chromatographic separation techniques are multi-stage separation methods in which the components of a sample are distributed between two phase, one of which is stationary, while the other is mobile. The stationary phase may packed in a column, spread as layer or distributed as a film, etc. The mobile phase may be gases or liquid or superficial fluid. The separation may be based on adsorption, mass distribution (partition), ionex change, etc. or may be based on differences in the physic-chemical properties of the molecule such as size, mass, volume, etc. Typically, column is a porous solid e.g. glass, silica or alumina that is packed into a glass or metal tube or that constitutes the walls of an open-tube capillary.

16.1 High Liquid Chromatographic Column

High performance liquid chromatography (HPLC) is a system that provides highly effective substance testing for the pharma, food and beverage, biological, and environmental industries. HPLC separates the compounds within samples so they can be individually tested, measured, or monitored. There many ways to classify liquid column chromatography. If this classification is based on the nature of the stationary phase and the separation process, three modes can be specified. In adsorption chromatography the stationary phase is an adsorbent (like silica gel or any other silica based packings) and the separation is based on repeated adsorption-desorption steps.

- In ion-exchange chromatography the stationary bed has an ionically charged surface of opposite charge to the sample ions. This technique is used almost exclusively with ionic or ionizable samples. The stronger the charge on the sample, the stronger it will be attracted to the ionic surface and thus, the longer it will take to elute. The mobile phase is an aqueous buffer, where both pH and ionic strength are used to control elution time.

- In size exclusion chromatography the column is filled with material having precisely controlled pore sizes, and the sample is simply screened or filtered according to its solvated molecular size. Larger molecules are rapidly washed through the column; smaller molecules penetrate inside the porous of the packing particles and elute later. Mainly for historical reasons, this technique is also called gel filtration or gel permeation chromatography although, today, the stationary phase is not restricted to a "gel".

- Concerning the first type, two modes are defined depending on the relative polarity of the two phases: normal and reversed-phase chromatography.

- In normal phase chromatography, the stationary bed is strongly polar in nature (e.g., silica gel), and the mobile phase is non-polar (such as n-hexane or tetrahydrofuran). Polar samples are thus retained on the polar surface of the column packing longer than less polar materials.

- Reversed-phase chromatography is the inverse of this. The stationary bed is non polar (hydrophobic) in nature, while the mobile phase is a polar liquid, such as mixtures of water and methanol or acetonitrile. Here the more non polar the material is, the longer it will be retained.

I. Types and Characteristics of HPLC Column

    A. *Normal Phase Columns:* The normal phase column uses polarity to separate the compounds of the sample. In normal phase chromatography, the stationary phase is polar and the mobile phase is non-polar. Once the sample reaches the column, its least polar compounds will separate first and its most polar compounds will separate last. Because each compound is extracted at different times, they can be individually recorded and tested by an HPLC lab technician. Normal phase columns are typically used to test organic acids, certain drugs, and a range of bio-molecules.

    B. *Reverse Phase Columns:* The reverse phase column is very similar to normal phase, but the polarities are reversed. In this type of column, the stationary phase is non polar and the mobile phase is polar. The reverse phase column is the most commonly used method and offers a range of testing options. Due to its great range of testing options, students can use the reverse phase column to test a wide variety of samples.

    C. *Ion Exchange Columns:* The ion exchange column uses either cationic or anionic charged ions. Cationic have a net positive charge of ions and anionic have a net negative charge. This will affect the way the sample responds and the lab technician will likely choose which charge to use based on the sample they are testing.

When the mobile phase and sample enter the column, the sample will begin to respond to the charged ions. The compounds will then begin to separate at different rates. This column type is typically used to separate carbohydrates, amino acids, and proteins. Therefore, students who choose to go into the food and beverage industry are most likely to use this type of column

D. **Size Exclusion Columns:** Unlike the other columns, size exclusion columns do not rely on the interaction of mobile phase, stationary phase, and sample to separate the sample. The sample is separated when it's filtered through a material that has different sized pores. The pores are made up of mesopores and micro pores. Mesopores have a diameter between two and 50 nanometres; while the micro pores have a diameter less than two nanometres.

Based on each compounds' ability to pass through the varying sized pores, they will separate at different rates. Larger compounds will pass more quickly, because they will avoid the holes, while smaller compounds that pass through the pores will take longer. Size exclusion columns are typically used to test proteins and carbohydrates.

16.2 Gas Chromatography Column

In gas chromatography, the column is the heart of the system where the separation of sample components takes place. They are classified in terms of tubing dimensions and type of packing material. Packed columns are generally 1.5 – 10m in length and 2 – 4mm id. These are generally made of stainless steel or glass. On the other hand capillary columns are 0.1 – 0.5 mm id and can be 10 – 100m long.

I. **Types of GC Column**

A. **Wall Coated Open Tubular (WCOT):** Internal wall of capillary is coated with a very fine film of liquid stationary phase.

B. **Surface Coated Open Tubular (SCOT):** Capillary tube wall is lined with a thin layer of solid support on to which liquid phase is adsorbed. The separation efficiency of SCOT columns is more than WCOT columns because of increased surface area of the stationary phase coating.

C. **Fused Silica Open Tubular (FSOT):** Walls of capillary fused silica tubes are strengthened by a polyimide coating. These are flexible and can be wound into coils.

II. Column Characteristics

A. *Column Materials:* Fused silica and stainless steel columns offer high degree of inertness and flexibility. When breakage is not of much concern fused silica is the best choice.

B. *Internal Diameter:* Sample concentration is the deciding factor for the internal diameter of the column. Loss of resolution, poor reproducibility and peak distortion result if sample concentration exceeds column capacity. Typical sample loading ranges around 10ng for 0.1mm id columns to up to 2,000ng for 0.53mm id columns.

C. *Length:* Longer columns provide greater resolution of sample components. However, increasing column length increases analysis time.

D. *Film Thickness:* Film thickness determines the retention and elution temperature of each sample component. Thick films increase the time a compound stays on the stationary phase and thinner films reduce retention time. Compounds having high volatility require more residence time for better separation and should be analyzed on thicker films. The commonly used film thickness in gas chromatography columns ranges from 0.1 to 5.0μm.

Columns are selected for use in a particular application based on column length and type of packing. Guidelines on selection of columns are provided in more detail in the certificate programme which will be launched in due course

16.3 Receipt, Issuance, Maintenance, Regeneration and Destruction of Column

After receipt of column from the manufacture or supplier check the Certificate of Analysis (COA) for correctness. On receipt of column details of the column should be recorded in "Column receipt and issuance record". Column receipt and issuance record should consist of Date of receipt of column, serial number, Type of column and unique identification number. Unique column number should be assign for each column for traceability and control. Column should be stored in designated place with its ends plugged. Performance qualification shall be performed as per manufacturer instruction or by system suitability parameter. If the column "Pass" then issuance the column to the product and enter in the Product column issuance log. In case of different column are used for the same product for same test or different test methods exists for same test, maintained column core code number. Number of injection and cumulative injection shall be recorded. If column does not meet the specification regeneration of column shall be performed as be manufacture recommendation and record in the column regeneration log. If the column does not meets the specification or analytical method, then discard the card with safety precaution and record in the destruction record.

16.4 Flow Chart of the Column Management

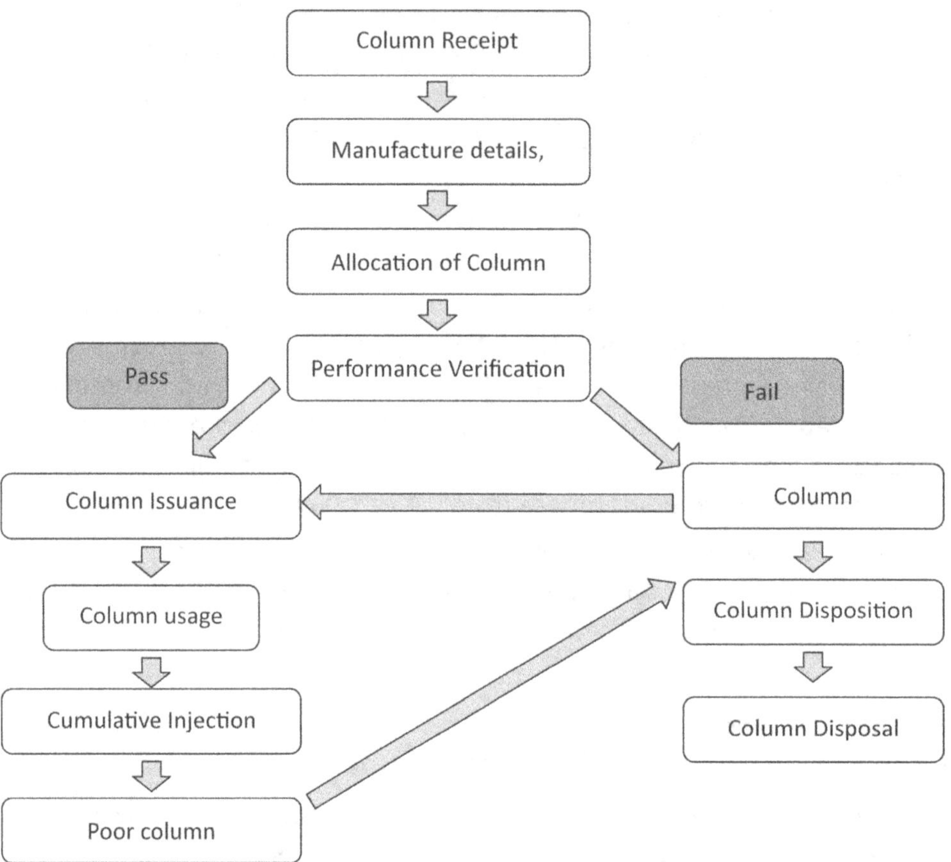

16.5 General Maintenance and Care

- Store column in the designated place.
- Do not store the column in water alone, as this may results in bacterial growth in the column
- Do not leave s column at elevated temperatures without eluent flow.
- Store the column in box with the compression screw.
- Flushing of the column before and after use with appropriate solvent and/or manufacturer recommendation.

16.6 Records Maintained for Column Management

A. Column Receipt and Qualification Record

Sr.no	Receipt					Signed and date	Qualification of column		Issued to Product/Test
	Date of receipt	Manufacture	Type of column	Serial No	Unique column number		Number of Injection injected	Results	
1									
2									

B. Column Issuance Record

Sr. No	Column no	Serial No	Type of Column	Name of Product to be Issuance	Name of Test for which column issued	Issued on date, sign and date of person	Discard on date, sign and date of person
1							
2							

C. Product Column Issuance Records

Product Name :

Test :

Sr.No	Column no	Serial No	Type of Column	Issued on date, sign and date of person	Analytical Record number	Number of Injection	Cumulative number of injection	Analyst Date and Sign	Discard on date, sign and date of person
1									
2									

D. Column Regeneration Record

Column Regeneration Record					
Product Name :				Test :	
Column Number :				Serial No.	
Generation Step	Solvent System	Flow rate	Time	Analyst Sign and date	Remark
1					
2					

E. Column Discard Record

Column Discard Record					
Product Name :				Test :	
Column Number:				Serial No.	
Number of Cumulative Injection:					
Last Analytical Report number :					
Sr.No.	Parameter	Results	Specification Limit	Result (Pass/Fail)	Analyst Sign and Date
1	Theoretical Plate				
2	Tailing Factor				
3	Resolution				
4	% RSD				
5	Any other				

GLP Standard Operating Procedure (SOP) Management

Introduction

- A Standard Operating Procedure (SOP) is a set of written instructions that document a routine or repetitive activity followed by an organization. The development and use of SOPs are an integral part of a successful quality system as it provides individuals with the information to perform a job properly, and facilitates consistency in the quality and integrity of a product or end-result. The term "SOP" may not always be appropriate and terms such as protocols, instructions, worksheets, and laboratory operating procedures may also be used. For this document "SOP" will be used. SOPs describe both technical and fundamental programmatic operational elements of an organization that would be managed under a work plan or a Quality Assurance (QA) Project Plan.

- SOPs detail the regularly recurring work processes that are to be conducted or followed within an organization. They document the way activities are to be performed to facilitate consistent conformance to technical and quality system requirements and to support data quality. They may describe, for example, fundamental programmatic actions and technical actions such as analytical processes, and processes for maintaining, calibrating, and using equipment. SOPs are intended to be specific to the organization or facility whose activities are described and assist that organization to maintain their quality control and quality assurance processes and ensure compliance with governmental regulations.

- If not written correctly, SOPs are of limited value. In addition, the best written SOPs will fail if they are not followed. Therefore, the use of SOPs needs to be reviewed and re-enforced by management, preferably the direct supervisor. Current copies of the SOPs also need to be readily accessible for reference in the work areas of those individuals actually performing the activity, either in hard copy or electronic format, otherwise SOPs serve little purpose.

17.1 SOP Process

The organization should have a procedure in place for determining what procedures or processes need to be documented. Those SOPs should then be written by individuals knowledgeable with the activity and the organization's internal structure. These individuals are essentially subject-matter experts who actually perform the work or use the process. A team approach can be followed, especially for multi-tasked processes where the experiences of a number of individuals are critical, which also promotes "buy-in" from potential users of the SOP.

SOPs should be written with sufficient detail so that someone with limited experience with or knowledge of the procedure, but with a basic understanding, can successfully reproduce the procedure when unsupervised. The experience requirement for performing an activity should be noted in the section on personnel qualifications. For example, if a basic chemistry or biological course experience or additional training is required that requirement should be indicated.

17.2 Writing Styles

SOPs should be written in a concise, step-by-step, easy-to-read format. The information presented should be unambiguous and not overly complicated. The active voice and present verb tense should be used. The term "you" should not be used, but implied. The document should not be wordy, redundant, or overly lengthy. Keep it simple and short. Information should be conveyed clearly and explicitly to remove any doubt as to what is required. Also, use a flow chart to illustrate the process being described. In addition, follow the style guide used by your organization, e.g., font size and margins.

17.3 Format and Structure

SOP is generally prepared in "Portrait", font to "Ariel/Calibre/Times New Roma", with font size at least 12 (readable) with line spacing to 1.5 the "alignment" to "justify" and font style to "Regular". Heading and subheading should be bold. Number clauses should restricted upto "3 digit", beyond that "alphanumerical or bullets" can be used.

A. SOP page set up

	A4 size paper or its equivalent	
Paper	Width	8.30
	Height	11.60
	Line Spacing	1.5
	Alignment	Justify
	Font style	Regular
Layout	Header	0.5"
	Fotter	0.25"
Margins	Top	0.5"
	Bottom	0.5"
	Left	1.0"
	Right	0.9
	Gutter	0
	Gutter Position	Left
	Orientation	Portrait
Apply to	Whole Document	

B. SOP Header

Header of SOP should be brief and self-descriptive in Bold with 1.5" spacing. Content should be bold and numbering should be alphanumerical or numerical. Company logo, Department and page number (X of Y pattern) should be part of Header. Version control of SOP should be start with "0" and increment serially upon subsequent revision.

Standard Operating Procedure			Log
Title and Version No		**Effective date**	
Department		**Page Number**	

C. SOP Footer

Footer number should be available on all pages which contained Prepared by, reviewed by and approved by with sign and date.

Prepared By	Checked By	Approved By
Name **Designee** **Employee ID**		
Sign and date	**Sign and date**	**Sign and date**

SOP numbering system and code should be defined in Quality unit for respective Business unit and should be approved by Quality Assurance. Printing of SOP shall be done by originating department and final soft copy of the SOP should be handover to QA for archival. Printed SOP signed should be done using a "blue or black ink pen". Whenever provision for electronic signature through validated platform is available. Author of SOP should sign and date in respective place of "Prepared by" section. In-charge or designated person should sign and date in the respective place of "Revision by" section. Once the signature of the author and reviewed are completed, the SOP shall be signed by Quality Unit.

17.4 SOP Review and Approval

SOPs should be reviewed (that is, validated) by one or more individuals with appropriate training and experience with the process. It is especially helpful if draft SOPs are actually tested by individuals other than the original writer before the SOPs are finalized.

The finalized SOPs should be approved as described in the organization's Quality Management Plan or its own SOP for preparation of SOPs. Generally the immediate supervisor, such as a section or branch chief, and the organization's quality assurance officer review and approve each SOP. Signature approval indicates that an SOP has been both reviewed and approved by management.

17.5 Frequency of Revisions and Reviews

SOPs need to remain current to be useful. Therefore, whenever procedures are changed, SOPs should be updated and re-approved. If desired, modify only the pertinent section of an SOP and indicate the change date/revision number for that section in the Table of Contents and the document control notation.

SOPs should be also systematically reviewed on a periodic basis, e.g. every 1-2 years, to ensure that the policies and procedures remain current and appropriate, or to determine whether the SOPs are even needed. The review date should be added to each SOP that has been reviewed. If an SOP describes a process that is no longer followed, it should be withdrawn from the current file and archived.

The review process should not be overly cumbersome to encourage timely review. The frequency of review should be indicated by management in the organization's Quality Management Plan. That plan should also indicate the individual(s) responsible for ensuring that SOPs are current.

Training should be provided (as necessary) when a document is revised or a new document is implemented.

17.6 Checklists

Many activities use checklists to ensure that steps are followed in order. Checklists are also used to document completed actions. Any checklists or forms included as part of an activity should be referenced at the points in the procedure where they are to be used and then attached to the SOP.

In some cases, detailed checklists are prepared specifically for a given activity. In those cases, the SOP should describe, at least generally, how the checklist is to be prepared, or on what it is to be based. Copies of specific checklists should be then maintained in the file with the activity results and or with the SOP.

Remember that the checklist is not the SOP, but a part of the SOP.

17.7 Document Control

Each organization should develop a numbering system to systematically identify and label their SOPs, and the document control should be described in its Quality Management Plan. Generally, each page of an SOP should have control documentation notation, similar to that illustrated below. A short title and identification (ID) number can serve as a reference designation. The revision number and date are very useful in identifying the SOP in use when reviewing historical data and is critical when the need for evidentiary records is involved and when the activity is being reviewed. When the number of pages is indicated, the user can quickly check if the SOP is complete. Generally this type of document control notation is located in the upper right-hand corner of each document page following the title page.

Short Title/ID #
Rev. #:
Date:
Page 1 of

17.8 SOP Document Tracking and Archival

The organization should maintain a master list of all SOPs. This file or database should indicate the SOP number, version number, date of issuance, title, author, status, organizational division, branch, section, and any historical information regarding past versions. The QA Manager (or designee) is generally the individual responsible for maintaining a file listing all current quality-related SOPs used within the organization. If an electronic database is used, automatic "Review SOP" notices can be sent. Note that this list may be used also when audits are being considered or when questions are raised as to practices being followed within the organization.

The Quality Management Plan should indicate the individual(s) responsible for assuring that only the current version is used. That plan should also designated where, and how, outdated versions are to be maintained or archived in a manner to prevent their continued use, as well as to be available for historical data review.

Electronic storage and retrieval mechanisms are usually easier to access than a hard-copy document format. For the user, electronic access can be limited to a read-only format, thereby protecting against unauthorized changes made to the document.

17.9 SOP General Format

SOPs should be organized to ensure ease and efficiency in use and to be specific to the organization which develops it. There is no one "correct" format; and internal formatting will vary with each organization and with the type of SOP being written. Where possible break the information into a series of logical steps to avoid a long list. The level of detail provided in the SOP may differ based on, e.g., whether the process is critical, the frequency of that procedure being followed, the number of people who will use the SOP, and where training is not routinely available. A generalized format is discussed next.

I. **Title Page**

The first page or cover page of each SOP should contain the following information: a title that clearly identifies the activity or procedure, an SOP identification (ID) number, date of issue and or revision, the name of the applicable agency, division, and/or branch to which this SOP applies, and the signatures and signature dates of those individuals who prepared and approved the SOP. Electronic signatures are acceptable for SOPs maintained on a computerized database.

II. **Table of Contents**

A Table of Contents may be needed for quick reference, especially if the SOP is long, for locating information and to denote changes or revisions made only to certain sections of an SOP.

III. **Text**

- Well-written SOPs should first briefly describe the purpose of the work or process, including any regulatory information or standards that are appropriate to the SOP process, and the scope to indicate what is covered. Define any specialized or unusual terms either in a separate definition section or in the appropriate discussion section. Denote what sequential procedures should be followed, divided into significant sections; e.g., possible interferences, equipment needed, personnel qualifications, and safety considerations (preferably listed in bold to capture the attention of the user). Finally, describe next all appropriate QA and quality control (QC) activities for that procedure, and list any cited or significant references.

- As noted above, SOPs should be clearly worded so as to be readily understandable by a person knowledgeable with the general concept of the procedure, and the procedures should be written in a format that clearly describes the steps in order. Use of diagrams and flow charts help to break up long sections of text and to briefly summarize a series of steps for the reader.

- Attach any appropriate information, e.g., an SOP may reference other SOPs. In such a case, the following should be included:

  (a) Cite the other SOP and attach a copy, or reference where it may be easily located.

  (b) If the referenced SOP is not to be followed exactly, the required modification should be specified in the SOP at the section where the other SOP is cited.

STANDARD OPERATION PROCEDURE			
Title			
SOP No. and Version No		Department	
Effective Date		Page No	
Objective Scope Responsibility Definition To be Reviewed Before Procedure Reference Abbreviations Annexure Format			
Prepared By (Sign and Date)	Reviewed By (Sign and Date)	Approved By (Sign and Date)	

17.10 Definitions

(i) *Appendix:* An appendix is a section added at the end of a document to provide additional information.

(ii) *Controlled Copy:* A formal copy of the latest, approved version of a document. A controlled copy must be systematically tracked, updated and stored for use.

(iii) *Uncontrolled Copy:* An informal copy of a document for which no attempt is made to update it after distribution. Copies of documents made by users (in paper or electronic form) are considered "uncontrolled copies". The responsibility of making sure a document is the most current approved document is with the user of the document. Documented Information:

information required to be controlled and maintained by an organization and the medium on which it is contained. (ISO 9000:2015(E))

(iv) ***Effectiveness:*** Extent to which planned activities are realized and planned results achieved. (ISO 9000:2015 (E))

(v) ***Effective Date:*** Date the documents are signed by the Chair person and it officially begins to be used.

(vi) ***Form:*** A document used to facilitate procedural implementation or document procedural objectives

(vii) ***Document Control:*** Ensuring that documents are reviewed for adequacy, approved for release, distributed to and used at the location where the prescribed activity is performed. Obsolete documents are to be retained.

Document Issue, Retrieval and Reconciliation Register									
Document Name :									
Issuance Detail						Retrieval Details			
Date of issuance	Document no/Format no/Checklist No	No. Copy issued	Receiving Department	Issued By (Sign and Date)	Issued to (Sign and Date)	Return By (Sign and Date)	Received By (Sign and Date)	Destroyed By (Sign and Date	Remark

(viii) ***Document Number:*** A unique identifier assigned by the firm to differentiate documents (Procedure, Form or Guidance) and their versions. All document numbers will include a standard prefix and a unique alpha-numeric identifier for a type of document and its version.

(ix) ***Form Number:*** A unique alpha-numeric identifier assigned by the firm used to differentiate forms and their versions. The Form Number consists of a prefix "F" (Form) followed by five digits. The first four digits designate the procedure the form is associated with. The next digit indicates the sequential number of the form.

(x) ***Implementation Date:*** Date that staff has received training and documents are posted to use.

(xi) ***Master Document List:*** A list of released documents maintained by the firm and including the Document Number, Version date, Revision Date and Publication Date.

(xii) ***Master file:*** This file includes all signed original documents and all electronic files.

(xiii) ***Document History Record:*** The Document History Record contains version and publication dates, document and version numbers, approval signatures, and revision records.

Document History Record			
SOP Name and Number	**Version No and Effective date**	**Description of changes**	**Changes done by**

Chapter - 18

GLP Specification Management

Introduction

A specification is defined as a list of tests, references to analytical procedures, and appropriate acceptance criteria which are numerical limits, ranges, or other criteria for the tests described. It establishes the set of criteria to which a drug substance, drug product or materials at other stages of its manufacture should conform to be considered acceptable for its intended use. "Conformance to specification" means that the drug substance and drug product, when tested according to the listed analytical procedures, will meet the acceptance criteria. Specifications are critical quality standards that are proposed and justified by the manufacturer and approved by regulatory authorities as conditions of approval.

Specifications are one part of a total control strategy designed to ensure product quality and consistency. Other parts of this strategy include thorough product characterization during development, upon which many of the specifications are based, adherence to Good Manufacturing Practices, a validated manufacturing process, raw materials testing, in-process testing, stability testing, etc.

Specifications are chosen to confirm the quality of the drug substance and drug product rather than to establish full characterization and should focus on those molecular and biological characteristics found to be useful in ensuring the safety and efficacy of the product.

"A specification is defined as a list of tests, references to analytical procedures, and appropriate acceptance criteria which are numerical limits, ranges, or other criteria for the tests described. It establishes the set of criteria to which a drug substance, drug product or materials at other stages of its manufacture should conform to be considered acceptable for its intended use. "Conformance to specification" means that the drug substance and drug product, when tested according to the listed analytical procedures, will meet the acceptance criteria. Specifications are critical quality standards that are proposed and justified by the manufacturer and approved by regulatory authorities as conditions of approval."

18.1 Procedures and Requirements

Documentation is an essential part of the Quality assurance system and, as such, shall be related to all aspects Good Manufacturing Practices (GMP). Its aim is to define the specifications for all materials, method of manufacture and control, to ensure that all personnel concerned with manufacture know the information necessary to decide whether or not to release a bath of drug for sale and to provide an audit trail that shall permit investigation of the history of any suspected defective batch. There shall be authorized and dated specifications for all materials, products, reagents and solvents including test of identity, content, purity and quality. These shall include specifications for water, solvents and reagents used in analysis.

Each specification for raw materials, intermediates, final products, and packing materials shall be approved by Quality Assurance and maintained by the Quality Control Department. Periodic revisions of the specifications shall be carried out wherever changes are necessary. Change history for each specification should be maintained. Typical flow for QC specification.

18.2 Types of the Laboratory Specifications

A. Raw materials and Packaging materials Specification should include

- The designated name and internal code reference;
- Reference, if any, to a pharmacopoeial monograph;
- Qualitative and quantitative requirements with acceptance limits;
- Name and address of manufacturer or supplier and original manufacturer of the material;
- Specimen of printed material;
- Directions for sampling and testing or reference to procedures;
- Storage conditions; and maximum period of storage before re-testing.

B. Product containers and closures should include

- All containers and closures intended for use shall comply with the pharmacopoeial requirements. Suitable validated test methods, sample sizes, specifications, cleaning procedure and sterilization procedure, wherever indicated, shall be strictly followed to ensure that these are not reactive, additive, absorptive, or leach to an extent that significantly affects the quality or purity of the drug. No second hand or used containers and closures shall be used. Whenever bottles are being used, the written schedule of cleaning shall be laid down and followed.
- Where bottles are not dried after washing, they should be rinsed with de-ionised water or distilled water, as the case may be.

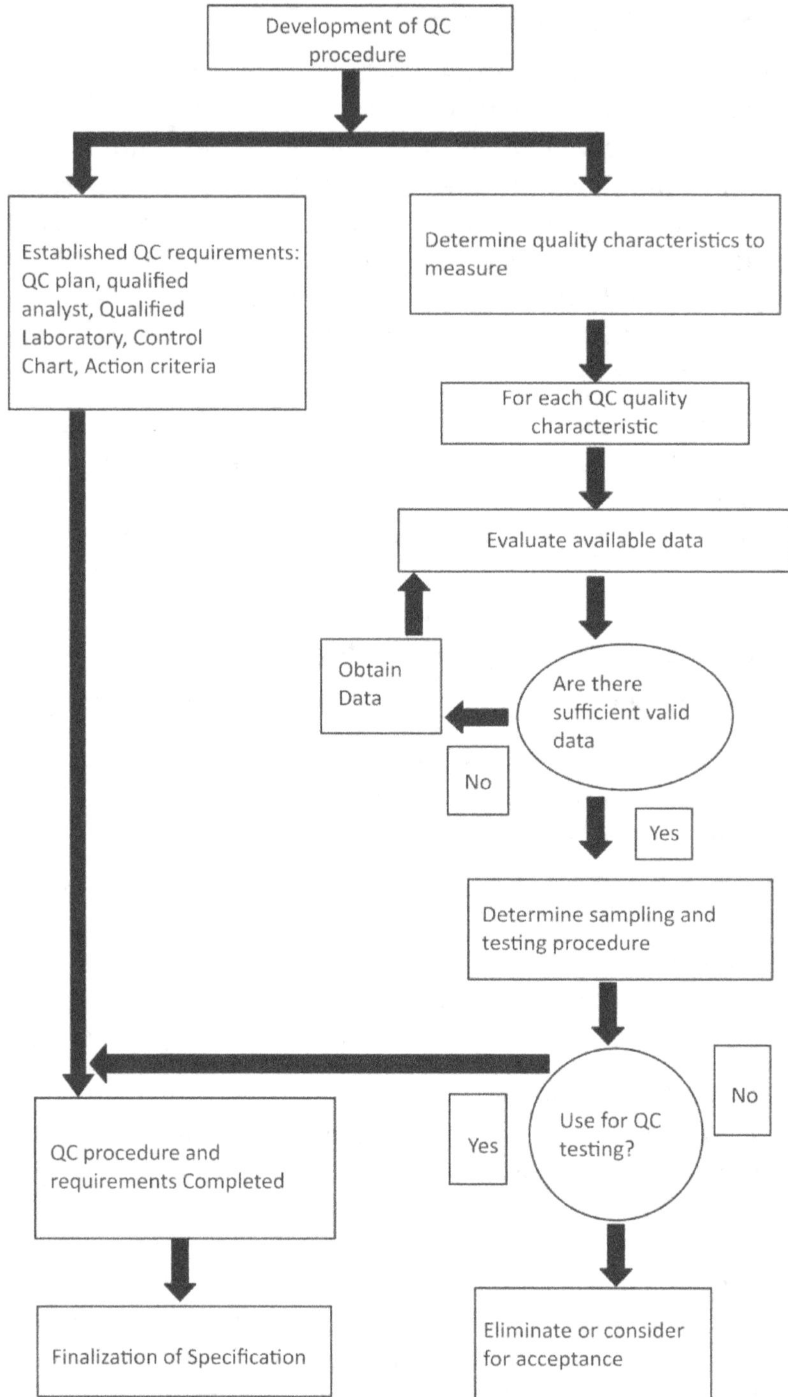

C. In-process and bulk products should include

- Specifications for in-process material, intermediate and bulk products shall be available. The specifications should be authenticated.

D. Finished products should include

- The designated name of the product and the code reference;
- The formula or a reference to the formula and the pharmacopoeial reference;
- Directions for sampling and testing or a reference to procedures;
- A description of the dosage form and package details;
- The qualitative and quantitative requirements, with the acceptance limits for release;
- The storage conditions and precautions, where applicable, and the shelf-life.

18.3 Specification Template

- Specification template is important to control the documents and gives mandatory requirements for testing the parameter for all test methods under the regulatory requirement.

A. Specification Format and Structure

Specification is generally prepared in "Portrait", font to "Ariel/Calibre/Times New Roma", with font size at least 12 (readable) with line spacing to 1.5 the "alignment" to "justify" and font style to "Regular". Heading and subheading should be bold. Number clauses should restricted upto "3 digit", beyond that "alphanumerical or bullets" can be used.

B. Specification page setup

	A4 size paper or its equivalent	
	Width	8.30
	Height	11.60
Paper	Line Spacing	1.5
	Alignment	Justify
	Font style	Regular
Layout	Header	0.5"
	Footer	0.25"
	Top	0.5"
	Bottom	0.5"
	Left	1.0"
Margins	Right	0.9
	Gutter	0
	Gutter Position	Left
	Orientation	Portrait
Apply to	Whole Document	

C. Specification Header

Header of specification should be brief and self-descriptive in Bold with 1.5" spacing. Content should be bold and numbering should be alphanumerical or numerical. Company logo, title, Document number and its version number, Category, Department and page number (X of Y pattern) should be part of Header. Version control of Speciation should be start with "0" and increment serially upon subsequent revision.

Specification and Method of Analysis			Log
Title		Document and Version No	
Material Code		Category	
Department		Effective date	
		Page Number	

D. Specification Footer

Footer number should be available on all pages which contained Prepared by, reviewed by and approved by with sign and date.

Prepared By	Checked By	Approved By
Name	Name	Name
Designee	Designee	Designee
Employee ID	Employee ID	Employee ID
Sign and date	Sign and date	Sign and date

E. Specification Structure

Specification numbering system and code should be defined in Quality unit for respective Business unit and should be approved by Quality Assurance. Printing of specification shall be done by originating department and final soft copy of the specification should be handover to QA for archival. Printed Specification signed should be done using a "blue or black ink pen". Whenever provision for electronic signature through validated platform is available. Author of Specification should sign and date in respective place of "Prepared by" section. In-charge or designated person should sign and date in the respective place of "Revision by" section. Once the signature of the author and reviewed are completed, the specification shall be signed by Quality Unit. Following is the format of the specification.

Specification and Method of Analysis			Log	
Title		**Document and Version No**		
Material Code		**Category**		
Department		**Effective date**		
		Page Number		

| Chemical name, formula, structure and weight |
| Safety Precaution |
| Storage condition |
| Quantity of sample for Chemical and Microbial test |
| Acceptance criteria for each test with reference |
| Procedure for Test and Methods of Analysis |
| Calculation of the results |
| Disposal of Sample |

Prepared By	Checked By	Approved By
Name	Name	Name
Designee	Designee	Designee
Employee ID	Employee ID	Employee ID
Sign and date	Sign and date	Sign and date

Chapter - 19

GLP Certificate of Analysis (COA) Management

Introduction

The COA is a legal document that certifies the quality that the batch conforms to the defined specifications, has been manufactured under GMP, and is suitable for use in pharmaceuticals and human dose. COA is provided to the user for each batch and/or delivery .The Certificate of Analysis is the document which lists the results of tests proving the purity of a pharmaceutical product. A Certificate of Analysis is a document issued by Quality Assurance that confirms that a regulated product meets its product specification. They commonly contain the actual results obtained from testing performed as part of quality control of an individual batch of a product. In accordance with GMP, the certificate can be used in lieu of testing by the manufacturer, provided that the reliability of the analysis. Certificates must be originals (not copies or duplicates) or their authenticity must otherwise be assured. There are Co As for excipients, APIs, packaging materials and finished products. It is important that all pages of the COA are numbered and include the total number of pages for document control and to assure the customer that all pages of the COA are present.

19.1 Process of Certificate of Analysis (COA)

The Certificate should list each test performed in accordance with compendial or customer requirements, including the acceptance limits, and the numerical results obtained (if test results are numerical). Information on the name of the intermediate or API or product including where appropriate its grade, the batch number, and the date of release should be provided on the Certificate of Analysis. For intermediates or APIs with an expiry date, the expiry date should be provided on the label and Certificate of Analysis. For intermediates or APIs with a retest date, the retest date should be indicated on the label and/or Certificate of Analysis. Certificates should be dated and signed by authorised personnel of the quality unit(s) and should show the name, address and telephone number of the original manufacturer. Where the analysis has been carried out by a repacker or reprocessor, the Certificate of Analysis should show the name, address and telephone number of the repacker/reprocessor and a reference to the name of the original manufacturer. If new Certificates are issued by or on behalf of repackers/reprocessors, agents or brokers, these Certificates should show the name, address and telephone number of the laboratory that performed the analysis. They should also contain a reference to the name and address of the original manufacturer and to the original batch Certificate, a copy of which should be attached.

19.2 Contents of COA

The certificate should include:

- The name and address of the laboratory performing the tests.
- The registration number of the certificate of analysis.
- Name of the material or product and the dosage form
- The name, description (i.e. grade, quantity received, type of container) and number (used by the original manufacturer and repacker/trader) of the batch for which the certificate is issued, the date of manufacture, and the expiry date (or retest date).
- Reference to the relevant specifications and testing procedures,
- The date on which the batch for which the certificate is issued was received.
- A reference to the test procedure used, including the acceptance criteria (limits).
- The results of all tests performed on the batch for which the certificate is issued (in numerical form, where applicable) and a comparison with the established acceptance criteria (limits).
- Any additional test results obtained on samples from the batch as part of a periodic statistically based testing programme.
- A statement indicating whether the results were found to comply with the requirements.
- The date(s) on which the test(s) was (were) performed.
- The signature of the head of the laboratory or an authorized person.
- The name, address, and telephone and fax numbers of the original manufacturer. If supplied by repackers or traders, the certificate should show the name, address, and telephone and fax numbers of the repacker/trader and a reference to the original manufacturer.

- A statement of the expected conditions of shipping, packaging, storage and distribution, deviation from which would invalidate the certificate.
- A copy of the certificate generated by the original manufacturer, if the sample is supplied by a repacked or trader.

19.3 Control on COA

Authorized person list should be available for compilation of the results, change in the specifications, enter the batch details in the predefined COA format. Review, approval and issuance of COA shall be performed by authorized person. In case of generation of COAs for different markets against the common specification or different Packing code for different markets, COA should be generated as per the request provided with different markets, COA should be generated as per the request provided with different markets names, after verification of particular product name and related approval for supply to the respective country or market through regulatory authority. Based on the request the number of copies shall be issued. In case of specific requirements from regulatory agency, COA's should be prepared accordingly.

19.4 Format of COA

<table>
<tr><td colspan="4" align="center">Firm Address and logo
CERIFICATE OF ANALYSIS</td></tr>
<tr><td>Product Name</td><td>Batch No.</td><td colspan="2">Retest/Expiry Date</td></tr>
<tr><td>Customer Name</td><td>Specification No.</td><td colspan="2">Analytical Report No.</td></tr>
<tr><td colspan="4">Storage Condition :</td></tr>
<tr><td align="center">Sr. No</td><td align="center">Test</td><td align="center">Specification</td><td align="center">Results</td></tr>
<tr><td></td><td></td><td></td><td></td></tr>
<tr><td></td><td></td><td></td><td></td></tr>
<tr><td colspan="4">The product Confirm to _________ specification.
Remark</td></tr>
<tr><td>Prepared by:
Sign and Date</td><td>Checked by:
Sign and Date</td><td colspan="2">Approved by:
Sign and Date</td></tr>
</table>

Chapter - 20

Analytical Rounding of Results

Introduction

When a number is obtained by calculations, its accuracy depends on the accuracy of the number used in the calculation. To limit numerical errors, an extra significant figure is retained during calculations, and the final answer rounded to the proper number of significant figures.

The observed or calculated values shall be rounded off to the number of decimal places that is in agreement with the limit expression. Number should not be rounded until final calculations for the reportable value have been completed. Intermediate calculation (e.g slope for linearity) may be rounded for reporting purposes, but the original (not rounded) value should be used for any additional required calculation. Acceptance criteria are fixed numbers and are not rounded. In case of multiple preparation, the rounding off should be employs for final average reported value.

20.1 Rules Involved in the Rounding of Results

(a) Identify the first two significant digits. Moving from left to right, the first non-zero number is considered the first significant digit. Zeros, which follow a decimal point, when there are only zeros ahead of the decimal point, are not considered significant figures.

(b) When the digit next beyond the one to be retained is greater than five, increase the retained figure by one. For example: 2.453 becomes 2.5 to two significant figures.

(c) When the digit next beyond the one to be retained is exactly five, and the retained digit is even, leave it unchanged; conversely if the digit is odd, increase the retained figure by one (even/odd rounding). Thus, 3.450 becomes 3.4 but 3.550 becomes 3.6 to two significant figures.

(d) Note: Even/odd rounding of numbers provides a more balanced distribution of results. Use of computer spreadsheets to reduce data typically follows the practice of rounding up.

(e) When two or more figures are to the right of the last figure to be retained, consider them as a group in rounding decisions. Thus, in 2.4(501), the group (501) is considered to be greater than 5 while for 2.5(499), (499) is considered to be less than 5.

20.2 Definitions

- *Specification:* A specification is defined as a list of tests, reference to analytical procedures and appropriate criteria which are numerical limits, ranges, or other criteria for the tests described.
- *Rounding:* Mathematical reduction of numerical values to digit required according to the significant digit or tolerance established for the particular calculation.
- *Tolerance:* the upper an lower limits of arrange include the two values themselves and all intermediate values outside the limits. The limits expressed in specifications, regardless of whether the values are expressed as percentages or as absolute number.
- *Significant digit:* Number of digits after decimal or whole number, to be considered as significant or representative of the real values, specification limit for a particular method.

Chapter - 21

Handling of Residual Solvent

Introduction

Residual solvents in pharmaceuticals are defined here as organic volatile chemicals that are used or produced in the manufacture of drug substances or excipients, or in the preparation of drug products. The solvents are not completely removed by practical manufacturing techniques. There is no therapeutic benefit from residual solvents; all residual solvents should be removed to the extent possible to meet product specifications, good manufacturing practices, or other quality-based requirements. Residual solvent should be monitored for all drugs substance, excipients and drugs product. If the calculation results, equal to or below of the limit specified, no testing requires for residual solvent of drugs. If the calculated level is above the recommend level, the drug product shall be tested to ensure that the product shall be tested to ensure that the process had reduced the relevant solvent level to within the acceptable amount. Drug product shall be tested if a residual solvent is used during its manufacturing process.

21.1 Classification of Residual Solvents by Risk Assessment

Based on the risk to human health, residual solvent are classified in three classes as follows:

A. **Class 1 solvents:** Solvents to be avoided as they are human carcinogens and environmental hazards.

B. **Class 2 solvents:** Solvents to be limited. Non-genotoxic animal carcinogens or possible causative agents of other irreversible toxicity such as neurotoxicity or teratogenicity. Solvents suspected of other significant but reversible toxicities.

C. **Class 3 solvents**: Solvents with low toxic potential. Solvents with low toxic potential to man; no health-based exposure limit is needed. Class 3 solvents have PDEs of 50 mg or more per day.

21.2 Analytical Procedures

Residual solvents are typically determined using chromatographic techniques such as gas chromatography. Any harmonised procedures for determining levels of residual solvents as described in the pharmacopoeias should be used, if feasible. Otherwise, manufacturers would be free to select the most appropriate validated analytical procedure for a particular application. If only Class 3 solvents are present, a nonspecific method such as loss on drying may be used.

21.3 Residual Solvent Level Reporting

- If Class 1 solvents are likely to be present they shall be identified and quantified.
- If only class 2 solvents are likely to be present and greater than the option 1 limits, then they shall be qualified
- If only class 3 solvents are likely to be present and greater than the option 1 limits or loss on drying is greater than 0.5%, then they shall be quantified.
- If only class 2 solvents and Class solvents are likely to present and grater than option 1 limits and 0.5% respectively, then they shall be quantified.

21.4 Limits of Residual Solvents

A. Solvents to be avoided

Solvents in Class 1 should not be employed in the manufacture of drug substances, excipients, and drug products because of their unacceptable toxicity or their deleterious environmental effect.

Table 1 Class 1 solvents in pharmaceutical products (solvents that should be avoided)

Solvent	Concentration limit (ppm)	Concern
Benzene	2	Carcinogen
Carbon tetrachloride	4	Toxic and environmental hazard
1,2-Dichloroethane	5	Toxic
1,1-Dichloroethene	8	Toxic
1,1,1-Trichloroethane	1500	Environmental hazard

B. Solvents to be limited

Solvents in Table 2 should be limited in pharmaceutical products because of their inherent toxicity. PDEs are given to the nearest 0.1 mg/day, and concentrations are given to the nearest 10 ppm. The stated values do not reflect the necessary analytical precision of determination. Precision should be determined as part of the validation of the method.

Table 2 Class 2 solvents in pharmaceutical products

Solvent	PDE (mg/day)	Concentration limit (ppm)
Acetonitrile	4.1	410
Chlorobenzene	3.6	360
Chloroform	0.6	60
Cumene	0.7	70
Cyclohexane	38.8	3880
1,2-Dichloroethene	18.7	1870
Dichloromethane	6.0	600
1,2-Dimethoxyethane	1.0	100
N,N-Dimethylacetamide	10.9	1090
N,N-Dimethylformamide	8.8	880
1,4-Dioxane	3.8	380
2-Ethoxyethanol	1.6	160
Ethyleneglycol	6.2	620
Formamide	2.2	220
Hexane	2.9	290
Methanol	30.0	3000
2-Methoxyethanol	0.5	50
Methylbutyl ketone	0.5	50
Methylcyclohexane	11.8	1180
Methylisobutylketone	45	4500
N-Methylpyrrolidone	5.3	530
Nitromethane	0.5	50
Pyridine	2.0	200
Sulfolane	1.6	160
Tetrahydrofuran	7.2	720
Tetralin	1.0	100
Toluene	8.9	890
1,1,2-Trichloroethene	0.8	80
Xylene*	21.7	2170
*usually 60% m-xylene, 14% p-xylene, 9% o-xylene with 17% ethyl benzene		

C. Solvents with low toxic potential

Solvents in Class 3 (shown in Table 3) may be regarded as less toxic and of lower risk to human health. Class 3 includes no solvent known as a human health hazard at levels normally accepted in pharmaceuticals. However, there are no long-term toxicity or carcinogenicity studies for many of the solvents in Class 3. Available data indicate that they are less toxic in acute or short-term studies and negative in genotoxicity studies. It is considered that amounts of these residual solvents of 50 mg per day or less (corresponding to 5000 ppm or 0.5% under Option 1) would be acceptable without justification. Higher amounts may also be acceptable provided they are realistic in relation to manufacturing capability and good manufacturing practice.

Table 3 Class 3 solvents which should be limited by GMP or other quality-based requirements

Acetic acid	Heptane
Acetone	Isobutyl acetate
Anisole	Isopropyl acetate
1-Butanol	Methyl acetate
2-Butanol	3-Methyl-1-butanol
Butyl acetate	Methylethyl ketone
tert-Butylmethyl ether	2-Methyl-1-propanol
Dimethyl sulfoxide	Pentane
Ethanol	1-Pentanol
Ethyl acetate	1-Propanol
Ethyl ether	2-Propanol
Ethyl formate	Propyl acetate
Formic acid	Triethylamine

D. Solvents for which no adequate toxicological data was found

The following solvents (Table 4) may also be of interest to manufacturers of excipients, drug substances, or drug products. However, no adequate toxicological data on which to base a PDE was found. Manufacturers should supply justification for residual levels of these solvents in pharmaceutical products.

Table 4 Solvents for which no adequate toxicological data was found

1,1-Diethoxypropane	Methyl isopropyl ketone
1,1-Dimethoxymethane	Methyl tetra hydro furan
2,2-Dimethoxypropane	Petroleum ether
Isooctane	Trichloro acetic acid
Isopropyl ether	Trifluoro acetic acid

21.5 Diagram Relating to the Identification of Residual Solvents and the Application of Limit Tests

A flow diagram of the procedure is shown in Figure. When a residual solvent (Class 2 or Class 3) is present at a level of 0.1 per cent or greater then the content may be quantitatively determined by the method of standard additions.

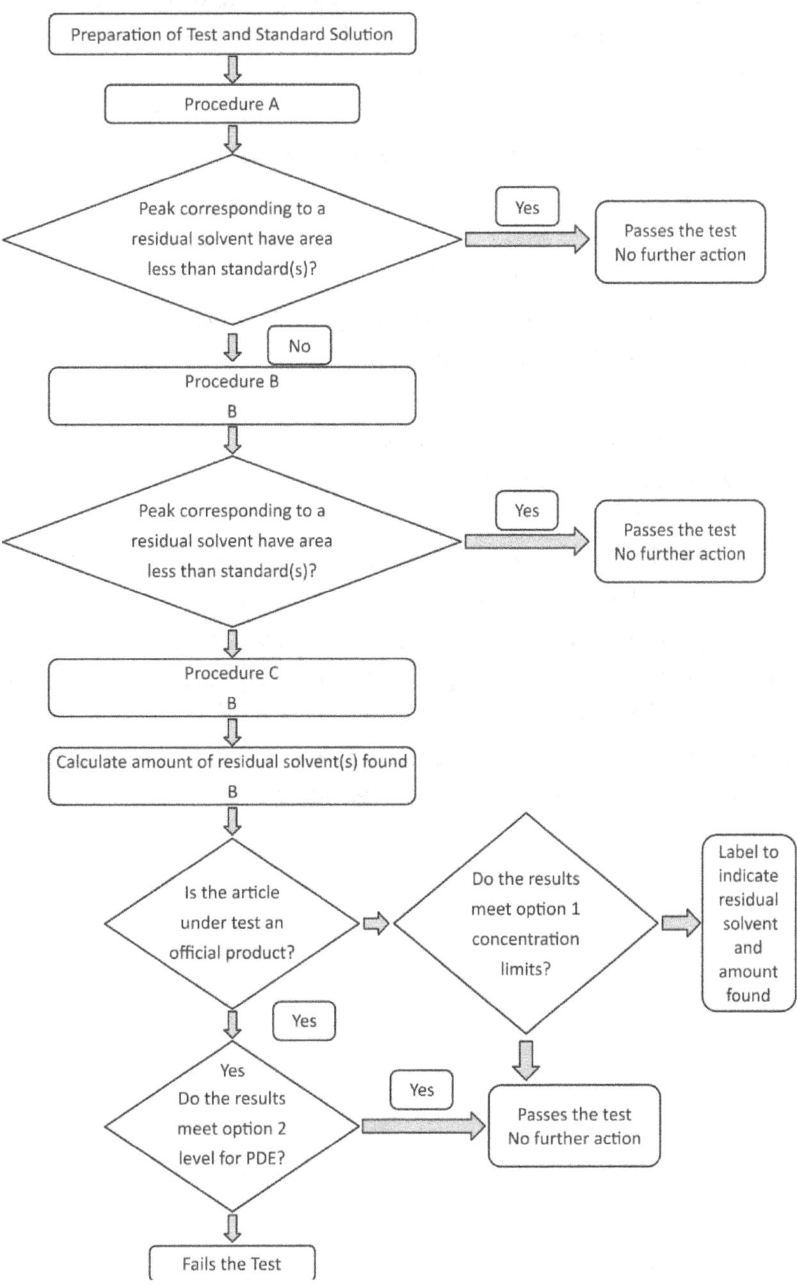

21.6 Glossary

1. ***Genotoxic Carcinogens:*** Carcinogens which produce cancer by affecting genes or chromosomes.

2. ***LOEL:*** Abbreviation for lowest-observed effect level.

3. **Lowest-Observed Effect Level:** The lowest dose of substance in a study or group of studies that produces biologically significant increases in frequency or severity of any effects in the exposed humans or animals.

4. ***Modifying Factor:*** A factor determined by professional judgment of a toxicologist and applied to bioassay data to relate that data safely to humans.

5. ***Neurotoxicity:*** The ability of a substance to cause adverse effects on the nervous system.

6. ***NOEL:*** Abbreviation for no-observed-effect level.

7. ***No-Observed-Effect Level:*** The highest dose of substance at which there are no biologically significant increases in frequency or severity of any effects in the exposed humans or animals.

8. ***PDE:*** Abbreviation for permitted daily exposure.

9. Permitted Daily Exposure: The maximum acceptable intake per day of residual solvent in pharmaceutical products.

10. ***Reversible Toxicity:*** The occurrence of harmful effects that are caused by a substance and which disappear after exposure to the substance ends.

11. ***Strongly Suspected Human Carcinogen:*** A substance for which there is no epidemiological evidence of carcinogenesis but there are positive genotoxicity data and clear evidence of carcinogenesis in rodents.

12. ***Teratogenicity:*** The occurrence of structural malformations in a developing fet us when a substance is administered during pregnancy.

Chapter - 22

Reserve Sample Management

Introduction

Reserve Sample s are the representative sample of each lot or batches of APIs, Excipient, Packing material, intermediates and finished product which are kept for purpose of future reference. Reserve sample is also called as control or retention sample. An appropriately identified reserve sample that is representative of each lot in each shipment of each active ingredient shall be retained. The reserve sample consists of at least twice the quantity necessary for all tests required to determine whether the active ingredient meets its established specifications, except for sterility and pyrogen testing.

22.1 Reserve Sample Collection Management

Reserve samples of drug product should be collected for each batch or lot, market order and each type of pack style. Reserve sample shall be collected and store in the same or stimulated or same immediate container-closure system s in which the drugs has been actually packed, labeled or marketed. Reserve sample should be collected throughout the packing time is start, middle and end of the batch, which gives representative of full batched. A list of quantity required quantity of reserve sample should be prepared and list should be periodically updated for addition or deletion or modification. Reserve sample list should be version control. The reserve sample quantity should be at least twice the quantity necessary for chemical analysis as per specification or as per any specific quantity is recommended by regulatory agencies. After collection of the sample, labeled it as "Reserve sample not for sale" and stored in the reserve sample room.

22.2 Reserve Sample Size and Retention Time

The reference sample should be of sufficient size to permit the carrying out, on, at least, two occasions, of the full analytical controls on the batch. Where it is necessary to do so, unopened packs should be used when carrying out each set of analytical controls. Any proposed exception to this should be justified to, and agreed with, the relevant competent authority. Where applicable, national requirements relating to the size of reference samples and, if necessary, retention samples should be followed.

Retention Time

A. Retention time for an active ingredient in a radioactive drug product, except for nonradioactive reagent kits:

- Three months after the expiration date of the last lot of the drug product containing the active ingredient if the expiration dating period of the drug product is 30 days or less; or

- Six months after the expiration date of the last lot of the drug product containing the active ingredient if the expiration dating period of the drug product is more than 30 days.

B. For an active ingredient in an OTC drug product shall be retained for 3 years after distribution of the last lot of the drug product containing the active ingredient.

C. Reserve samples of compressed medical gases need not be retained.

D. For a drug product other than above section, the reserve sample shall be retained for 1 year after the expiration date of the drug product.

E. materials should be retained for the duration of the shelf life of the finished product concerned

F. No need to retain the reserve samples of Rejected samples or volatile solvents/gases/water used in manufacturing process or hazardous materials (acid, alkalies etc.), flavors, liquid raw materials & excipients.

G. The period of retention time may be shortened if the period of stability of the material, as indicated in the relevant specification, is shorter.

22.3 Storage Management for Reserve Sample

After collection and labelling of the reserve sample "Reserve Sample, Not to Sale", details of reserve sample should be enter in "Reserve Sample Record" for each and every sample traceability. Records of Traceability of samples to be maintained & should be available for review by competent authorities Reserve sample shall be maintained for at least one year after the expiration of the drug product or as mentioned in Reserve Sample Retention Section. The reserve sample shall be store as per product storage requirements. Environmental temperature monitoring shall performed and recorded periodically. Any discrepancy in maintaining the environment conditions shall be investigated under quality notification. Access to these samples limited to authorized people.

22.4 Periodic Visual Inspection of Reserve Sample

Statistical approach shall be used for examination of reserve sample. Visual inspection reserve sample schedule shall be prepared for all representative samples. Visual inspection shall be done at least once a year for pack condition and description. All inspection results shall be recorded and maintained. Any evidence of reserve sample deterioration shall be investigated and documented.

22.5 Reserve Sample Issuance and Retrieval Management

Reserve sample are required for investigation of non-conformance. Withdraw of any reserve sample shall be approved by designee person by mentioning the reason of the removal of the reserve sample and quantity required.

22.6 Destruction of Reserve Sample

Every month Quality Control shall review the Reserve sample control register if any sample crossed the retention period. On completion of the retention period the Quality control should take approval for the destruction of sample. Upon destruction the details shall be record in "Reserve sample Control Register" with date and time

22.7 Format Used in Reserve Sample Management

A. Reserve sample Quantity records

Reserve sample record shall be maintained for the quantity to be with draw during packing run. It acts as ready reference on the shop floor for collection of the sample quantity for the Reserve sample Management.

Reserve Sample Quantity Register			
Version number:			Page Number
Sr. no.	Product Name	Specification number	Quantity of Reserve Sample
1			
2			
Prepared By: Sign and Date		Checked By: Sign and Date	Approved By: Sign and Date

B. Reserve Sample Records

<table>
<tr><td colspan="3" align="center">Reserve Sample Record</td></tr>
<tr><td colspan="2">Version number:</td><td>Page Number</td></tr>
<tr><td colspan="3">

Part A

Records should consist of following information

- Name of product
- Batch number
- Pack Type
- Market
- Reserve sample collected
- Manufacturing and Expiry Date
- Retain till date
- Location of Racks
- Responsible person sign and fate

Part B

- Issuance details
- Request number
- Quantity to be issued
- Balance quantity in Reserve sample management

Part C

- Disposed by date and sign

</td></tr>
<tr><td>Prepared By:
Sign and Date</td><td>Checked By:
Sign and Date</td><td>Approved By:
Sign and Date</td></tr>
</table>

C. Reserves Sample Request Management Log

Sr. No	Request No	Date	Product and Batch Number	Reason For Withdraw	Requested By Date and Sign	Quantity Issued	Quantity Issued by Sign and Date	Received by Sign and Date	Destruction or Returned Detail with Date and Signed
Reserve Sample Management log									

Sr. No	Request No	Date	Product and Batch Number	Reason For Withdraw	Requested By Date and Sign	Quantity Issued	Quantity Issued by Sign and Date	Received by Sign and Date	Destruction or Returned Detail with Date and Signed
1									
2									

D. Reserve Sample Visual Inspection Record

Reserve sample Visual Inspection Record							
Product Name		Batch No	Manufacturing Date	Expiry date	Packed Style	Physical Appearance	Description
Resting Frequency	Test ➡						
	Results						

Pharmaceutical Product Stability Management

Introduction

Stability testing is an important step to establish a re-test period for the drug substance or a shelf life for the drug product and recommended storage conditions. The objective of a stability study is to determine the shelf-life, namely the time period of storage at a specified condition within which the drug product still meets its established specifications .Stability studies are representative of quality of material, which provide evidence on how the quality of a drug substance or drug product varies with time under the influence of a variety of environmental factors, such as temperature, humidity, and light, and to establish a retest period for the drug substance or a shelf life for the drug product and recommended storage conditions. Three Key attributes of Drugs Product or substance are: Safety, Efficacy and Quality.

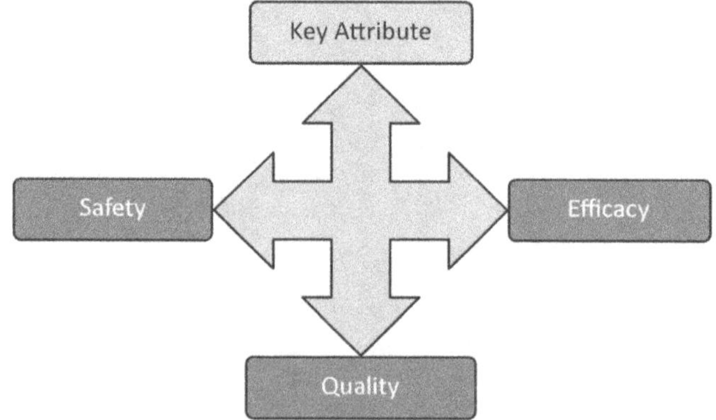

23.1 Stability Testing Condition and Testing Condition

In order to be able to reduce the amount of stability testing required, the number of different long-term testing conditions must be reduced to a sufficient extent. This approach was proposed by Paul Schumacher in 1972 (1) and by Wolfgang Grimm in 1986 (2), and in 1998 (3) when they defined four different long-term testing conditions, which match with the climatic conditions of the target markets categorized in just four different climatic zones.

Stability Testing Zone and Testing Condition			
WHO and ICH Q1A(R2) guidelines Long-term testing conditions (Temp/Humidity)	Testing conditions	Climatic zones	Long-term stability testing which could be realized in order to cover all climatic zones (Temp/Humidity)
21°C/45% RH	Long-term	I	25°C/60%RH
25°C/60% RH	Long-term	I and II	
30°C/35% RH	Long-term	III	30°C/65% RH
30°C/65% RH	Intermediate Long-term	I and II IVa	
30°C/75% RH	Long-term	IVb	30°C/75% RH

23.2 Testing of Stability Sample

Stability study should be conducted on market packed. The storage conditions and the lengths of studies chosen should be sufficient to cover storage and shipment. In case of Brazil market, stability studies for the drug product shall be conducted on the primary packed only.

Stability Study	Testing frequency (months)
Long Term*	0,3,6,9,12,18,24,36,48,60
Intermediate**	0,3,6,9,12
Accelerated***	0,1,2,3,6

- *Based on requirements, the long term study may be extended beyond and up to 60 months, as needed to ensure the expected or intended shelf life interval has been tested. Such successful long term study data may be useful for shelf life extensions via appropriate regular submission.

- **In case of any significant changes/failure in accelerated study, the intermediate study shall be performed.

- ***For accelerated study minimum three time points, including the initial and final points shall be performed (e.g: 0,1,2,6 or 0,1,3,6).

A. Photo stability testing

Photo stability testing is an integral part of stress testing, and should be conducted on at least one primary batch of the finished packed product as per approved protocol. The intrinsic photo stability characteristics of new drug substances and products should be evaluated to demonstrate that, as appropriate, light exposure does not result in unacceptable change. Decision flow for photo stability is as below.

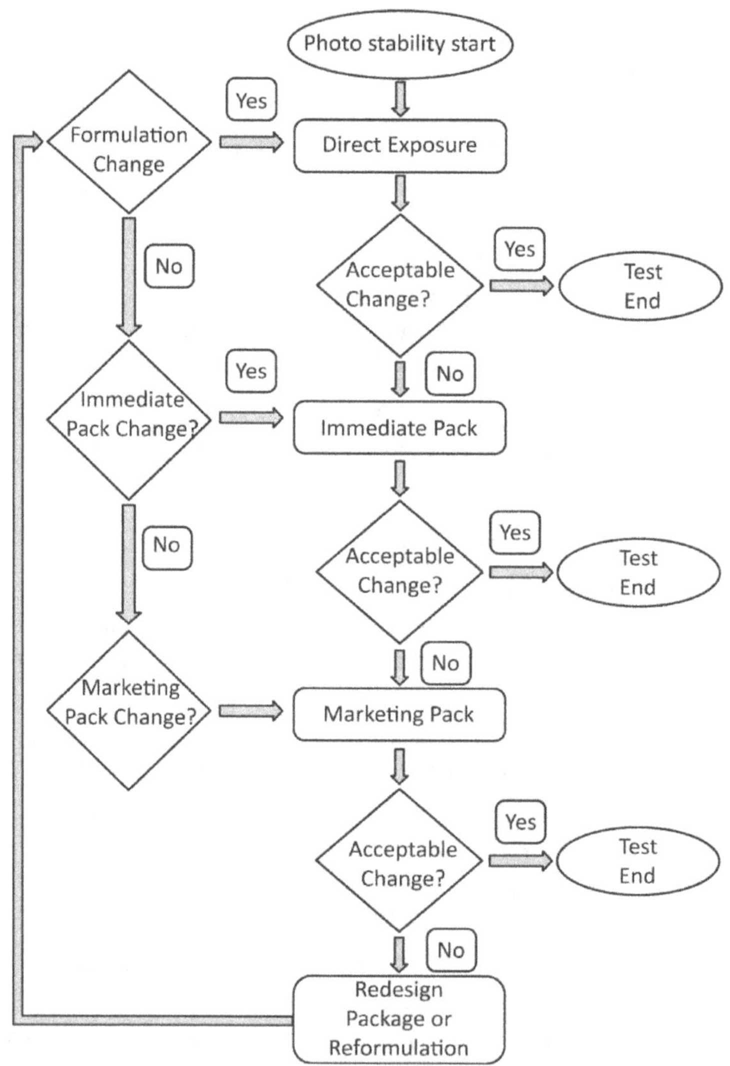

B. Orientation Product stability

The orientation of the product during storage, i.e. upright versus inverted, may need to be included in a approved protocol where contact of the product with the closure system may be expected to affect the stability of the products contained, or where there has been a change in the container closure system.

C. In-used stability and "Freezing, Thawing and overheating effect

In-used stability and "Freezing, Thawing and overheating effect and shelf life study shall be performed for drug product, wherever required as per country requirement through an approved protocol.

D. Bulk Product Stability

In case of drug product stored for long period before being packed and /or shipped from manufacturing site to packing site ,the stability data shall be generated on bulk product,.

E. Continuing Stability Testing

A Continuing Stability Programme (i.e. ongoing stability studies) should be implemented to ensure compliance with the approved shelf life specifications. A minimum of one batch of every strength of the drug product should be enrolled into the continuing stability programme each year. Bracketing and martyring can be applied, if scientifically justified. One batch (preferably first batch of the year) of drug substance or product shall be added to the stability monitoring program.

F. Cases when Stability study to be initiated

- New product
- Change in the manufacturing process
- Change in the composition of the drug product
- Change in packing profile and market
- Change in vendors of critical raw materials, or significant changes in existing vendor manufacturing for critical raw material
- Change in batch size

23.3 Selection of Batches for Stability Sample

Stability data should be provided for batches of the same formulation and dosage form in the container closure system proposed for marketing at least three primary batches of the drug substance.

23.4 Stability Sample Quantity

The total sample quantity shall be collected based on the number of storage conditions and testing frequencies for the respective study. The quantity of the stability sample shall also include sufficient extra samples which shall be used in case of any investigation. Stability studies shall be executed a per approved procedure/protocol.

Before charging the sample for stability, the sample shall be label with different colors of label for long term, intermediate, accelerated and refrigerator condition

Condition	Stick label
40 ± 2°C/75 % ±5 RH	Blue
25 ± 2°C/60 % ±5 RH	Yellow
30 ± 2°C/65 % ±5 RH	Green
30 ± 2°C/75 % ±5 RH	Orange
5 ± 3°C	Purple

23.5 Stability Protocol and Report

Before initiating the stability study, stability protocol shall be prepared and same shall be reviewed and approved.

Protocol shall have a unique number Stability protocol shall be revised for any additional or deletion of any test parameter, testing frequency/additional sample by change control and version control management.

23.6 Analysis of Stability Sample

The stability sample shall be analysed within specified time period for all test parameters as per approved protocol/validated procedure/validated specification. In general, accelerated study time periods should be tighter than long term time period for completion and long term completion interval should be targeted well less than one month. Any significant changes shall be assessed and investigated though quality notification.

23.7 Evaluation of Stability Data

A systematic approach to evaluate and determine the critical quality attributes of product. Evaluation of stability data include results from the physical, chemical, biological, and microbiological tests with supporting document .The basic concepts of stability data evaluation are the same for single- versus multi-factor studies and for full- versus reduced-design studies. Each attribute should be assessed separately, and an overall assessment should be made of the findings for the purpose of proposing a retest period or shelf life. The retest period or shelf life proposed should not exceed that predicted for any single attribute.

Evaluation of stability data for estimating the retest period/shelf life for the products proposed to be stored at storage conditions(other than storage in refrigerator and freezer) shall be follow as per below table.

A. Evaluation of stability data to establish retest period or shelf life, If the accelerated stability data for 6 months is satisfactory

Evaluation of stability data to establish retest period or shelf life		
If the accelerated stability data for 6 months is satisfactory		
Stability	**X**	**Y**
Accelerated Stability for 6 month found satisfactory	Long term(6 months)	Y=2x Retest date is 12 months
	Long term(9 months)	Y=2x Retest date is 18 months
	Long term(12 months)	Y=2x Retest date is 24 months
	Long term(18 months)	Y=x + 12 Retest date is 30 months
	Long term(24 months)	Y=x + 12 Retest date is 36 months
	Long term(36 months)	Y=x Retest date is 36 months

B. Evaluation of stability data to establish retest period or shelf life, If the accelerated stability data for 6 months is not satisfactory

Evaluation of stability data to establish retest period or shelf life		
If the accelerated stability data is not satisfactory within 6 months		
Stability	**X**	**Y**
Significant change in test results at Accelerated condition	Long term(6 months) Intermediate (6 months)	Y=x+3 Retest date is 9 months
	Long term(9 months) Intermediate (9 months)	Y=x+3 Retest date is 12 months
	Long term (12 months) Intermediate (12 months)	Y=x+3 Retest date is 15 months
Extrapolation of data shall not be permitted, if there is a significant change observed test results at intermediate storage condition.		

C. Retest period /shelf life for the products proposed to be stored at refrigerator

Evaluation of stability data to establish retest period or shelf life		
If the accelerated stability data for 6 months is found satisfactory		
Stability	**X**	**Y**
Significant change in test results at Accelerated condition	Long term(6 months)	Y=1.5x Retest date is 9 months
	Long term(9 months)	Y=1.5x Retest date is 13.5 months
	Long term(12 months)	Y=1.5x Retest date is 18 months
	Long term(18 months)	Y=x+6 Retest date is 24 months
	Long term(24 months)	Y=x No extrapolation beyond 24 months
Extrapolation of data shall not be permitted, if there is a change at accelerated condition within 3 months.		

D. Freezer conditions Products

While evaluating stability data for estimating the retest period/shelf life for the products proposed to be stored at freezer at freezer conditions, no extrapolation of data is permitted and the retest period assigned shall be based on the available ling term data.

23.8 Extrapolation of Stability Data

Extrapolation is the practice of using a known data set to infer information about future data. Extrapolation to extend the retest period or shelf life beyond the period covered by long-term data can be proposed in the application, particularly if no significant change is observed at the accelerated condition. Whether extrapolation of stability data is appropriate depends on the extent of knowledge about the change pattern, the goodness of fit of any mathematical model, and the existence of relevant supporting data. Any extrapolation should be performed such that the extended retest period or shelf life will be valid for a future batch released with test results close to the release acceptance criteria.

23.9 Significant Change at Accelerated Condition

If a "significant change" occurs between 3 and 6 months' testing at the accelerated storage condition, the proposed shelf-life should be based on the long term data available at the long term storage condition.

In general, "significant change" for a drug product is defined as:

1. A 5% change in assay from its initial value, or failure to meet the acceptance criteria;
2. Any degradation product exceeding the acceptance criterion;
3. Failure to meet the acceptance criteria for appearance, physical attributes, and functionality tests (e.g. colour, phase separation, re-suspend ability, caking, hardness, dose delivery per actuation); however, some changes in physical attributes (e.g., softening of suppositories, melting of creams) may be expected under accelerated conditions and as appropriate for the dosage form.
4. Failure to meet the acceptance criteria for pH;
5. Failure to meet the acceptance criteria for dissolution for 12 dosage units (capsule or tablet).

If the "significant change" occurs within the first 3 months testing at the accelerated storage condition, a discussion should be provided to address the effect of short term excursions outside the label storage condition, e.g., during shipping or handling. This discussion can be supported, if appropriate, by further testing on a single batch of the drug product for a period shorter than 3 months but with more frequent testing than usual. It is considered unnecessary to continue to test a drug product through 6 months when a "significant change" has occurred within the first 3 months.

This can be applied to products such as ointments, cream or suppositories that are impossible to test at accelerated condition where only long term testing is required.

Note: The following physical changes can be expected to occur at the accelerated condition and would not be considered significant change that calls for long term testing if there is no other significant change:

(a) Softening of a suppository that is designed to melt at 37 ºC, if the melting point is clearly demonstrated, failure to meet acceptance criteria for dissolution for 12 units of a gel at in capsule or gel-coated tablet if the failure can be unequivocally attributed to cross-linking.

(b) However, if phase separation of a semi-solid dosage form occurs at the accelerated condition, testing at the long term condition should be performed. Potential interaction effects should also be considered in establishing that there is no other significant change.

23.10 Destruction of Stability Samples

After compilation and approval of entire stability report, left over stability samples shall be discarded as per individual destruction Procedure.

23.11 Stability Storage Chamber Management

Stability chamber is also called as incubator ot environmental control chamber. It is an enclosed controlled environment for the storage of stability samples and is as such continuously monitored for temperature and relative humidity. Capacity of a drug substance or a drug product to comply with specification laid down for the duration of the shelf life assigned to it, when stored under the conditions stated on the label of the containers.

- Stability storage chamber, room equipment must appropriately mapped, qualified and calibrated.
- Stability chamber should be well designed, installed and qualified at its working range. During qualification and requalification of chamber empty chamber performance

verification, Door open recovery, Power failure recovery, load chamber performance check, sensor calibration, determination of Hot and cold point of chamber, Chamber shall be clean regular and details are recorded.

- Actual temperature and relative humidity shall be monitored daily. When the data loggers/software systems are used for monitoring, data shall be recorded/printed at the time of interval for 30 or 60 minute for all sensors.

- In case chamber break down or insufficient place in the stability chamber or failure, stability sample shall be transfer to other chamber of same condition in same or other units. All sample withdraw and loaded shall be recorded.

- Preventive maintenance of stability chamber shall be performed by considering cleaning of reservoir tank/boiling tank/heater coils, functioning of float valve, polishing of sensor and verification of alarm system.

- Back of stability environment data shall be done on periodic basis. Data shall be reviewed randomly for it legibility, accuracy and archival.

- Excursion handling: Short terms spikes sue to opening of door of the storage facility are acceptable. Excursion above validated limits shall be investigated though the quality notification.

23.12 Mean Kinetic Temperatures

Mean kinetic temperature is a useful tool in pharmaceutical stability studies .It helps to calculate the degradation of stability sample. Mean Kinetic Temperature (MKT) is a simplified way of expressing the overall effect of temperature fluctuations during storage or transit of product.

MKT is an expression of cumulative thermal stress experienced by a product at varying temperatures during storage and distribution. In other words, MKT is a calculated, single temperature that is analogous to the effects of temperature variations over a period of time. MKT is not a simple weighted average. The calculation of MKT gives the higher temperatures a greater weight when computing the average than would a simple numerical average or an arithmetic mean. This weighting is determined by a geometric transformation--the natural logarithm of the absolute temperature.

The International Conference on Harmonization (ICH) stability testing guidelines define MKT as "a single derived temperature, which, if maintained over a defined period, would afford the same thermal challenge to a pharmaceutical product as would have been experienced over a range of both higher and lower temperatures for an equivalent defined period".

By using this unequal weighting of the higher temperatures in a temperature series, MKT takes into consideration the accelerated rate of thermal degradation of materials at these higher temperatures. Therefore, MKT provides for the non-linear effect of temperature.

MKT is expressed as:

$$\dfrac{\Delta H/R}{-ln\left(\dfrac{e^{-\Delta H/RT1}+e^{-\Delta H/RT2}+e^{-\Delta H/RTn}}{n}\right)}$$

Where.

DH = activation energy (typically from 60 to 100 KJ/mol for solids and liquids)

R = 8.314472 J/mol-K (Universal gas constant)

T = temperature in degrees K

n = the number of sample periods over which data is collected

Note: In is the natural log and e_x is the natural log base.

Calculation of Mean Kinetic Temperature Data

Collect the temperature data of each stability chamber monthly through software and calculate the mean temperature and mean kinetic temperature. The highest and lowest temperature shall be considered for calculating the average temperature for each day. Calculate the mean kinetic as per formula. If calculated manually validated excel shall be used.

23.13 Format used for Stability Study Management

    A. Stability Protocol and Report Numbering record

    B. Stability Product Protocol

    C. Stability sample inward and outward register

    D. Stability data compilation report

    E. Stability Sample Schedule

A. Stability Protocol and Report Numbering record

Stability Protocol and Report Numbering Record						
Sr.no	Product Name	Strength	Protocol number	Report Number	Allocated by/Date	Authorized by/Date

B. Stability Product Protocol

<table>
<tr><td colspan="3">Stability Product Protocol</td></tr>
<tr><td colspan="3">Product Name :
Strength :
Market :
Batch /Lot No.:
Stability charging location :
Stability Conditions:</td></tr>
<tr><td colspan="3">Protocol Approval :</td></tr>
<tr><td>Prepared By :
Sign and Date</td><td>Checked By :
Sign and Date</td><td>Approved By :
Sign and Date</td></tr>
<tr><td colspan="3">

Table of content

1. Objective
2. Responsibility
3. Stability Plan
- Reference documents :
    - Batch manufacturing record
    - Batch Packing record
    - Stability Testing Specification
- Active Pharmaceutical Ingredient and Manufacture name with item code number
- Batch size
- Packing details
- Label claim
- Storage condition
- Number of Sampling unit for stability testing
4. Procedure
- Collection of sample for stability study
- Stability collection schedule
- Quantity for analysis
- Stability sample loading date
- Stability data compilation
</td></tr>
</table>

C. Stability Sample Inward and Outward Records

colspan Stability Sample Inward and Outward Records

<table>
<tr><td colspan="12" align="center">Stability Sample Inward and Outward Records</td></tr>
<tr><td colspan="6">Equipment Number:</td><td colspan="6">Stability Storage condition :</td></tr>
<tr><td rowspan="2">Sr. no</td><td rowspan="2">Sample Batch number</td><td colspan="2">Sample Loading</td><td rowspan="2">Done By Sign and Date</td><td rowspan="2">Checked By Sign and Date</td><td colspan="3">Sample Withdrawing</td><td rowspan="2">Recon-ciliation</td><td rowspan="2">Done By Sign and Date</td><td rowspan="2">Checked By Sign and Date</td><td rowspan="2">Remark</td></tr>
<tr><td>Quantity</td><td>Date Time</td><td>Testing interval</td><td>Quantity</td><td>Date Time</td></tr>
<tr><td></td><td></td><td></td><td></td><td></td><td></td><td></td><td></td><td></td><td></td><td></td><td></td></tr>
</table>

D. Stability compilation Report

Product Name and label claim :
Strength :
Market :
Batch /Lot No.:
Stability charging location :
Stability Conditions:
Testing interval :
Packing description:
Protocol No.

Interval	Specification No.						Compilation and Reviewed		
	Parameter	Description	Assay	RS	Water content	Remark	Prepared By Sign/Date	Checked By Sign/Date	Remark
Initial									
1 month									
2 month									

E. Stability Sample Schedule

Stability Sample Schedule for Year __																	
Product Name	Bach No	Interval	Quantity	Jan	Feb	Mar	April	May	June	July	Aug	Sep	Oct	Nov	Dec	Remark

Prepared By : Sign and Date	Checked By : Sign and Date	Approved By : Sign and Date

23.14 Definition

1. *Accelerated testing:* Studies designed to increase the rate of chemical degradation or physical change of a drug substance or drug product by using exaggerated storage conditions as part of the formal stability studies. Data from these studies, in addition to long-term stability studies, can be used to assess longer term chemical effects at no accelerated conditions and to evaluate the effect of short-term excursions outside the label storage conditions such as might occur during shipping. Results from accelerated testing studies are not always predictive of physical changes.

2. ***Intermediate testing:*** Studies conducted at 30°C/65% RH and designed to moderately increase the rate of chemical degradation or physical changes for a drug substance or drug product intended to be stored long-term at 25°C.

3. ***Long-term testing:*** Stability studies under the recommended storage condition for the retest period or shelf life proposed (or approved) for labelling.

4. ***Stress testing (drug product):*** Studies undertaken to assess the effect of severe conditions on the drug product. Such studies include photo stability testing (see ICH Q1B) and specific testing of certain products (e.g., metered dose inhalers, creams, emulsions, refrigerated aqueous liquid products).

5. ***Formal stability studies:*** Long-term and accelerated (and intermediate) studies undertaken on primary and/or commitment batches according to a prescribed stability protocol to establish or confirm the retest period of a drug substance or the shelf life of a drug product

6. ***Dosage form:*** A pharmaceutical product type (e.g., tablet, capsule, solution, cream) that a drug substance generally, but not necessarily, in association with excipients.

7. ***Drug product:*** The dosage form in the final immediate packaging intended for marketing.

8. ***Drug substance:*** The unformulated drug substance that may subsequently be formulated with excipients to produce the dosage form.

9. ***Retest date:*** The date after which samples of the drug substance should be examined to ensure that the material is still in compliance with the specification and thus suitable for use in the manufacture of a given drug product.

10. ***Expiration date:*** The date placed on the container label of a drug product designating the time prior to which a batch of the product is expected to remain within the approved shelf life specification, if stored under defined conditions, and after which it must not be used.

11. ***Retest period:*** The period of time during which the drug substance is expected to remain within its specification and, therefore, can be used in the manufacture of a given drug product, provided that the drug substance has been stored under the defined conditions. After this period, a batch of drug substance destined for use in the manufacture of a drug product should be retested for compliance with the specification and then used immediately. A batch of drug substance can be retested multiple times and a different portion of the batch used after each retest, as long as it continues to comply with the specification. For most biotechnological/biological substances known to be labile, it is more appropriate to establish a shelf life than a retest period. The same may be true for certain antibiotics.

12. ***Shelf life (also referred to as expiration dating period):*** The time period during which a drug product is expected to remain within the approved shelf life

specification, provided that it is stored under the conditions defined on the container label.

13. ***Storage condition tolerances:*** Defined as the acceptable variations in temperature and relative humidity of storage facilities for stability studies.

14. ***Bracketing:*** The design of a stability schedule such that only samples on the extremes of certainde sign factors (e.g., strength, package size) are tested at all time points as in a full design. There sign assumes that the stability of any intermediate levels is represented by the stability of the extremes tested. Where a range of strengths is to be tested, bracketing is applicable if the strengths are identical or very closely related in composition (e.g., for a tablet range made with different compression weights of a similar basic granulation, or a capsule range made by filling different plug fill weights of the same basic composition into different size capsule shells).Bracketing can be applied to different container sizes or different fills in the same container closure system.

15. ***Matrixing:*** The design of a stability schedule such that a selected subset of the total number of possible samples for all factor combinations is tested at a specified time point. At a subsequent time point, another subset of samples for all factor combinations is tested. The design assumes that the stability of each subset of samples tested represents the stability of all samples at a given time point. The differences in the samples for the same drug product should be identified as, for example, covering different batches, different strengths, different sizes of the same container closure system, and, possibly in some cases, different container closure systems

16. ***Container closure system:*** The sum of packaging components that together contain and protect the dosage form. This includes primary packaging components and secondary packaging components if the latter are intended to provide additional protection to the drug product. A packaging system is equivalent to a container closure system.

17. ***Impermeable containers:*** Containers that provide a permanent barrier to the passage of gases or solvents (e.g., sealed aluminium tubes for semi-solids, sealed glass ampoules for solutions).

18. ***Mean kinetic temperature:*** A single derived temperature that, if maintained over a defined period of time, affords the same thermal challenge to a drug substance or drug product as would be experienced over a range of both higher and lower temperatures for an equivalent defined period. The mean kinetic temperature is higher than the arithmetic mean temperature and takes into account the Arrhenius equation. When establishing the mean kinetic temperature for a defined period, the formula of J. D. Haynes (*J. Pharm. Sci.,* 60: 927-929, 1971) can be used.

19. ***Supporting data:*** Data, other than those from formal stability studies, that support the analytical procedures, the proposed retest period or shelf life, and the label storage statements. Such data include (1) stability data on early synthetic route batches of drug substance, small scale batches of materials, investigational formulations not proposed for marketing, related formulations, and product presented in containers and closures other than those proposed for marketing; (2) information regarding test results on containers; and (3) other scientific rationales.

Chapter - 24

Good Chromatographic Integration Practice in GLP

Introduction

Integration is a fundamental task that all chromatographers perform on a daily basis. Due to faster chromatography, which is creating more data every day, the amount of time spent on integration is ever increasing. Peaks of varying symmetries and size, overlapping peaks, valleys, shallow peak rises, and declines are just some of the different variants and effects that a chromatographer has to face. Other considerations include peak splitting, extreme fronting or tailing, shifting apexes and valleys of unresolved peaks, and baselines with large sloping background absorption and background noise.

24.1 Steps Involved Chromatographic Integration

Upon completion of analysis sample set or batch sequence, the processing method shall be set up to peak(s) of interest are identified and quantifies correctly. The integration parameters shall remain the same for all the chromatograms in the same sequence for the same type of test.

24.2 Integration of Well Resolved Peaks and Properly Integrated Single Peak

Integration parameters such as peak tailing, retention time and other parameters shall be set appropriately to detect peak(s) of interest."Base to Base" shall be the preferred type of integration for symmetrical peaks having major responses without co-eluting/overlapping peak(s) and where the baseline is stable.

24.3 Acceptable Practice of Peak Integration for a Single Peak during Assay Test

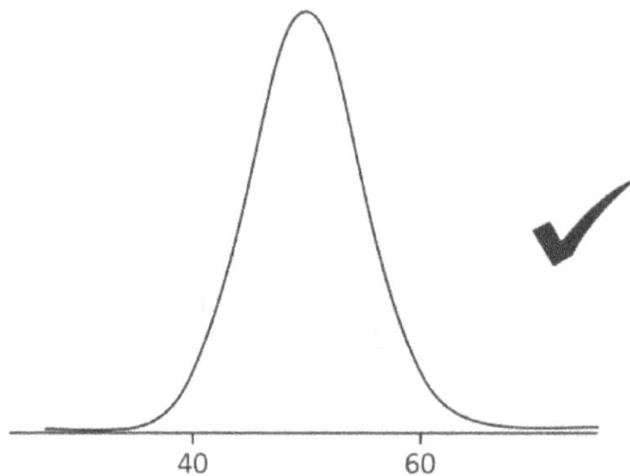

The peak is symmetrically shaped and exhibits no indication of coelution. The baseline is stable and returns to the same level (i.e., the baseline is flat). This is an example of baseline-to-baseline integration. Peaks of this nature are usually appropriately integrated automatically by the software. On occasion, the analyst must integrate a peak of this nature manually due to a retention time shift that causes the data system to incorrectly determine that the peak is not a target analyte.

A. Unacceptable practice of peak integration for a single peak during assay test

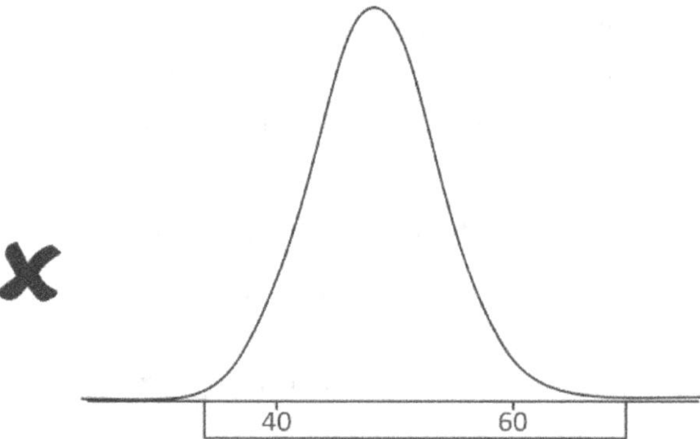

(i) ***Peak enhancing:*** Extending the area which does not belong to subject peak, this is over-estimates the peak area.

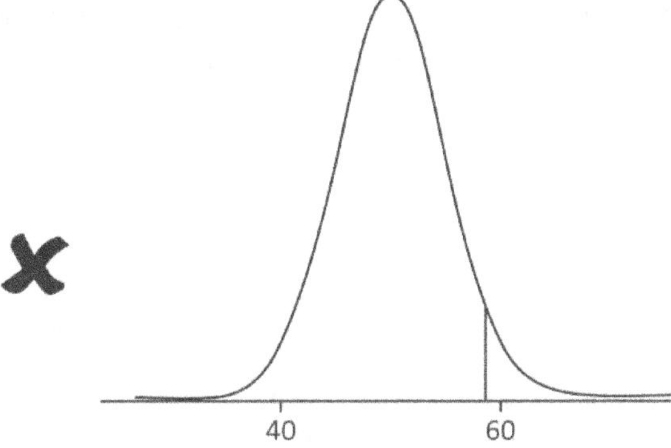

(ii) *Peak Shaving:* Eliminating a part of subjected peak area, this underestimates the peak area.

(iii) *Peak shaving by removing tail:* This is an example of an improperly integrated peak. The at ailing side of the peak has been removed eliminating significant area that should be included in the peak. This is not an example of removing an excessive area due to peak tailing because the Gaussian shape of the peak has clearly not been preserved in the integration.

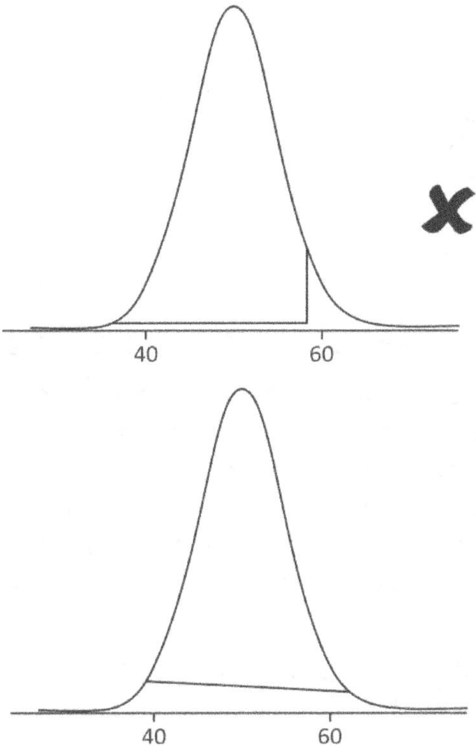

(iii) *Peak shaving through elevating the baseline:* This is an example of an Improperly elevated baseline. This clearly excludes a large area of the peak that a baseline-to-baseline integration would correct.

(iv) ***Gross peak shaving:*** This is an improperly integrated peak that includes both elevating the baseline and eliminating the leading and tailing edge of the peak.

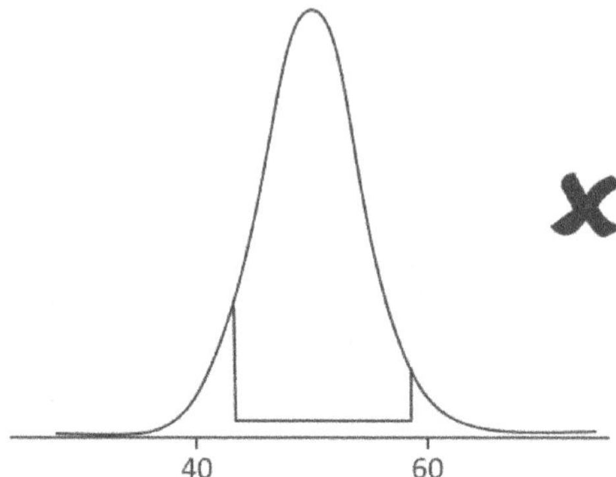

(v) ***Baseline above the original baseline:*** Baseline above the original baseline, this under-estimates the peak area.

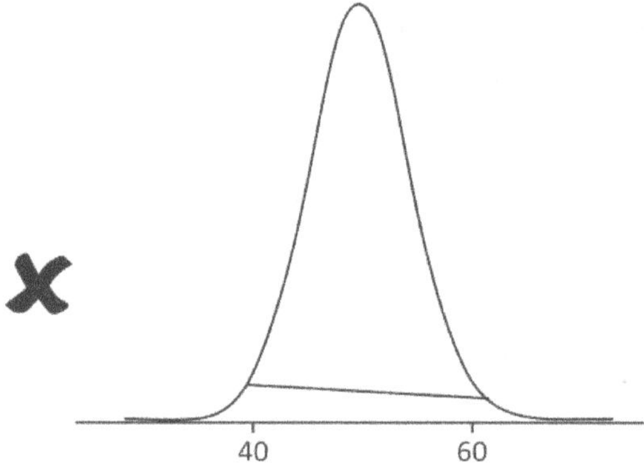

24.4 Chromatographic Integration of Unresolved Peaks

In the case of unresolved peak, "Base to Base" integration shall be used for symmetrical peaks having minimal co-eluting/overlapping peaks(s) and where the baseline is stable. Us of "Valley to Valley" integration may also be appropriate, if required by the sample chromatogram. For example, "valley to valley" could be used if a known baseline disturbance is present under a set of eluted peak. Performing "base to base" integration in this situation would over-estimate the area of the peaks by taking into account the baseline disturbance.

Another type of integration that could be used if the above types are unsuitable is a perpendicular drop to baseline. This can be used when peaks are significantly unresolved (but peak shape clearly defined) and where the baseline is stable. A vertical line is drawn from the valley between the unresolved peaks to the baseline and the outer sides of the peaks are integrated to the baseline as normal.

In the event of unresolved peaks where one peak has almost completely co-eluted with the other peak (undefined peak shape), using a perpendicular drop would significantly over-estimate the area of the peak. In this case it is acceptable to treat the undefined peak as a shoulder or rider peak and skim the peak .A peak may be considered a rider peak if it has an area <10% the area of the main peak.

I. Unresolved Peak

A. ***Acceptable practice of peak integration of unresolved peaks:*** Proper integration of several peaks which are not completely resolved (i.e., the response does not return to the baseline between peaks). In this instance the lowest point between the two peaks, the valley, is selected as the appropriate end point for the peaks

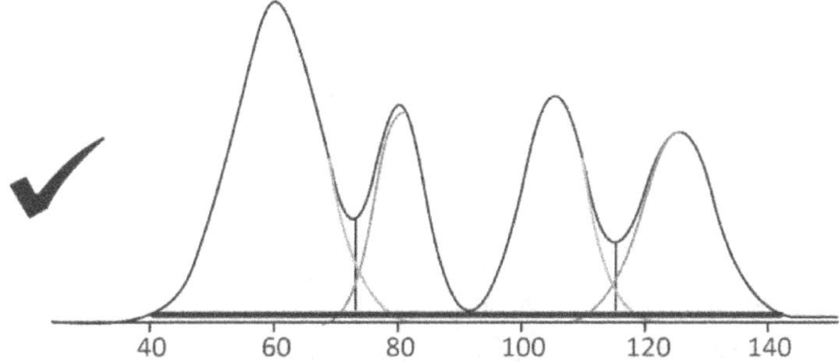

(i) ***Acceptable practice of Perpendicular drop to baseline:*** This includes the area of the each peak to the base line and gives the best representation of each peak area based on a steady baseline..mm

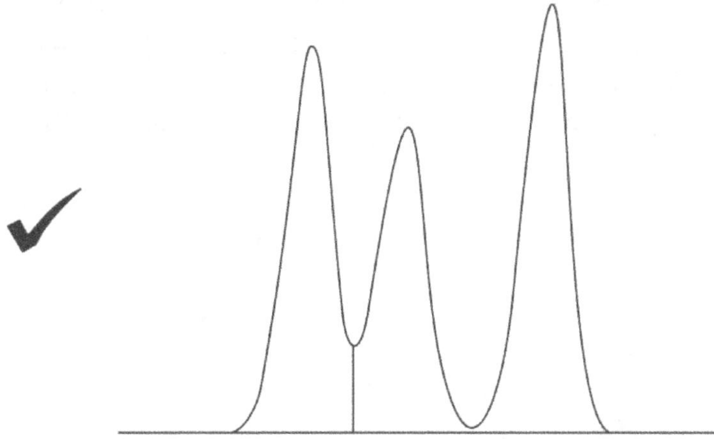

(ii) ***Acceptable practice of Proper integration to remove interfering peaks:*** Peaks with slight interferences either just prior to or immediately after the target peak. These interfering peaks are not resolved and may be included in the automatic integration.

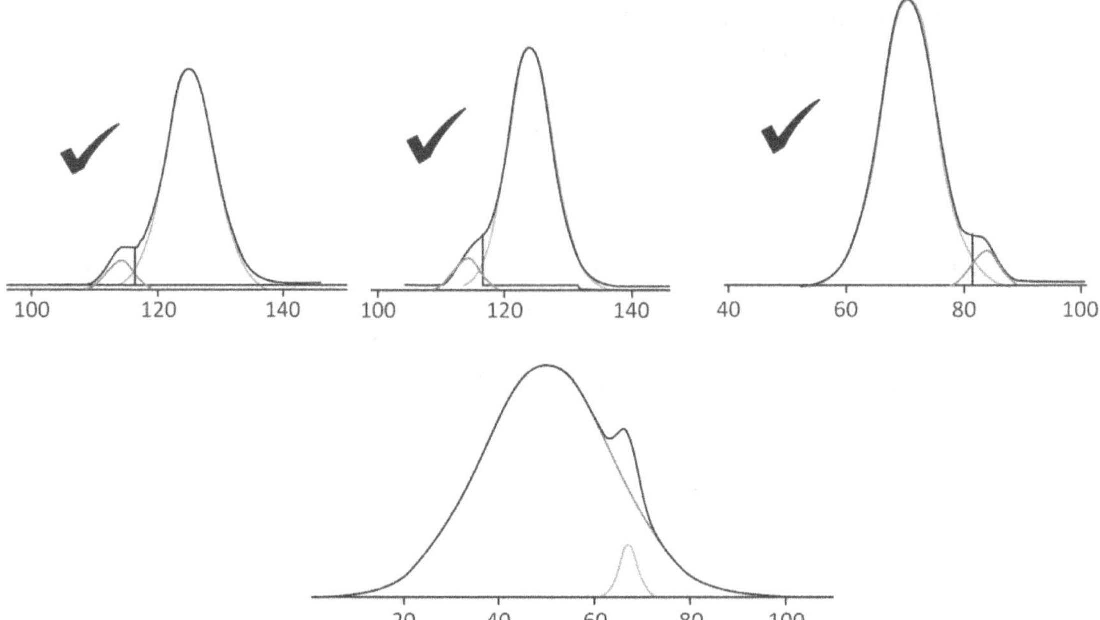

(iii) ***Acceptable practice of Peak shape requiring use of peak skimming:*** A peak that may require the use of more sophisticated software to remove the area due to a co eluting peak. Depending on the sophistication of the data system, it may be possible to remove the additional area. It is necessary that the resulting integration area preserve the Gaussian peak shape.

B. Unacceptable practice of peak integration of unresolved peaks:

(i) ***Valley to Valley:*** A tangent is drawn from then the baseline starts to rise, to the valley between the unresolved peaks. This is ignoring the area beneath the valley to the baseline, under estimating the area of the both peaks.

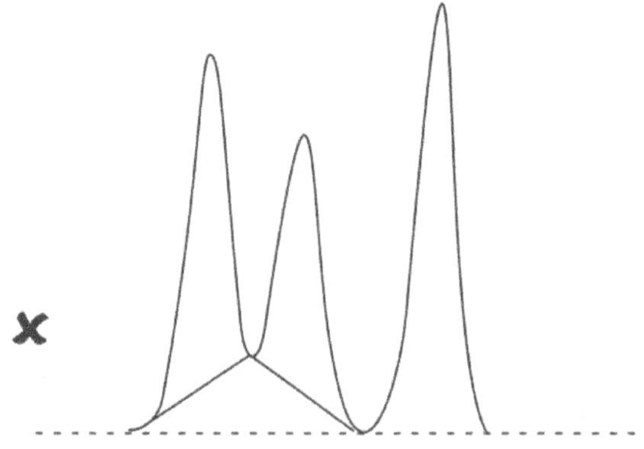

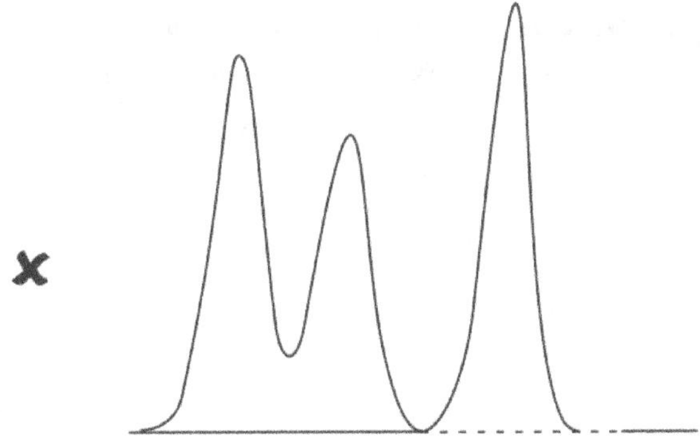

(ii) *Over-estimated Peak:* Improper integration which causes the area of the peak to be over-estimated.

II. Addition of co eluting peak or Shouldering with large peak

This is an example of co elution of the tailing edge of a peak resulting in additional area being included in the automated integration. The manual integration is performed to preserve the peak shape and eliminate the additional area.

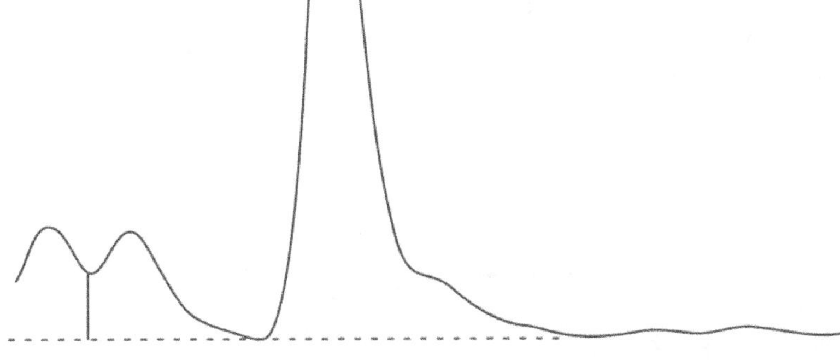

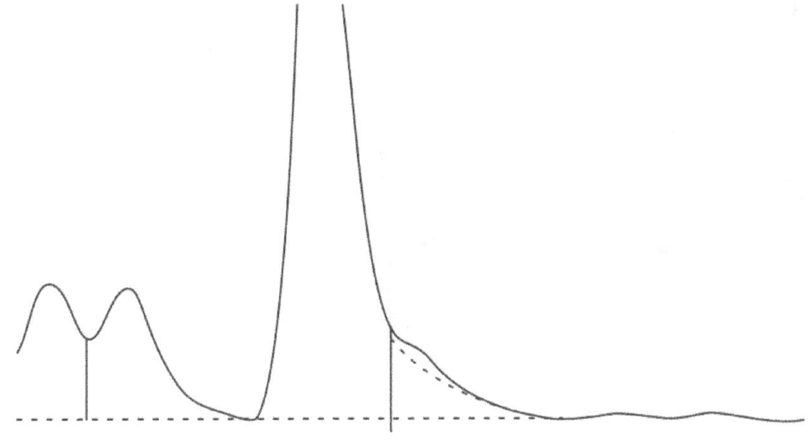

A. ***Acceptable Shouldering or Co-elution with large peak:*** Integrate form when the signal first starts to rise, to where the signal falls and meets the baseline disturbance.

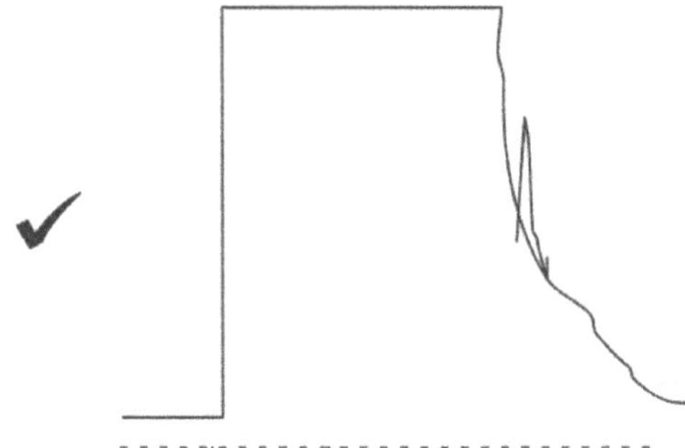

B. ***Unacceptable Shouldering or Co-elution with large peak:*** Using a perpendicular drop to baseline over-estimates the peaks Shouldering or co-elution with large peak

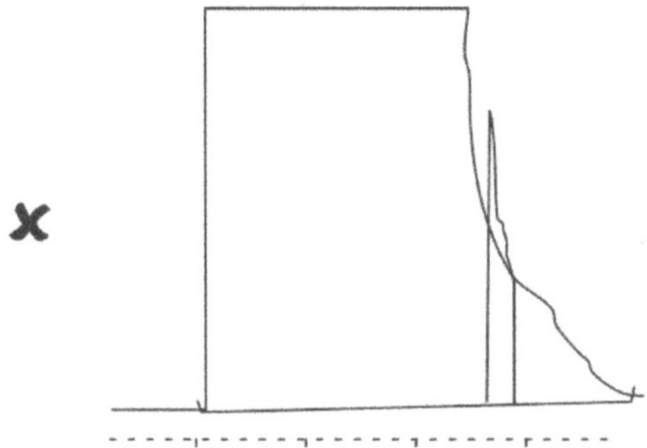

III. Spike or Baseline ripple

A. ***Acceptable Practice of integration of Spike or Baseline ripple:*** Start integration from the expected baseline and end when the peak drops and meets the baseline.

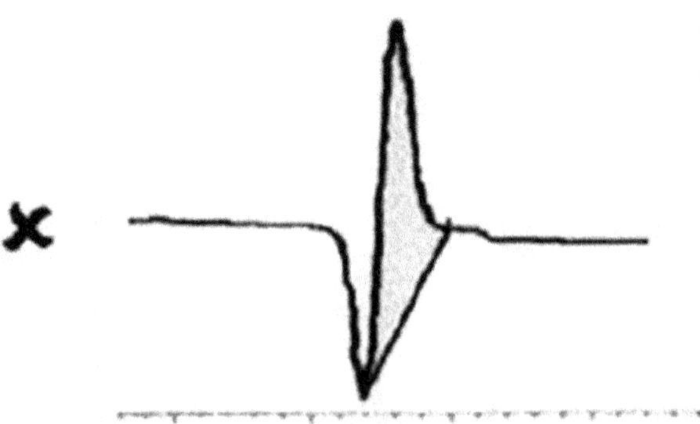

B. ***Unacceptable Practice of integration of Spike or Baseline ripple:*** Improper integration including baseline drop in peak which over-estimates peak area.

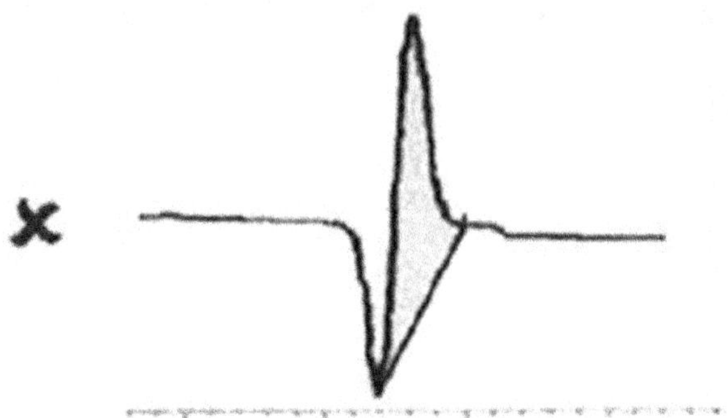

24.5 Noisy Baseline

For noisy baselines it may be applicable to inhibit integration of disregard particular peaks, such as blank/diluents peaks. On occasion, there may be a blank peak, or baseline shift that is present in both the blank and sample chromatogram. In the event that this causes interference near the peak of interest, it may be necessary to carry out a blank subtraction using the blank injection. This automatically substrate the blank chromatogram from all other chromatograms in the sequence, hence removing any interference from the peak/baseline shift question.

A. ***Noisy baseline:*** This is an example of a noisy baseline resulting in poor integration by the data system that is attempting to integrate using a valley-to-valley integration procedures. The appropriate integration (shown in the second figure) of the peak eliminates area associated with baseline changes and integrates only the target peak.

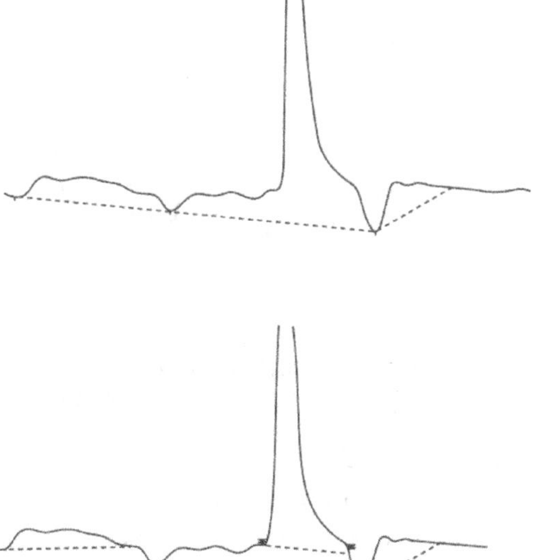

B. *Including area due to a noisy baseline:* This is an example of a peak tail that had not been included in the integration due to a noisy baseline. The tail of the peak should be integrated assuming a Gaussian peak shape as clearly indicated by the shape of the secondary ion trace.

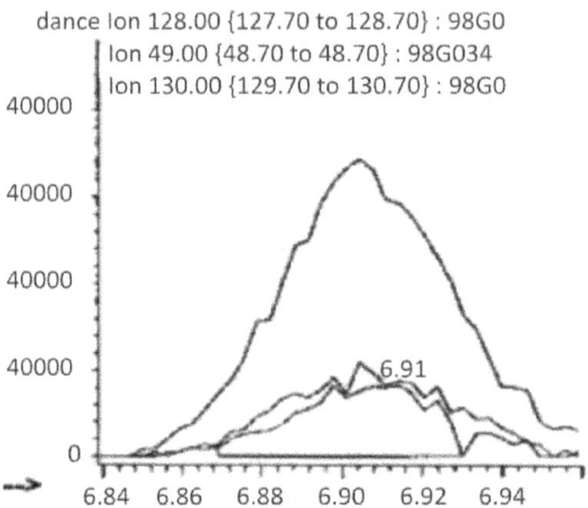

C. *Peak splitting:* This is an example of automated integration that can occur when detector response is noisy. The baseline is integrated at an excessively high level and the peak is split due to the noise observed at the top. (The peak is shown superimposed over a normal peak for comparison purposes.) The manual integration of this peak includes all the area reasonably attributable to the peak while excluding the noisy baseline

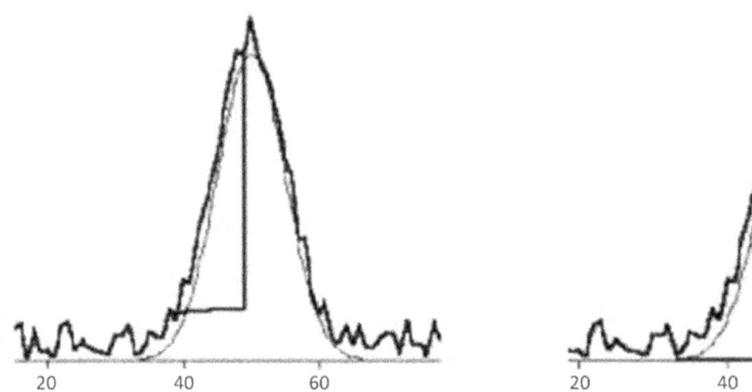

24.6 Integration for Tests Involving Qualification of Minor Peaks in Combination with Principle Peak(s)

The processing method shall be defined as per the test procedure, which will also include specific integration events such as inhibits integration or disregard limits. Naming conventions for processing methods shall be as per test procedure for components/peak e.g. main peak, known impurities, unknown impurities dilute peaks etc.

The preferred method for naming known impurities shall be based on their RRT values as stated in the test procedure .In exception to this, when the known impurities are incorporated into the test as part of SST criteria, the naming of the known impurities will be primarily based on the retention time (RT).This is done by comparing the retention of the known impurity in the SST to that of the peak present in the sample chromatogram with a matching chromatogram. Known impurities can still be named as per their RRT values if a comparison between the SST and test sample chromatograms is not relevant e.g there is a shift in retention time between the two injections.

When 3D spectral data is available, this can also be used as guidance in determining if the identification of a known impurity should be based on RRT or RT.

Where "disregards limit" are applicable, a note will be made in the relevant document printout indicating impurities that fall below the disregard limit, this will be accompanied by a manual calculation totalling the known/unknown impurities above the disregard limit.

Inhibit integration requirements will be instructed as per the test procedure and will clearly state the retention time where integration is to be inhibited e.g inhibit integration up to 3 minutes. They shall only be utilized to remove interference of HPLC/GC blank peaks/Diluent peaks. Inhibit integration may also be used at the analysts discretion in cases of noisy baseline, but only to remove interference from the peak types mentioned above.

24.7 Chromatographic Inhibit integration

Inhibit limits shall be designed in such a way that the retention times of known, unknown impurities and analyte peaks of interest are not suppressed. Actual "inhibit integration" timed events should only utilized to remove the interference of blank/Placebo peaks that are present in the test sample chromatograms as recommended in the respective specification. Any test sample peak that is showing substantially higher area or% area >0.05% response than the blank peak at the same retention time, shall be investigated. An overlay of chromatograms of the blank and sample shall be attached to the processing method to demonstrate the inhibit integration event.

24.8 Chromatographic Manual integrations

Manual integration is a perfectly acceptable and expected procedure to be performed on chromatographic data has not been integrated appropriately using the software's automatically integration procedures. Manual integration shall not routinely take place. It will only be performed in the event suitable chromatogram cannot be obtained by system or software.

"No manual Integration policy is a problem. Software aren't always right"

Manual integration is used to provide accurate quantisation of peak area where the original integration or identification provided by the data system is in error. This integration must only include the area attributable to the specific target compound. The area integrated must not include baseline background noise. Manual integration may only be used when peaks are distinct with a valley between the peaks. The area integrated must not extend past the point where the sides of the peak intersect with the baseline noise. Manual integration relies solely upon the experience of the analyst to determine proper integration for each peak and must be employed only by experienced analysts thoroughly trained in the chromatographic software. Integration parameters (both automated and manual) must adhere to valid scientific chromatographic principles. All data must be integrated consistently and appropriately in standards, samples and quality control samples.

An SOP or guideline for sample data reintegration should be established. This SOP or guideline should explain the reasons for reintegration and how the reintegration is to be performed. The rationale for the reintegration should be clearly described and documented. Original and reintegration data should be reported. For chromatogram where manual integration has occurred, a zoomed in chromatogram will be included to clearly show al minor peaks that have been integrated. In the event where some peaks have been inhibited due to diluents or placebo peaks, the sample chromatogram will be over-layed with the blank and included in the document so that the retention time and size of the peaks in question can be compared between sample and blank.

All computerized data reduction must be reviewed carefully by the analyst to determine the accuracy and appropriateness of the quantitation performed by the data system. Any errors must be corrected using this document as a guideline to define appropriate integration. The document must describe procedures for completing and documenting corrections to analytical results.

Documentation for reintegrated data. Documentation should include the initial and repeat integration results, the method used for reintegration, the reported result, assay run identification, the reason for the reintegration, the requestor of the reintegration, and the manager authorizing reintegration. Reintegration of a clinical or preclinical sample should be performed only under a predefined SOP.

Some of the "exceptional" cases where manual integration is needed are listed below including, but not limited to the following scenarios.

Type	Scenarios	Description
Limitations due to Software integration	Missed Peak	Peak is not detected
	Wrong Peak	Incorrectly detects a peak attribution
Complicated chromatography due to	Split peaks	Integration results in dividing the peak at the inflection point

Table *contd...*

Type	Scenarios	Description
sample and/or matrix interference	Shouldering or co-elution/Unresolved	Where there is no perpendicular drop for unresolved peaks or other improper integration of this type
	Baseline noise	Where baseline is either rising/falling usually due to gradient elution, due to which integration results in error.
	Negative peak	Where there is sudden drops in base line, usually integration results in error
	Excessive peak tailing	Where failure of instrument response to return to baseline usually due to excessive amount of sample load and concentration.
Fluctuation in the chromatograph	Spike	Where a sudden di/spike up of baseline due to vibration or electrical noise, usually expected in chromatography

24.9 Flow Chart of Good Chromatographic Practice in GLP

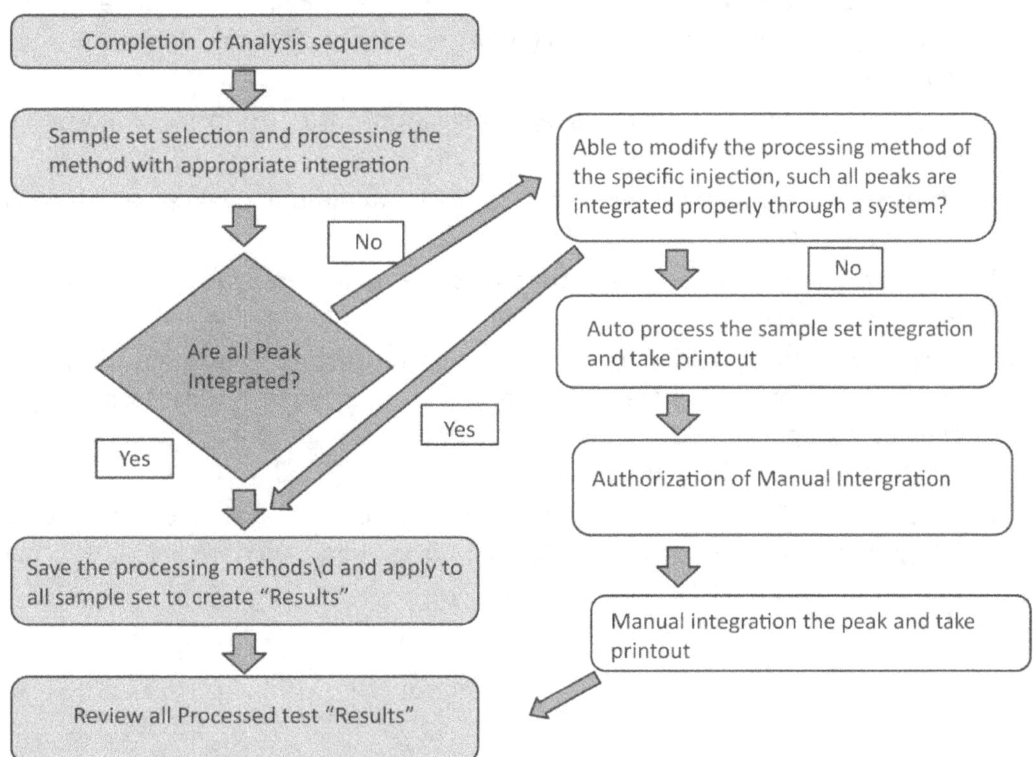

24.10 Definition

(a) ***Peak:*** Graphical display of a detector signal when the bad of separated component passes through the detector recorded as a deflection from the baseline.

(b) ***Baseline:*** The highest point on a positive chromatographic peak or the lowest point on a negative chromatographic peak.

(c) ***Chromatographic integration:*** A process to detect peaks I order to calculate responses of different peaks (i.e) Area/height and retention time (RT).

(d) ***Integration:*** Integration is the process of calculating an area that is bounded in part or in whole by a curved line. The goal of chromatographic peak integration is to obtain retention times, heights, and areas of these peaks

(e) ***Auto integration:*** A mathematical process (employs by the software) directed by a defined set of integration parameters to detect peak and to calculate peak response.

(f) ***Manual Integration:*** A process employed by the date user to integrate the peaks responses

(g) ***Manual integration:*** A process employed by the data user to integrate the peaks by using the cursor.

(h) ***Reporting threshold:*** A limit above an impurity is to be reported.

(i) ***Disregard limit:*** In chromatographic tests, the nominal content at or below which peak/signals are not taken into account for calculations a sum of impurities.

(j) ***Inhibit integration:*** Prevents peak integration during the time window of the event.

(k) ***Minimum area:*** A parameter that specifies minimum peak area required for detecting a peak.

(l) ***Minimum height:*** A parameter that specifies minimum peak height required for detecting a peak.

(m) ***Processing:*** Processing is the manipulation of data to determine the identities and/or amounts of separated components. It most often involves integrating chromatographic peaks to calibrate standards and generate a calibration curve, and to quantitate the source components.

(n) ***Processing Methods:*** Processing methods define how software detects, integrates, calibrates, and quantitates unprocessed, raw data from a 2D channel or a 2D-derived channel.

Chapter - 25

Management of Unknown and Extraneous Peaks in Chromatographic Tests

Introduction

Impurities may be introduced into the drug product though different potential cause pathways such as described in the root cause approach using the fishbone of Ishikawa.

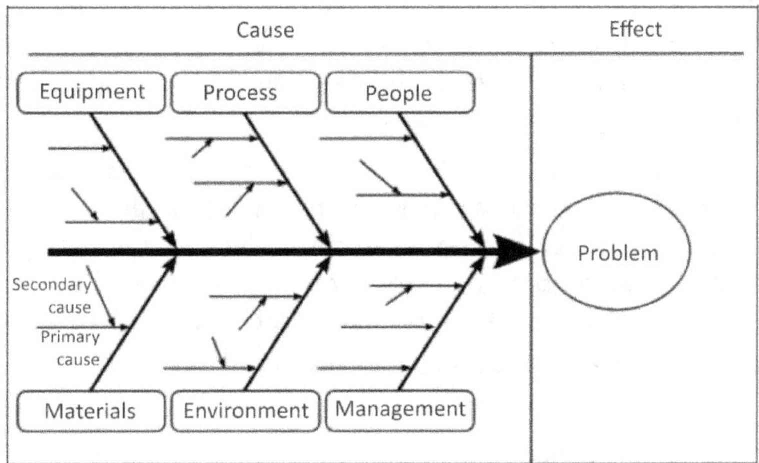

Impurities can originate from:

- Unexpected degradation of Active Ingredient.
- Genotoxic impurities, present as a result of the synthesis of the active ingredient's solvent residue, catalysts reaction products in synthesis, intermediate synthesis compounds,...)
- Chemical compound introduced in the drug product as a results of the contact between the primary packaging and the drug product (leachable)

- Chemical compounds, introduced in the drug product as a results of the contact between processing materials and the product stream(Storage bags, filters, tubing materials)
- Secondary Leachable, being formed as a result of a chemical reaction between a leachable and drug product components (Active Ingredient, Excipient, Adjuvants, Preservatives..)

25.1 Impurities as per ICH Q3A

- Drug substance impurities addressed in 2 ways:
- Chemistry aspects-classification/identification, report generation, setting specifications, analytical procedures;
- Safety aspects-guidance for qualifying impurities either not present or present in substantially higher levels in drug batches used in safety/clinical studies. Thresholds given below which qualification not required.

25.2 Classification of Impurities

- ***Organic(process and drug related):*** includes starting materials, by-products, intermediates, degradation products, reagents, ligands and catalysts
- ***Inorganic:*** **includes** reagents, ligands and catalysts, heavy metals or other residual metals, inorganic salts, other materials (eg. filter aids, charcoal etc.)
- ***Residual solvents:*** organic or inorganic
- Unexpected impurities, suddenly arising in quality control analysis. The sudden appearance of an unknown peak during an chromatographic analysis of a pharmaceutical product can be a critical finding causing delays and requiring considerable resource (both time and money) to resolve. The immediate need to investigate an unknown peak is a common problem.
- The source of an unknown peak can be attributed to:
- Laboratory sample contamination
- Instrument related peaks
- Method resolution (e.g. a vendor modification to HPLC column packing altering the column performance revealing a previously undetected entity)
- Raw material impurities or contamination
- Product or aliquot degradation (a material stability issue or reaction with another excipient)
- Leachable/extractable (from a manufacturing step or a container closure system)
- Manufacturing process cross-contamination
- Some other unknown source

Regardless, once an impurity analysis on a drug product reveals that an anomalous peak causing a failure of a specification, it jeopardizes the release of material into production, or even has the potential to initiate a product recall.

As a first step, one needs to assess if the peak is real and not an artefact. Once confirmed, it is critical to determine the identity and the source of the unknown peak in order to ascertain the impact this peak may have on the affected product and ultimately the patient.

Prior to making the decision to begin the isolation and identification process, some in-lab investigations and standard OOS practices should be performed.

- Re-perform the sample preparation in scrupulously pre-rinsed glassware to rule out laboratory contamination
- Re-analyze the sample on a different instrument
- Analyze the raw materials used in the manufacturing process individually using the drug product impurity methodology in an attempt to identify the source of the peak
- Review historical data to determine if the unknown entity had been present (maybe at lower levels) in prior batches
- If cross contamination is suspected, review batch and equipment records to assess potential candidate components or

Once the decision is made to initiate the unknown peak investigation, the first data that should be generated in an unknown peak investigation by: Diode a aryanalys is, Use of volatile buffer to achieve an LC-MS compatible method, LC-MS/MS and High resolutions MS analysis or/by Peak isolation and NMR Analysis.

25.3 Unknown or Extraneous Peaks Observed during a Chromatographic Test

A. Unknown peaks in purity methods

All unknown peaks above their reporting threshold when calculated as directed by the purity method will be evaluated against their identification thresholds, approved specification limits as well as historical batch and product data for drug substance and drug products.

Unless otherwise specified by the test method, unknown peak in drug products with multiple API's, will be calculated versus the API that yields the highest % estimate for the peak, using the calculations defined in the test method. If the method does not provide a calculation for degradation compounds, the unknown peak will be calculated as an area%.

If the origin or identity of the peak is determined, e.g. through spectral analysis,

(a) It may either be calculated against the relevant API,

(b) Disregarded if the origin is determined to be placebo related/synthesis impurity (drug product only),

(c) Invalidated if the origin is lab error.

B. Extraneous peaks in method not including API component

If the chromatographic test for drug product does not include a peak for the API component or if the API peak present has interference from the chromatography (e.g., blank interference, peak off-scale, or peak is too small for use in calculations), which may be the case in examples such as preservatives assay testing, then the suspect extraneous peak should be calculated against the main peak of interest in the test to estimate its level in terms of % w/w relative to the product (see Appendix-2 for example calculation). In this case, the Investigation trigger for extraneous peaks will be >0.05% w/w relative to the product. If there are multiple components of interest (e.g., multiple preservatives), perform the calculation against the smallest main peak of interest. (e.g., if assaying for multiple preservatives). Use the smallest preservative peak as basis of calculation of the potential extraneous peak.

C. Unknown/Extraneous peaks in non-purity methods :

When testing by non-purity chromatographic methods (e.g., as may be the case for Assay, content uniformity, dissolution etc.):

(a) If the non-purity test method being used is not the validated method for purity:

(i) Only the peak(s) intended to be quantified, will be integrated, except when an unexpected peak is observed at a level that exceeds the investigation trigger. In this case, such peak should be integrated as a potential extraneous peak and investigated.

(ii) Also, if a peak other than the analyte of interest is observed and it can be shown to be part of the historical chromatographic profile for that product (as determined by review of the historical chromatograms, of the historical data, or of the other relevant reports, such as method validation reports), then the peak should be disregarded as being typical (even if that historical peak is observed to exceed the investigation trigger), such peaks should not be integrated.

(b) If the non-purity test method being used is validated for purity, unknown peaks will not be treated as extraneous peaks, as per the definitions in this SOP (they should be evaluated as "unknown"); however consideration should be given to the differences between the purity sample preparation and the sample preparation of the non-purity test sample (e.g., in Content Uniformity testing, if only a few dosage units show a peak slightly above Identification threshold, then further investigation likely would not be needed. However, the decision to not investigate further should be justified, documented and approved by QA)

25.4 Reporting, Identification and Qualification Thresholds for Unknown Peaks

(a) The ICH reporting thresholds are based on MDD. The ICH identification thresholds are estimated based on the MDD or TDI calculated for each drug product, i.e., whichever is stringent.

(b) The MDD of the API in a drug product can be calculated based on the Highest Daily Dose (HDD) of the drug product as described on the instruction leaflet and the concentration of the API in the product; alternatively it can be obtained from the stability owner.

   - MDD = HDD X concentration.

(c) Refer current issued list for ICH Reporting and Identification Thresholds.

(d) The maximum daily doses (MDD) for most of the Drug Substances are ≤2 g/day. Therefore, the reporting and identification thresholds for unknown impurities in Drug Substances (except those listed in current issued list for ICH Reporting and Identification Thresholds) are as follows:

   - Reporting threshold = >0.05%

   - Identification threshold = >0.10%

(e) The exceptions for the above thresholds are:

   - Drug Substances with a maximum daily dose > 2 g/day, and

   - Drug Substances where Total Daily Intake (TDI) values were used.

(f) For Veterinary products, the criteria for RT and IT are in accordance with the VICH guidelines (VICH GL 11-Quality):

   - Reporting Threshold: 0.3%

   - Identification Threshold: 1.0%

A. Determination of Reporting, Identification and Qualification Thresholds for Unknown Peaks in drug Products

The ICH reporting, identification and qualification thresholds for unknown impurities in drug products are summarized in Tables 1, 2 and 3.

Table 1 Reporting thresholds for Drug Products per ICH Q3B (R2)

Maximum Daily Dose[1]	Reporting Threshold[2,3]
≤ 1 g	0.1%
> 1 g	0.05%

Table 2 Identification Thresholds for Drug products Per ICH Q3B (R2)

Maximum Daily Dose[1]	Reporting Threshold[2,3]
< 1 mg	1.0% or 5 µg TDI, whichever is lower
1 mg - 10 mg	0.5% or 20 µg TDI, whichever is lower
>10 mg - 2 g	0.2% or 2 mg TDI, whichever is lower
> 2 g	0.10%

Table 3 Identification Thresholds for Drug Products per ICH Q3B (R2)

Maximum Daily Dose[1]	Reporting Threshold[2,3]
< 10 mg	1.0% or 50 µg TDI, whichever is lower
10 mg - 100 mg	0.5% or 200 µg TDI, whichever is lower
>100 mg - 2 g	0.2% or 3 mg TDI, whichever is lower
> 2 g	0.15%

Notes :

- [1]The amount of drug substance administered per day
- [2]Thresholds for degradation products are expressed either as a percentage of the drug substance or as total daily intake (TDI) of the degradation product. Lower thresholds can be appropriate if the degradation product is unusually toxic.
- [3]Higher thresholds should be scientifically justified.

  To verify if a threshold is exceeded, a reported result has to be evaluated against the thresholds as follows: when the threshold is described in %, the reported result rounded to the same decimal place as the threshold should be compared directly to the threshold. When the threshold is described in TDI, the reported result should be converted to TDI, rounded to the same decimal place as the threshold and compared to the threshold e.g. an amount of 0.18% degradation level corresponds to a TDI of 3.4 mg impurity (absolute amount) which is then rounded down to 3 mg; so the qualification threshold expressed in TDI (3 mg) is not exceeded.

❖ **Illustration of Reporting Degradation Product Results for Identification and Qualification in Drug product**

Example-1: 50 mg Maximum Daily Dose

Example-1: 50 mg Maximum Daily Dose

Reporting threshold: 0.1%

Identification threshold: As per Table-2: If MDD >10 mg - 2 g, Identification threshold is 0.2% or 2mg TDI, whichever is lower

For the given MDD, conversion of % Identification Threshold in terms of drug Substance to "mg" for comparison with TDI of degradation product is performed as follows:

50 mg X 0.2% = 0.1 mg

0.1mg (0.2%) is more stringent than 2 mg TDI hence, **Identification Threshold is 0.2%.**

Qualification threshold: As per Table-2: If MDD >10 mg - 100 mg, Qualification threshold is 0.5% or

200µg TDI, whichever is lower For the given MDD, conversion of % Qualification Threshold in terms of drug Substance to "µg" for comparison with TDI of degradation product is performed as follows:.

50 mg X 0.5% = 250 µg

200 µg TDI is more stringent than 250 µg (0.5%) hence, Qualification Threshold is 200 µg

Raw' Result (%)	Reported Result (%) (Reporting Threshold =0.1%)	Total Daily Intake (TDI) of the Degradation Product (rounded result in µg)	Action	
			Identification	Qualification Threshold 200 µg TDI Exceeded?
0.04	Not reported	20	None	None
0.2143	0.2	100	None	None
0.349	0.3[1]	150	Yes	None[1]
0.550	0.6[1]	300	Yes	Yes[1]

Notes :

[1] After identification, if the response factor is determined to differ significantly from the original assumptions, it can be appropriate to re-measure the actual amount of the degradation product present and re-evaluate against the qualification threshold.

Example 2

<table>
<tr><td colspan="5">Example 2: 1.9 grams Maximum Daily Dose</td></tr>
<tr><td colspan="5">Reporting threshold: 0.05%
Identification threshold: 2 mg
Qualification threshold: 3 mg</td></tr>
<tr>
<td rowspan="2">'Raw' Result (%)</td>
<td rowspan="2">Reported Result (%) (Reporting Threshold = 0.05%)</td>
<td rowspan="2">Total Daily Intake (TDI) of the Degradation Product (rounded result in mg)</td>
<td colspan="2">Action</td>
</tr>
<tr>
<td>Identification Threshold 2 mg TDI exceeded?</td>
<td>Qualification Threshold 3 mg TDI exceeded?</td>
</tr>
<tr><td>0.049</td><td>Not reported</td><td>1</td><td>None</td><td>None</td></tr>
<tr><td>0.079</td><td>0.08</td><td>2</td><td>None</td><td>None</td></tr>
<tr><td>0.183</td><td>0.18[1]</td><td>3</td><td>Yes</td><td>None[1]</td></tr>
<tr><td>0.192</td><td>0.19[1]</td><td>4</td><td>Yes</td><td>Yes[1]</td></tr>
<tr><td colspan="5">Notes :
[1] After identification, if the response factor is determined to differ significantly from the original assumptions, it can be appropriate to re-measure the actual amount of the degradation product present and re-evaluate against the qualification threshold.</td></tr>
</table>

B. Determination of Reporting, Identification and Qualification Thresholds for Unknown Peaks in drug Substance

The ICH reporting, identification and qualification thresholds for unknown impurities in drug substances are summarized in Tables 1

Table 1 Reporting and Identification Thresholds for APIs per ICH Q3A (R2)

<table>
<tr>
<td>Maximum Daily Dose**1</td>
<td>Reporting Threshold2,3</td>
<td>Identification Threshold 3</td>
<td>Qualification Threshold 3</td>
</tr>
<tr>
<td>≤ 2g/day</td>
<td>> 2g/day</td>
<td>0.10% or 1.0 mg per day intake (whichever is lower)</td>
<td>0.15% or 1.0 mg per day intake (whichever is lower)</td>
</tr>
<tr>
<td>> 2g/day</td>
<td>0.03%</td>
<td>0.05%</td>
<td>0.05%</td>
</tr>
<tr><td colspan="4">** The amount of drug substance administered per day</td></tr>
<tr><td colspan="4">Notes
1. The amount of drug substance administered per day
2. Higher reporting thresholds should be scientifically justified
3. Lower thresholds can be appropriate if the impurity is unusually toxic</td></tr>
</table>

Table 1 *Contd...*

To verify if a threshold is exceeded, a reported result has to be evaluated against the thresholds as follows: when the threshold is described in %, the reported result rounded to the same decimal place as the threshold should be compared directly to the threshold. When the threshold is described in TDI, the reported result should be converted to TDI, rounded to the same decimal place as the threshold and compared to the threshold. For example the amount of impurity at 0.12% level corresponds to a TDI of 0.96 mg (absolute amount) which is then rounded up to 1.0 mg; so the qualification threshold expressed in TDI (1.0 mg) is not exceeded

❖ **Illustration of Reporting Impurity Results for Identification and Qualification in Drug Substance**

Example 1: 0.5 g Maximum Daily Dose				
Reporting threshold = 0.05%				
Identification threshold = 0.10%				
Qualification threshold = 0.15%				
'Raw' Result (%)	**Reported Result (%) (Reporting Threshold = 0.05%)**	**Calculated Total Daily Intake (TDI) (mg) of the impurity (rounded result in Mg**	**Action**	
			Identification (Threshold 0.10% exceeded?)	**Qualification (Threshold 0.15% exceeded?)**
0.044	Not reported	0.2	None	None
0.0963	0.10	0.5	None	None
0.12	0.12[1]	0.6	Yes	None[1]
0.1649	0.16[1]	o.8	Yes	Yes[1]
Notes				
[1] After identification, if the response factor is determined to differ significantly from the original assumptions, it can be appropriate to re-measure the actual amount of the degradation product present and re-evaluate against the qualification threshold.				

Example 2

Example 2: 0.8 g Maximum Daily Dose
Reporting threshold = 0.05%
Identification threshold = 0.10%

Qualification threshold = 1.0 mg TDI				
'Raw' Result (%)	Reported Result (%) (Reporting Threshold = 0.05%)	Calculated Total Daily Intake (TDI) (mg) of the impurity (rounded result in mg	**Action**	
			Identification (Threshold 0.10% exceeded?)	Qualification (Threshold 1.0 mg TDI exceeded?)
0.066	0.07	0.6	None	None
0.124	0.12	1.0	Yes	None[1]
0.143	0.14	1.1	Yes	Yes[1]
Notes				
[1]After identification, if the response factor is determined to differ significantly from the original assumptions, it may be appropriate to re-measure the actual amount of the impurity present and re-evaluate against the qualification threshold.				

25.5 Investigation Trigger for Extraneous Peaks

The Investigation trigger for Extraneous Peaks is based on the ICH Identification Thresholds of the product, except in the following situations. Identification Threshold, the individual unspecified impurities shelf-life specification will be used as:

- If the regulatory shelf-life specification for individual unspecified impurities is greater than the Identification Threshold, the individual unspecified impurities shelf-life specification will be used as the basis of the investigation trigger for extraneous peaks, rather than the Identification Threshold.
- For HPLC dissolution and drug release testing, an internal investigation trigger has been established based on the evaluation of different drug products groups depending on their API content, dissolution characteristics, and types of dissolution apparatus used.
  (a) For extraneous peaks observed during the HPLC dissolution analysis of immediate-release drug products, an internal investigation trigger of 10.0% will be used based on the lowest API content.
  (b) For extended-release drug products, the investigation trigger of 10.0% will be applied to the cumulative HPLC peak areas of the extraneous peak and the API in the drug release solution. Thus, the areas of the extraneous peak in the HPLC chromatograms of the individual drug release solutions will be added together and compared with the total API areas in all the solutions.

25.6 Investigating "Unknown" or "Extraneous" Peaks

A suspect OOT, Phase I laboratory investigation must be initiated in the following cases:

(a) Unknown atypical/OOT peaks detected above their reporting thresholds. i.e. results that are inconsistent with the product trend or approaching the Identification Threshold/specification Limit.

(b) Unknown typical peaks detected above their identification thresholds.

(c) Extraneous peaks estimated to be **above** their investigation trigger.

- In case the unknown is above its specification limit, the investigation will be classified as a suspect OOS investigation.
- The objectives of the investigation will be to determine if the source of the peak may be attributable to laboratory error (e.g., contamination of sample / solutions the laboratory), or if it may have potential product impact.
- The Phase I investigation may include, but is not limited to the following activities.

(a) Reanalysis of original solutions

(b) Review of the available chromatographic assay and purity method development and validation reports for potential insight into possibilities for the identity of the unknown impurity or extraneous peak

(c) Evaluation of placebo or known process impurities or degradants for retention time match.

If an unknown peak is confirmed to be above its identification thresholds or specification limit or if an extraneous peak is estimated to be above its investigation trigger and the initial laboratory investigation cannot attribute the presence of the suspect peak to laboratory error, then the Phase I investigation may be complete.

As a part of Phase II Investigation, a retest can be initiated.

Expanded Investigation for Identification of Confirmed Unknown or Extraneous Peaks can be performed utilizing a laboratory with the appropriate expertise for determining peak identity.

A. Investigation should to make the identification process can be done by following steps

- (a) Understanding what is known already
- (b) Impurity profile
- (c) Linking results of the impurities
- (d) Linking the impurities to their source
- (e) Confirmation of the identity of the impurity and its Quantification.
- (f) Performing a toxicological assessment

B. Approach for unknown impurity identification

I.   Investigation Process Flow

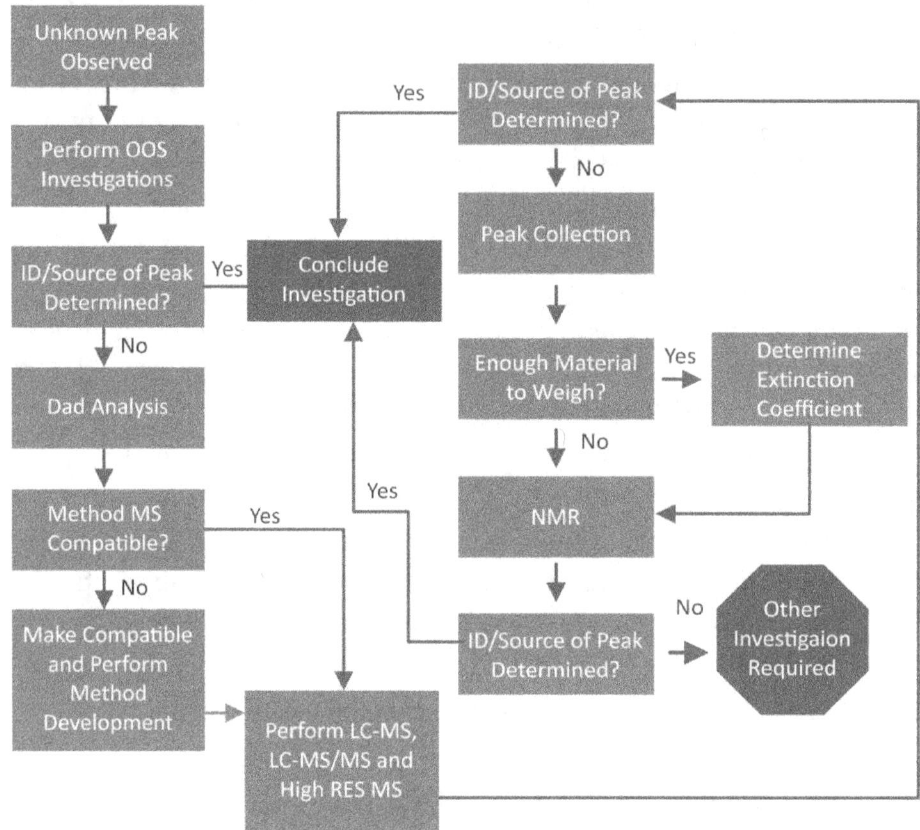

II. Method used for Investigation Process Flow

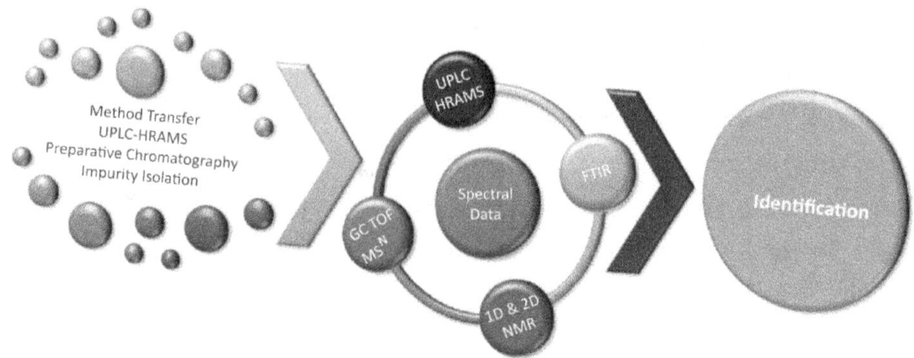

25.7 Definition

- *API:* Active Pharmaceutical Ingredient.
- *TDI:* Total Daily Intake
- *Extraneous peak:* Any unexpected/atypical peak in a non-purity, chromatographic test, e.g., Assay, Content Uniformity (CU), Dissolution, Release Rate, Preservatives Assay, Residual Solvents, Ion Chromatography that is observed above the established investigation trigger for extraneous peaks, as defined in this SOP.
- *Unknown peak:* A peak of unknown origin that is detected during any testing that uses a purity test method. Note: Distinction is made between "extraneous" and "unknown" peaks in this procedure, since there are differences in managing these two situations in the laboratory (as outlined in this SOP).
- *Investigation Trigger:* A limit above (>) which a peak should be investigated.
- *Reporting Threshold (RT):* A limit above (>) which an impurity should be reported.
- *Identification Threshold (IT):* A limit above (>) which an impurity should be identified.
- *Qualification:* The process of acquiring and evaluating data that establishes the biological safety of an individual degradation product or a given degradation profile at the level(s) specified.
- *Qualification Threshold (QT):* A limit above (>) which a degradation product should be qualified.
- *Purity test methods:* Purity test method is intended to mean any chromatographic method validated for the determination of impurities/ degradation products, even when the method is not being employed for the determination of impurities. For example, Content Uniformity testing using an HPLC Assay/Purity method.

25.8 Annexure

A. Appendix-1: Process flow for handling Unknown and Extraneous Peaks

Appendix 1 Process flow for handling Unknown and Extraneous Peaks

Appendix 2 Example of How to Estimate Level of Potential Extraneous Peak in Test When the Calculation is Performed Relative to the Analyte, which is not the API

Product Components of Interest	% w/w in Product Formulation	Peak Area	Area % Relative to Main Analyte[a]	Assay Result for Analyte	% w/w of Analyte Relative to Product[b]
Analyte (e.g., preservative, residual solvent, ion)	0.2%	500,000	N/A	80%	0.16%
Extraneous Peak	N/A	50,000	10%[c]	N/A	0.02%[d]

Analyte = smallest peak of interest in the chromatogram

Amount in the formulation, for preservative assay the estimate of % w/w in the product is in terms of % labelclaim of the preservative in the formulation, corrected for the assay result

(Area of Extraneous Peak / Area of Analyte) × 100% = 50,000 / 500,000 × 100 = 10%

Area% Rel. to Analyte × % Analyte Rel. to Product (w/w)] / 100% = (10 × 0.16) / 100 = 0.02%. No investigation needed, the investigation trigger of >0.05 w/w% was not exceeded.

Chapter - 26

Quality Agreements in Laboratory

Introduction

Quality agreement is legally binded and mutual negotiation between two parties that is Contract Giver and Contract Acceptor .It define, in a formalized manner, clear role and responsibilities relative to quality tasks to assure the manufacture and supply of safe materials "A quality agreement is a comprehensive written agreement between parties involved in the contract manufacturing of drugs that defines and establishes each party's manufacturing activities in terms of how each will comply with CGMP."

Quality agreements is based on cGMP principles by defining, establishing, and documenting their activities in drug manufacturing operations, including processing, packing, holding, labeling operations, testing, and quality control operations.Quality Agreementshould not cover general business terms and conditions such as confidentiality, pricing or cost issues, delivery terms, or limits on liability or damages.

26.1 Elements of a Quality Agreement

Quality agreements should made in formalized documents with clear well-written language. It define key roles and responsibilities communication mode, services,expectations, and approval of varies activities of both parties. Element of a Quality agreement are as follows, but not limited to;

- ***Purpose/Scope:*** To cover the nature of the contract manufacturing services to be provided
- ***Definitions:*** To ensure that the owner and contract facility agree on precise meaning of terms in the quality agreement
- ***Resolution of disagreements:*** To explain how the parties will resolve disagreements about product quality issues or other problems

- ***Manufacturing activities:*** To document quality unit and other activities associated with manufacturing processes as well as control of changes to manufacturing processes Life cycle of, and revisions to, the quality agreement

26.2 Activities in Quality Agreements

(a) ***Quality unit:*** Following activities are involved;
- Quality activity to ensure the product meets cGMP activities.
- Approving ,rejection and dispatches of product
- Communication between each other
- Audits, inspections, and communication of findings
- Communicating inspection observations and findings

(b) ***Facilities and Equipment:*** Following activities are involved;
- Facility as per cGMP
- Qualification, Validation, Calibration, Preventive maintenance
- Technology and automated control systems, environmental monitoring and room classification, utilities, and any other equipment and facilities.
- It should indicate how the parties will communicate information about preventing cross-contamination and maintaining traceability when a contract facility processes drugs for multiple owners.

(c) ***Materials management:*** Following activities are involved;
- Auditing, qualifying, and monitoring component suppliers.
- Sampling and testing in compliance with CGMP
- Inventory management, including labeling, label printing, inventory reconciliation, and product status identification
- Facility prevent mix-ups and cross-contamination.
- Define responsibility for physical control of materials at different points in the manufacturing process
- Each party's roles in storage and transport
- Each party's roles in Environmental conditional monitoring during storage and transport

(d) ***Product-specific considerations:*** Following activities are involved;
- Product/component specifications
- Defined manufacturing operations, including batch numbering processes

- Responsibilities for expiration/retest dating, storage and shipment, and lot disposition
- Responsibilities for process validation, including design, qualification, and ongoing verification and monitoring
- Provisions to allow owner personnel access to the contract facility when appropriate
- Product knowledge, Product development report technology transfer.

(e) *Laboratory controls:* Following activities are involved;
- Procedures delineating controls over sampling and testing samples
- Protocols and procedures for communicating all laboratory test results conducted by contract facilities to the owner for evaluation and consideration in final product disposition decisions
- Procedures to verify that both owner and contract facilities accurately transfer development, qualification, and validation methods when an owner uses a contract facility for laboratory testing
- Routine auditing procedures to ensure that a contract facility's laboratory equipment is qualified, calibrated, and maintained in a controlled state in accordance with CGMP.
- Designation of responsibility for investigating deviations, discrepancies, failures, out-of-specification results and out-of-trend results in the laboratory, and for sharing reports of such investigations

(f) *Documentation:* Following activities are involved;
- Review and approval of documents
- Changes made to standard operating procedures, manufacturing records, specifications, laboratory records, validation documentation, investigation records, annual reports, and other documents related to products or services provided by the contract facility.
- Roles in making and maintaining original documents or true copies in accordance with CGMP
- Records readily available during audit, inspection or whenever required.
- Electronic records :storage, periodic review and retrieval
- Record-keeping time frames

26.3 Disqualification

Contract should be disqualified if it does not meets the approved agreement.

26.4 Template used for Quality Agreement of Contract Laboratory

Quality Agreement	
Version : No :	**Page No.**

<u>Quality Agreement</u>

This document specifies the Quality Agreement with respect to quality and technical requirements between XXXXX here-in after referred as 'Client'/Contract giver in this agreement and M/s. XXXXX. Here-in after referred to as Testing Laboratory/contract acceptor in this agreement.

XXXXX Testing Laboratory	**Address :**
Effective Date:	This agreement is valid for 4 years from the effective date.

Introduction

This Quality agreements to come an agreement between Contract Giver CONTRACT GIVER Address ,Email ,and between Contract Acceptor CONTRACT GIVER Address ,Email relating to the compliance with cGMPs & GLP (as hereinafter defined) in the FACILITY (as hereinafter defined) of Testing Services listed in Attachment-C.

The purpose of this Quality Agreement is to assign responsibilities among the CONTRACT ACCEPTOR and CONTRACT GIVER as to assure compliance with cGMPs and GLP. Nothing in this Quality Agreement shall limit CONTRACT ACCEPTOR's obligation to comply with cGMPs, GLP and under APPLICABLE LAWS (as hereinafter defined) including, without limitation, the guidance and directives set forth in: ICH : Q7 Good Manufacturing Practice Guidance for Active Pharmaceutical Ingredients, Chapter-11, 21 CFR part 211, subpart-I and 21 CFR part 11. Eudralex GMP Rules, Volume-4, Chapter-6, Chapter-7 and Annexure 11.

 This Quality Agreement takes the form of a detailed checklist of the activities associated with testing of samples. Responsibility for each activity is assigned to either CONTRACT GIVERand/or CONTRACT ACCEPTOR in the appropriate box in the Delegation Responsibility Checklist that follows. For each responsibility listed, the respective party is required to put into effect all applicable procedures and to take all necessary actions to effectuate that responsibility in accordance with cGMPs, APPLICABLE LAWS and the MARKETING AUTHORIZATIONS (as hereinafter defined).To facilitate routine communications between the parties, a list of contacts is provided in Attachment-A.

The responsibilities set forth in this Quality Agreement with all related definitions shall survive the completion of performance, expiration or termination of this Quality Agreement.

This Quality Agreement shall be reviewed annually for its compliance with APPLICABLE LAWS and quality specification with trend analysis of the samples (tested in previous year). Quality Agreement shall be amended to fulfill the requirement (if required). The information to be updated in Annexure and shall be modified with version number and effective date.

Table *Contd...*

<table>
<tr><td colspan="2" align="center">Quality Agreement</td></tr>
<tr><td colspan="2">Contents</td></tr>
</table>

Definition **4**

Term and Termination..**5**

RESPONSIBILITY DELEGATION CHECKLIST **6**

1.   Regulatory Authorizations and Communications **6**

2.   GMP Compliance..**6**

    2.1   Organization and Personnel **6**

    2.2   Buildings and Facilities..**7**

3.   Laboratory: **7**

    3.1   Analytical Methods 7

    3.2   Testing and OOS Investigations..**7**

    3.4   Disposition/Rejection/Destruction....................................**9**

    3.5   Records Retention **9**

4.   CONTRACT GIVER Oversight **9**

    4.1 Audits..**9**

    4.2 Regulatory & Customer Inspections/Notifications **10**

    4.3 Review & Revision of Quality Agreement.......................**11**

5.   Dispute Resolution: **11**

SIGNATURES..**12**

ATTACHMENT – A..**14**

ATTACHMENT – B..**15**

ATTACHMENT – C..**16**

Definitions :

The following list provides definitions of related terms used in this Quality Agreement. Such terms, when used in this Quality Agreement, will be shown in all capital letter fonts.

"AUTHORITY" shall mean any applicable local, national or supranational government regulatory authority involved in granting approvals for the MANUFACTURE of DRUG SUBSTANCE/DRUG PRODUCT at the FACILITY or A COMMERCIAL LABORATORY.

"APPLICABLE LAWS" shall mean all laws, ordinances, rules and regulations applicable to the MANUFACTURE of DRUG SUBSTANCE/DRUG PRODUCT and the obligations of CONTRACT ACCEPTOR and/or Contract Giver, as the context requires under this Quality Agreement, including, without limitation, the cGMPs& GLP covered under laws and regulations of Indian Drugs & Cosmetic Acts and various countries i.e. U.S, EU guidelines.

"cGMP" shall mean all APPLICABLE LAWS and regulations relating to the MANUFACTURE of SUBSTANCE/DRUG PRODUCT, including, without limitations, Current Good Manufacturing Practices as specified in the ICH Guidelines for Active Pharmaceutical Ingredients (Q7) and EU GMP regulations, Country of origin and Destination country.

"EMERGENCY COMPLAINT" shall mean a complaint that represents a possible widespread or serious production error or a serious health risk to the patient. (Examples include, but

Table Contd...

Quality Agreement
are not limited to, widespread or critical manufacturing errors, cross-contamination or product mixes, product/packaging tempering, suspected counterfeits, and production incidents possibly resulting in product misuse). "FACILITY" shall mean the CONTRACT ACCEPTOR's facilities located, Listed in Attachment-C. "MANUFACTURE/MANUFACTURING/MANUFACTURED"/ANALYZED shall mean all operations performed by CONTRACT ACCEPTOR, as specified in the relevant specifications & test methods, including, without limitation, storage, testing & releasing of raw materials, packaging materials and ofPRODUCTs. "MARKETING AUTHORIZATIONS" shall mean and include all licenses, permits, authorizations and approvals of, all pricing agreements and price reimbursement agreements with, and all registrations, filings and other notifications to, any AUTHORITY or department necessary, appropriate or useful for testing of raw materials, packaging materials and Products. "MATERIALS" shall mean all raw materials, components, and other items necessary for the MANUFACTURE or SAMPLES HANDLED of Products. QUALITY STANDARDS" shall mean an official compilation of specifications, analytical methods, and standardized assays used to insure the identity, strength, quality, and purity of product. Raw data means all original records and documentation, or verified copies thereof, which are the result of the original observations and activities in a study or analysis that are necessary for the reconstruction and evaluation of an analysis or study. Raw data includes record of analysis/data sheets, work sheets, chromatographs photographs, computer readable media, recorded data from automated instruments, or any other data storage medium that has been recognized as capable of providing secure storage. In the case of basic electronic equipment which does not store electronic data, or provides only a printed data output (e.g. balance or pH meter), the printout constitutes the raw data. Primary data (terms from regulatory): The analytical data which is primarily generated which will be the source for the further calculation and reporting of results. Eg. Sequence in the chromatographic analysis. "STANDARD OPERATING PROCEDURES" shall mean the procedures in effect at the time of TESTING of Raw Material, Packaging Material and Product samples, which have been approved by CONTRACT ACCEPTOR's Quality Unit
Term and Termination:
This Agreement shall be effective on the date the last Party has signed this Agreement and shall remain effective for a period of Three (3) years upon completion of which parties may through mutual consent expressed in writing extend the term of this Agreement. Either party may terminate this Agreement by giving 6 months written notice to the other party. After such termination and if so requested by CONTRACT ACCEPTOR will negotiate with CONTRACT GIVER in good faith a subsequent Quality Agreement

Table Contd…

Quality Agreements		
RESPONSIBILITY DELEGATION CHECKLIST **(Column marked with "*" and Bold letters are mandatory.)**		
Compliance	**Contract Giver**	**Contract Acceptor**
Regulatory Authorizations and Communications:		
1.1* Will maintain a valid relevant license(s)/ certification required by APPLICABLE LAWS. A copy of license/certification will be provided to CONTRACT GIVER for submission to Local AUTHORITY.		X
1.2 Will adhere to approved registration documentation/marketing authorization (as applicable) for the Testing of Raw Materials (RM), Packaging Materials (PM) and finished Product.		X
1.3 Will notify CONTRACT ACCEPTOR of any communications received from AUTHORITIES related to the Dossier of the Drug Product.	X	
1.4* Will notify CONTRACT GIVER of any significant communications by AUTHORITIES related to the potential discontinuance and/or withdrawal of the relevant license/certifications.		X
1.5 Is not debarred or does not employ or use any service of any individual who is debarred or who has engaged in activities that could lead to being debarred by AUTHORITY.		X
1.6 Will provide Certified copies of all documentation necessary for CONTRACT GIVER to respond to inquiries by AUTHORITIES relating to Contracted activities.		X
1.7* Will provide a current (no older than three years) relevant certificate and will keep update with every renewal of those Certificates/Certificate from applicable marketing countries/territories.		X

Table Contd...

Quality Agreements			
RESPONSIBILITY DELEGATION CHECKLIST **(Column marked with "*" and Bold letters are mandatory.)**			
Compliance		**Contract Giver**	**Contract Acceptor**
2. GMP Compliance			
2.1 Organization and Personnel			
2.1.1	Will maintain an Independent Quality Unit having the responsibility and authority to approve or reject samples.		X
2.1.2	Will ensure that the use of consultants is supported by documented verification that the consultant(s) has(have) the sufficient education, training, experience, or any combination thereof, to advise on the subject for which they are retained.		X
2.1.3	Will document initial, on-going cGMPs, and analysis, specific training of employees and the personnel involved in Contract Activities.		X
2.2 Buildings and Facilities			
2.2.1	Will provide testing services for the materials and product at the sites located at the address listed in Attachment-C.		X
2.2.2	Will maintain and operate the areas of the FACILITIES related to the storage & testing in compliance with the cGMP and GLP.		X
2.2.3	Will be responsible for ensuring that the FACILITY owned, Contracted and operated by the CONTRACT ACCEPTOR, prior to the effective date of this Agreement, operates in compliance with the cGMPs and GLP.		X
2.2.4*	Will notify CONTRACT GIVER immediately of any changes to the ownership of FACILITY or any sub-contract.		X

Table *Contd...*

Quality Agreements		
RESPONSIBILITY DELEGATION CHECKLIST **(Column marked with "*" and Bold letters are mandatory.)**		
Compliance	**Contract Giver**	**Contract Acceptor**
3. Laboratory		
3.1 Analytical Methods		
3.1.1 Will ensure that analytical methods are established at testing facility before it is used for sample analysis.	X	
3.1.2 Will ensure test methods used for sample analysis finished product testing shall be validated and determine necessary method transfer requirements to the testing Laboratory.	X	
3.2 Testing and OOS Investigations		
3.2.1 Will conduct tests/method validation according to approved Testing Procedures and Specifications provided by CONTRACT GIVER and will use qualified, calibrated, 21CFR Part 11 compliant equipment and staff.		X
3.2.2 Samples shall be handled as per respective sampling procedures.		X
3.2.3 Will provide the updated approved test methods for testing of material/products with each sample.		X
3.2.4 Control identity of all activities to perform with respect to chemicals, columns, instruments etc.		X
3.2.5 Will use valid impurities and Reference Standards from the compendia source (with necessary COA) mentioned in the Testing Procedures or provide by Contract Giver.		X
3.2.6 Will store the samples according to storage conditions defined on sample label/analytical request sheet e.g. 5 ± 3°C, NMT 25°C etc.		X
3.2.7 Will prepare and maintain all documents and records concerning the material/products given for testing but not limited to raw data, documents related to test, equipment usage log, backup printouts, system suitability, calibration,		X

Table Contd...

	Quality Agreements		
	RESPONSIBILITY DELEGATION CHECKLIST **(Column marked with "*" and Bold letters are mandatory.)**		
	Compliance	**Contract Giver**	**Contract Acceptor**
	handling and reporting of Out Of Specifications Results, Planned and Unplanned deviations (incidents) and issuing Certificate of Analysis in accordance with current Good Laboratory Practices.		
3.2.8	Will review the raw data and results of the samples tested, before sharing the analytical results to the client.		X
3.2.9	Responsible to approve or reject the batch based on the analytical results review against the specification.		X
3.2.10	Will provide a Certificate of Analysis for each batch/lot of each sample. Certificate of Analysis" means a certificate which sets forth: Relevant product, batch information, a statement certifying that the products were tested in accordance with the spec/Method of Analysis and whether it complies to the specification or not.		X
3.2.11	Will immediately notify within 24 hrs to client for any unanticipated findings of deviations, Out-of-Specifications Results. Will perform investigation in the presence of Client's representative, as applicable.		X
3.2.12	Will store the solutions, reagents, chemicals etc till the results of investigation are declared.		X
3.2.13	Will maintain a system to assure the proper acceptance and identification of Products, Test Solutions, Volumetric Solutions, Reagents, Reference Standards, throughout the testing of the Product.		X
3.2.14	Records will be maintained to allow for the Traceability of the specific batches/lots of Products Tested, Volumetric/Test Solutions used		X

Table Contd...

Quality Agreements			
RESPONSIBILITY DELEGATION CHECKLIST **(Column marked with "*" and Bold letters are mandatory.)**			
	Compliance	**Contract Giver**	**Contract Acceptor**
	while testing a particular batch/lot of Product supplied by Client		
3.2.15	Laboratory will support the client with best efforts (with expertise, investigation, documentation as appropriate) in case of any complaints or issues related to tests.		X
3.2.16	Will provide the copies of raw data/primary along with supporting documents pertaining to the test to CONTRACT GIVER upon request.		X
3.2.17	For any batch of PRODUCT analyzed, will notify CONTRACT GIVER within three (3) business days upon discovery of any deviations from STANDARD OPERATING PROCEDURES, SPECIFICATIONS, or TEST METHODOLOGY that may have a direct or indirect impact on PRODUCT quality or market suitability.		X
3.2.18	Will investigate and document via written report the resolution of any significant deviations.		X
3.3 Data Integrity			
3.3.1*	Will maintain processes to ensure all testing data (electronic, logbooks, notebooks, etc.) demonstrate the characteristics of ALCOA: Attributable, Legible, Contemporaneous, Original and Accurate, in addition to, Complete, Consistent, Enduring and Available.		X
3.3.2	Will have systems in place to prevent the alteration or loss of data. If data is lost or altered, systems/processes must be in place to detect the occurrence.		X
3.3.3	Self-Inspection process will include an assessment of the site Data Integrity Program to check for effectiveness and traceability.		X

Table *Contd...*

Quality Agreements		
RESPONSIBILITY DELEGATION CHECKLIST **(Column marked with "*" and Bold letters are mandatory.)**		
Compliance	**Contract Giver**	**Contract Acceptor**
3.3.4* Will notify to "Contract Giver" immediately (within 3 business days) if data integrity is found compromised and reported.		X
3.3.5 Will ensure their Data Governance systems include imparting training to staff on importance of data integrity principles.		X
3.3.6* Any incidence of integrity observed during "Contract Giver" inspection will be defined critically and may lead immediate suspension of services and will provide impact assessment within 30 days		X
3.4 Disposition/Rejection/Destruction		
3.4.1 Sample disposition will be made based on the analytical data generated during the analysis.		X
3.4.2 Left over samples destroyed as per the approved in-house procedures.		X
3.4.3 Will provide the certificate of analysis with the details as specified in Attachment-B.		X
3.5 Records Retention		
3.5.1 Will store all the original documentation generated for Storage, Testing that is required to be maintained under cGMPs in a fireproof environment with limited/controlled access for 5 years or till one year after the expiration period.		X
3.5.2 Will retain the original testing records for analytical method, transfer and verification for the entire term.		X
4. CONTRACT GIVER Over sight		
4.1 Audits		
4.1.1* Will conduct internal audits of all relevant activities to ensure compliance with the cGMPs & GLP and applicable STANDARD OPERATING PROCEDURES.		X

Table *Contd...*

Quality Agreements			
RESPONSIBILITY DELEGATION CHECKLIST **(Column marked with "*" and Bold letters are mandatory.)**			
	Compliance	**Contract Giver**	**Contract Acceptor**
4.1.2*	Will allow and support routine Laboratoryqualification audit of FACILITY as per CONTRACT GIVER policy and frequency.		X
4.1.3	Will provide reasonable prior notice to CONTRACT ACCEPTOR for scheduling quality audits.	X	
4.1.4*	Will allow and support additional (Cause base) audits if requested by CONTRACT GIVER to address specific quality problems related to the SERVICES or in response to AUTHORITY requirements.		X
4.1.5	Will provide reasonable access to all areas within the FACILITY during audits.		X
4.1.6	Will provide CONTRACT ACCEPTOR with a written report detailing observations and conclusions arising from each quality audit.	X	
4.1.7	Will provide CONTRACT GIVER with a written response within stipulated timelines on audit observations and conclusions.		X
4.1.8	Will discuss and agree on corrective action plans to address audit observations and conclusions.	X	X
4.1.9	Will allow CONTRACT GIVER to share a summary of CONTRACT GIVER audit of FACILITY if requested by an AUTHORITY.		X
4.2 Regulatory & Customer Inspections/Notifications			
4.2.1	Will allow inspectors from AGENCIES to perform required inspections.		X
4.2.2	Will notify CONTRACT GIVER within three (3) business days of learning of any proposed inspection or visit of the FACILITY by any governmental authority, including, without limitation, any AUTHORITY or any environmental regulatory authority or any customer (in case of		X

Table *Contd…*

	Quality Agreements		
	RESPONSIBILITY DELEGATION CHECKLIST **(Column marked with "*" and Bold letters are mandatory.)**		
	Compliance	**Contract Giver**	**Contract Acceptor**
	Critical observations) that is directly related to, or which could potentially impact, supply of Intended SERVICES.		
4.2.3	Will notify CONTRACT GIVER as soon as reasonably possible, but in any case within three (3) calendar days, after the commencement of any unannounced visit or inspection of the FACILITY by any governmental authority, including, without limitation, any AUTHORITY or any environmental regulatory authority that is directly related to, or which could potentially impact, supply of intended SERVICES.		X
4.2.4	Will notify CONTRACT GIVER within three (3) business day of any requests for information, notices of violation or other communication from any AUTHORITY relating to quality, environmental, occupational health and safety compliance, relating directly to intended PRODUCT, including without limitation, any information CONTRACT ACCEPTOR receives regarding any threatened or pending action by any AUTHORITY or any AUTHORITY non-approval or regulatory action.		X
4.3 Review & Revision of Quality Agreement			
4.3.1	As a mandatory revision, Quality Agreement shall be revised every 4th year.	X	
4.3.2	Will review this Quality Agreement on a yearly basis for compliance of requirements and material, if required this can be revised.	X	X
4.3.3	Will obtain written approval from both parties for any amendments to this Quality Agreement.	X	X
4.3.4	Will provide revision control of this Quality Agreement.	X	

Table Contd...

Quality Agreements			
RESPONSIBILITY DELEGATION CHECKLIST **(Column marked with "*" and Bold letters are mandatory.)**			
Compliance		**Contract Giver**	**Contract Acceptor**
4.3.5	In between two revisions, changes in Attachment/Annexure to the Quality Agreement can be updated through signing off by both parties.	X	*X*
5 Dispute Resolution			
5.1	Will attempt to amicably resolve any dispute arising out of or relating to this Quality Agreement by escalating such disputes to senior management within the Quality organizations of both parties.	X	X

SIGNATURES		
IN WITNESS WHEREOF, the parties hereto have caused this Quality Agreement to be executed by their duly authorized representatives as of the effective date.		
	For and on behalf of "Contract Giver "	**For and on behalf of "Contract Acceptor"**
Name:		
Title:		
Signature & Company seal:		
Date:		

Quality Agreement			
CHANGE CONTROL			
Version	**Supersedes**	**Date**	**Description of the Changes made in the quality agreement**
Contract Taken (signature & date) and Seal		Contract Giver (signature & date) and Seal	

Table Contd...

ATTACHMENT – A

Contact Personnel

Note: This page to be amended, whenever contact personal details revised

Revision No.	00

1.0 For External testing laboratory (Contract acceptor):

Department	Name & Designation	Contact No.& Email ID
Quality control		
Quality Assurance		
Head – Quality		

2.0 For XXXXXXLABORATORIES LTD.: (Contract giver)

Department	Name & Designation	Contact No.& Email ID

Note: Any changes in List of product shall be controlled by version number (starts from 00, 01, 02, 03. etc.)

ATTACHMENT – B

<u>**Release Documentation**</u>

Release documentation to be provided to XXXXXX should include

A Certificate of Analysis shall contain the following details:

- Name, address, and contact phone number of the CONTRACT ACCEPTOR's FACILITY where SUBSTANCE is being TESTED.
- SUBSTANCE name and Grade (Pharmacopoeia)
- CONTRACT ACCEPTOR's batch / lot number & Vendor Lot Number
- Date of MANUFACTURE
- Date of Release
- Date of Expiry.
- Re-Test date
- A list of each test performed, the acceptance limits as indicated in the SPECIFICATIONS, and the results obtained.
- The CoA should document actual values, where specifications are quantitative, and maintain the significant figures and rounding of numbers defined in the SPECIFICATIONS.
- A remarks on COA "This batch meets specifications and analyzed as per, applicable laws, DMF (as appropriate), and the controlled validated process.

Table *Contd...*

- Release statement signed and dated by a Responsible Person, Qualified Person or Pharmacist representing the CONTRACT ACCEPTOR Quality Unit indicating that the batch: meets SPECIFICATIONS; is TESTED in accordance with cGMP's, MARKETING AUTHORIZATIONS, APPLICABLE LAWS, DMF (as appropriate), and the controlled validated process.

Contract Taken (signature & date) and Seal	Contract Giver (signature & date) and Seal

ATTACHMENT – C

List of Analytical techniques / Tests

Revision No.	00

Note: This page to be amended, whenever any additional site/test is added or removed.

Sr. No.	Name of Service	Site Name and Full Address (where Services taken)
1.	Identification, Related substance, OVI, and residual solvent by GC.	
2.	Assay (chemically, Instrumental, microbiology and MLT).	
3.	Water analysis as per EPA norms.	
4.	Viscosity, IR.	
5.	Elemental analysis by AAS/ICP.	
6.	Sterility testing.	

Note: Any changes in List of product shall be controlled by version number (starts from 00, 01, 02, 03…etc)

Contract Taken (signature & date) and Seal	Contract Giver (signature & date) and Seal

Chapter - 27

Sampling and Testing in Laboratory Management

Introduction

Sampling contains the operations intended to select a portion of a product for a defined purpose. The sampling procedure should be appropriate to the purpose of sampling. The written procedure should be available. All operations related to sampling should be performed with care, using proper equipment and tools. Care shall be taken during sampling to guard against contamination or mix up of, or by, the material being sampled. All sampling equipment that comes into contact with the material shall be clean. Some particularly hazardous or potent materials may require special precautions. Samples shall be representative of the batches of material from which they are taken in accordance with the approved written procedure. It is good practice to prepare a checklist so that nothing is missing or forgotten before a sampling expedition is undertaken.

Personnel who will collect samples must be fully trained in both sampling techniques and field test procedures. Quality control personnel must have access to production areas for sampling and investigation as appropriate.

27.1 Sampling Responsibility

The samplers need to be adequately trained in the practical aspects of sampling, qualified to perform the sampling operation, and should have sufficient knowledge of pharmaceutical substances to allow them to execute the work effectively and safely. Given that the sampling technique itself can introduce bias, it is important that personnel carrying out the sampling should be suitably trained in the techniques and procedures used. The training should be documented in the individual's training records. Sampling records should clearly indicate the date of sampling, the sampled container and the identity of the person who sampled the batch.

27.2   Different Types of Sampling in the Pharmaceutical Industry

Following material are sampled in the pharmaceutical industry:

- Starting materials for use in the manufacture of finished pharmaceutical products;
- Intermediates in the manufacturing process (e.g. bulk granule);
- Pharmaceutical products (in-process as well as before and after packaging);
- Primary and secondary packaging materials; and
- Cleaning and sanitizing agents, compressed gases and other processing agents.

    Containers from which samples are withdrawn shall be opened carefully and subsequently reclosed. They shall be marked to indicate that a sample has been taken

A. Incoming materials sampling and testing

At least one test to verify the identity of each batch of material shall be conducted, with the exception of the materials. A supplier's certificate of analysis can be used in place of performing other tests, provided that the manufacturer has a system in place to evaluate suppliers.

Supplier approval shall include an evaluation that provides adequate evidence (e.g. past quality history) that the manufacturer can consistently provide material meeting specifications. Full analyses shall be conducted on at least three batches before reducing in-house testing. However, as a minimum, a full analysis shall be performed at appropriate intervals and compared with the certificates of analysis. Reliability of certificates of analysis shall be checked at regular intervals. Processing aids, hazardous or highly toxic raw materials, other special materials or materials transferred to another unit within the company's control do not need to be tested if the manufacturer's certificate of analysis is obtained, showing that these raw materials conform to established specifications. Visual examination of containers, labels and recording of batch numbers shall help in establishing the identity of these materials. The lack of on-site testing for these materials shall be justified and documented.

Samples shall be representative of the batch of material from which they are taken. Sampling methods shall specify the number of containers to be sampled, which part of the container to sample, and the amount of material to be taken from each container. The decision on the number of containers to sample and the sample size shall be based upon a sampling plan that takes into consideration the criticality of the material, variability of the material, past quality history of the supplier and the quantity needed for analysis.

Sampling shall be conducted at defined locations and by procedures designed to prevent contamination of the material sampled and contamination of other materials.

B. Re-evaluation of incoming material

Materials shall be re-evaluated as appropriate to determine their suitability for use (e.g. after prolonged storage or exposure to heat or humidity).

C. In-process sampling and controls

Written procedures shall be established to monitor the progress and control the performance of processing steps that cause variability in the quality characteristics of intermediates and APIs. In process controls and their acceptance criteria shall be defined based on the information gained during the development stage or historical data.

The acceptance criteria and type and extent of testing can depend on the nature of the intermediate or active pharmaceutical ingredient being manufactured, the reaction or process step being conducted and the degree to which the process introduces variability in the product's quality. Less stringent in-process controls may be appropriate in early processing steps, whereas tighter controls may be appropriate for later processing steps (e.g. isolation and purification steps). Critical in-process controls (and critical process monitoring), including the control points and Methods, shall be stated in writing and approved by the quality unit(s). In-process controls can be performed by qualified production department personnel and the process adjusted without prior quality unit(s)' approval if the adjustments are made within pre-established limits approved by the quality unit(s). All tests and results shall be fully documented as part of the batch record. Written procedures shall describe the sampling methods for in-process materials, intermediates and APIs. Sampling plans and procedures shall be based on scientifically sound sampling practices.

In-process sampling shall be conducted using procedures designed to prevent contamination of the sampled material and other intermediates or APIs. Procedures shall be established to ensure the integrity of samples after collection. Out of specification investigations are not normally needed for in-process tests that are performed for the purpose of monitoring and or adjusting the process.

D. Cleaning validation

Sampling shall include swabbing, rinsing or alternative methods (e.g. direct extraction), as appropriate, to detect both insoluble and soluble residues. The sampling methods used shall be capable of quantitatively measuring levels of residues remaining on the equipment surfaces after cleaning. Swab sampling may be impractical when product contact surfaces are not easily accessible due to equipment design and/or process limitations (e.g. inner surfaces of hoses, transfer pipes, reactor tanks with small ports for handling toxic materials and small intricate equipment such as micronizers and micro fluidizers). Printed packaging materials are considered critical to the conformity of the pharmaceutical product to its labelling and special attention shall be paid to sampling and the safe and secure storage of these materials. There shall normally be a separate sampling area for starting materials. If sampling is performed in the storage area, it shall be conducted in such a way as to prevent contamination or cross contamination.

E. Weighing areas

The weighing of starting materials and the estimation of yield by weighing shall be carried out in separate weighing areas designed for that use, for example, with provisions for dust control. Such areas may be part of either storage or production areas.

F. Handling of material and its storage

Materials shall be handled and stored in such a manner as to prevent degradation, contamination and cross-contamination. Materials stored in fibre drums, bags or boxes shall be stored off the floor and, when appropriate, suitably spaced to permit cleaning and inspection.

Materials shall be stored under conditions and for a period that will have no adverse effect on their quality and shall normally be controlled so that the oldest stock is used first. Certain materials in suitable containers can be stored outdoors, provided identifying labels remain legible and containers are appropriately cleaned before opening and use. Rejected materials shall be identified and controlled under a quarantine system designed to prevent their unauthorized use in manufacturing.

27.3 Different Sampling Category

The category of analysis for sample should be defined according to the requirement for regulatory or monitoring purpose:

- *Chemical analysis:* Required chemical tests to prove the safety and quality of the product. Nutritional tests required if product exhibits a claim
- *Microbial analysis:* Test for Absence of pathogens and safety in microbial counts
- *Physical analysis:* Test for extraneous matter, damaged product
- *Sensory analysis:* Test for retention of original characteristics including flavour, texture etc. and other expected characteristics Verifying the identity;
- Performing complete pharmacopoeia or analogous testing; and performing special or specific tests.

27.4 Sampling Process

Sampling process consist of the following steps:

1. Sampling procedure: Approved written procedure should be available in brief.
2. Trained personnel: Personnel should be trained on the approved procedure and material safety data sheet
3. Preparation of sampling : Sampling Kit and gowning should be followed
4. Sampling facility: Sampling should be carried out in appropriate facility to avoid cross contamination.
5. Sampling operation: Appropriate sampling tool shall be used.
6. Sampling on receipt: Sampling document shall be maintained. Sampling is done for ; Starting material, Intermediate and bulk product, Finished product and Packaging material
7. Sampling plan: Appropriate sampling plan should be used
8. Sampling tools: Cleaned and appropriate sampling tool shall be used
9. Storage and retention of sampled material: Sample shall be stored in as per storage direction.

27.5 Sampling Plan

Samples should be representative of the batch of material from which they are taken. Sampling methods should specify the number of containers to be sampled, which part of the container to sample, and the amount of material to be taken from each container. The number of containers to sample and the sample size should be based on a sampling plan that takes into consideration the criticality of the material, material variability, past quality history of the supplier, and the quantity needed for analysis. Sampling should be conducted at defined locations and by procedures designed to prevent contamination of the material sampled and contamination of other materials. Containers from which samples are withdrawn should be opened carefully and subsequently reclosed. They should be marked to indicate that a sample has been taken.

Sampling plans and procedures should be based on scientifically sound sampling practices. The choice of a sampling plan should always take into consideration the specific objectives of the sampling and the risks.

27.6 Sampling Facility used in Pharmaceutical Industry

The flow of materials and personnel through the building or facilities should be designed to prevent mix-ups or contamination.

Sampling facilities should be designed to:

- prevent contamination of the opened container, the materials and the operator;
- Prevent cross-contamination by other materials, products and the environment; and protect the individual who samples (sampler) during the sampling procedure.

  The sampling should be performed in an area or booth designed for and dedicated to this purpose. The area in which the sample was taken should be recorded in the sampling record and a sequential log should be kept of all materials sampled in each area.

There should be defined areas or other control systems for the following activities:

- Receipt, identification, sampling, and quarantine of incoming materials, pending release or rejection
- Quarantine before release or rejection of intermediates and APIs
- Sampling of intermediates and APIs
- Holding rejected materials before further disposition (e.g., return, reprocessing or destruction)
- Storage of released materials
- Production operations
- Packaging and labelling operations
- Laboratory operations

27.7 Sampling Tools

A sampling tool should be:

- uncontaminated
- approximately uniform in cross section to the desired depth
- provide reproducible sampling units
- they must be robust enough to withstand handling operations;
- they must be easy to clean;
- all parts must be made of materials resistant to: the effects of the goods being sampled (e.g. fruit acid or chemicals); and the cleaning agents (e.g. bleach or surfactants);
- they must also conform to safety requirements.

  Sampling tools must always check their cleanliness before use. After use, you must clean the sampling tools according to specific instructions.

I. Types of Sampler generally used in pharmaceutical industry

    A. *Sampling liquids:* The following tools may be used: vacuum pumps, dipping vessels, pipette-type samplers, sampling scoops, piston-tube samplers etc.

B. ***Sampling solids in powder or granulated form:*** The following tools may be used: spear samplers, tube-type samplers, zone samplers, sampling trowels, spiral samplers, samplers for frozen goods, hand-drill samplers, etc.

C. ***Sampling gases:*** Sampling requires a metal cylinder (sample cylinder) for collection and transport. Sample cylinder is an assembly that includes a pressure rated cylinder and inlet and outlet isolation valves.

II. Storage of Sampler

Separate the used and clean tools accordingly. Keep the used sampling tools to wash down area of quality control department. Wash the tools as per SOP for cleaning of sampling tools. From the hot air oven transfer dry tools on the fixed place of Q.C. Laboratory. Now wrap all the tools individually in butter papers or Aluminium foil. Ensure that there is no exposure to sampling tool after wrapping with butter paper. Seal the butter paper by transparent cellulose paper. Affix the tag 'CLEAN' label duly signed with the date on the sampling tools. Kept clean tools in the cupboard duly locked.

III. Pharmaceutical Sampling Tools

A. ***Scoops:*** Scoops are used for powders or blend and should be made of an inert material, such as polypropylene or stainless steel.

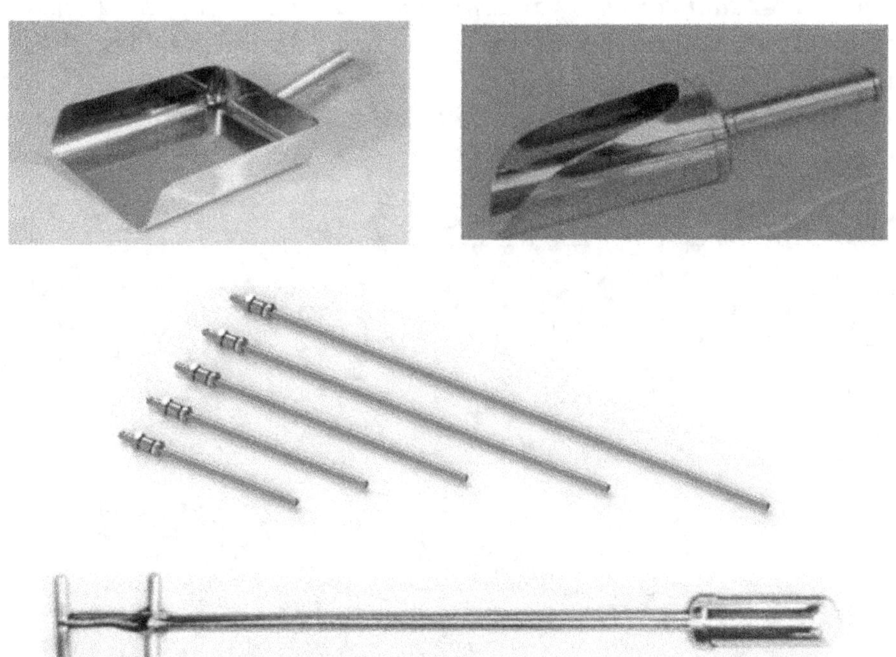

B. ***Dip tubes:*** Dip tubes should be used for sampling liquid and topical products and should be made of an inert material, such as polypropylene or stainless steel.

C. ***Weighted containers:*** For taking samples from large tanks and storage vessels, a container in a weighted carrier can be used. Container should be made of an inert material, such as polypropylene or stainless steel.

D. ***Thieves:*** Sample thieves should be used when taking samples from deep containers of solids.

Multilevel powder sampler with variable sample volumes: The Multi Thief is ideal for taking multiple bulk samples from free flowing powders and granules. The high quality sampler is manufactured to the highest standard in 316 stainless steel. The versatile design enables the sample volume to be quickly and easily changed by changing the removable sample cells.

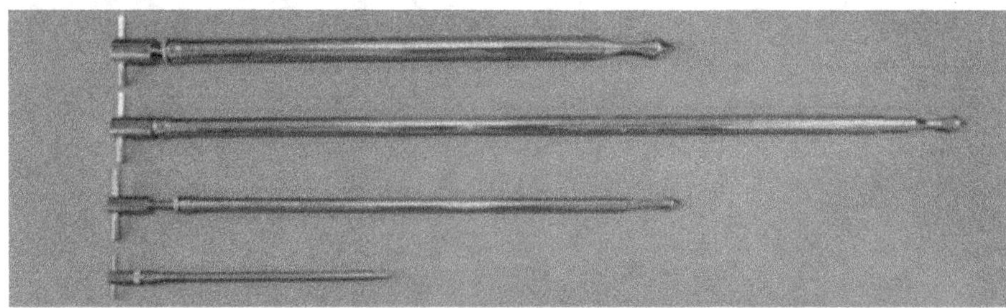

27.8 Sampling Container and Labelling

The sample container shall be used for storage of sample. The containers that have been sampled shall be marked accordingly and carefully resealed after sampling. Each sample container shall bear a label indicating:

- The name of the sampled material;
- The batch or lot number;
- The number of the container from which the sample has been taken;
- The number of the sample;
- The signature of the person who has taken the sample;
- The date of sampling.

27.9 Storage for Sampled Material

Materials should be handled and stored in a manner to prevent degradation, contamination, and cross-contamination. Materials stored in fibber drums, bags, or boxes should be stored off the floor and, when appropriate, suitably spaced to permit cleaning and inspection. Materials should be stored under conditions and for a period that have no adverse effect on their quality, and should normally be controlled so that the oldest stock is used first. Certain materials in suitable containers can be stored outdoors, provided identifying labels remain legible and containers are appropriately cleaned before opening and use. Rejected materials should be identified and controlled under a quarantine system designed to prevent their unauthorized use in manufacturing.

27.10 Golden Rules of Sampling

I. **Control of Sampling activity**

   The material shall be sampled in a closed room (Sampling/dispensing room). A list of materials shall be compiled and approved at firm level. Ensure the required air pressure within the sampling/dispensing room is achieved prior to sampling. Record the air pressure as read from the magnahelic gauge prior to and after completion of sampling.

   (a) Ensure that the other materials surrounding the material to be sampled are in closed condition.

   (b) Material shall be sampled from a small opening and for a minimum time to avoid undue exposure of material to environment.

   (c) Ensure the cleaning of the floor if any spillage occurs during the sampling before proceeding for next container/material sampling.

Sample material one batch at a time. Open the container or bag one at a time, close it properly after sampling and paste the sampled labels before proceeding for the next container. For materials, samples shall be collected from top unless otherwise stated in the respective material method of analysis to follow a specific sampling method. An annexure shall be prepared and approved. In case of sampling of materials due for retesting, sampling person shall ensure that previous Sampled & Approved labels are removed or defaced.

II. Sampling documentation Practice

Print Good receipt note using as per approved procedure. Get an record of analysis issued from authorized quality control personnel. Verify record of analysis against the Good receipt note for the confirmation of details namely material code, material name, In-house Batch Number, analytical reference number of the inspection lot. Verify Vendor Certificate of analysis for confirmation of the analytical results against the specification limits in the Certificate of analysis. Raise an incident report in case of discrepancy, investigate and proceed for sampling only after taking necessary corrective action and preventive action. The number of containers or bags to be sampled shall follow the procedure outlined under 'sampling plan'. Prepare 'Sample for analysis' labels for pasting on the sample bags. The sample bags and sampled label shall be numbered as 1 of x, 2 of x and so on based on the number of containers to be sampled. Collect and carry the required number of 'sampled' labels. Take the appropriate sampling tools [stainless steel Spatula/spoon], cleaned and dried. Check visually for cleanliness before use. Read MSDS (Material safety data sheet) and understand the precautions to be observed for the safe handling of material. Collect and carry necessary safety wares and sampling tools needed to open the packages, containers and drums (like knives, pliers, hammers & wrenches) and material to re-close the packages (nylon threads etc.). The sample shall be collected using a clear polythene bag unless otherwise specific need for particular sample bags or sample containers (E.g. light sensitive materials requiring black colored poly bag etc.) is specified under respective method of analysis. Calculate the minimum amount of sample to be collected from each container using the formula 'Total quantity specified/No. of containers to be sampled'.

The minimum quantity to be sampled from each container or bag shall be one spatula full or one spoon full.   Sampling additional number of times from each container or bag shall be considered, if the need is determined (if the one time collection of samples from containers/bags may fall short of the minimum total quantity specified). An annexure shall be prepared and approved at plant following the template given below to standardize the sampling tool requirement.

Standard Sampling Tools			
Sr. No.	Sampler Type (Spatula / Spoon)	Size (Small/ Medium/ Large)	Approx. Qty that can be withdrawn using one full spatula / spoon

III. Sample register and analytical report number

A unique "Analytical Report Number'' shall be assigned for all samples. After generating the "Analytical report Number'' in the Inward Register, it shall be entered on the "Analytical Test Requisition'. New "Analytical Report number'' system shall start from 1st January of every year.

IV. Sample Testing

Sample testing shall be carried out as per respective Specification &method of analysis. Record the raw data in Record of analysis during analysis. In case of out of specification results reported, investigate with appropriate corrective action and preventive action. Assign retest period for the material based on the vendor certificate of analysis or from stability data. Print the analytical test report .The analytical test report shall be verified against the record of analysis/raw data and duly signed. Quality control person shall generate the approved/rejected labels and paste these labels with the help of warehouse person on yellow portion of the Quarantine labels available on all containers/ags. Upon pasting the labels, the stock shall be posted to unrestricted or blocked stock. Discard the remaining sample after completion of the analysis to the 'Sample for disposal container'.

V. Sample record of analysis

Record of analysis shall be printed by authorized quality control persons .In case of electronic system failure, quality assurance person shall take photo copy from the master record of analysis and mention the Batch no. and analytical reference number on the entire page the record of analysis. Quality assurance shall issue the record of analysis by stamping with Quality assurance issuance stamp with sign & date on first page (Figure 1) and without sign & date on rest of the pages (Figure 2).

Figure 1

Figure 2

Quality assurance shall record the issuance details of the record of analysis in 'Issue, Retrieval and Reconciliation of Batch Records & Record of Analysis' register. In case a Reprint of record of analysis is required, quality control shall request quality assurance and record the details in "Document issue Request" as per current version. Authorized personnel in quality assurance shall issue the   record of analysis to quality control by mentioning the reason for reprint. Unexecuted record of analysis shall be handled as per procedure. List of quality control and quality assurance persons authorized to give record of analysis print shall be maintained at plant level and same shall be version controlled. Changes in the list of quality control and quality assurance persons authorized for record of analysis printing shall be made through change notification.

VI. Disposal of samples

Left over samples after analysis: A quantity of the sample remaining after completion of analysis and on its approval/ rejection, these samples shall be disposed off. These samples shall be collected into a separate container and sent to the incinerator/effluent treatment plant (ETP). (Usually the samples analyzed on the previous day) whenever it is required. For this no records to be maintained. Expired chemicals, waste solutions, hazardous chemicals, dilute or concentrated acids or alkalis and out dated working standards shall be sent to incinerator/effluent treatment plant (ETP).

VII. Vendor Assessment for sampling

Three vendor batches/lots (different batches/lots) supplied in bulk quantities for the first time from a new vendor source shall be assessed for sample variability to establish the sampling plan for the routine sampling. Sampling shall be done in all (100 % sampling) containers or bags of each batch. Sample shall be taken from three layers of each container and transferred to individual sample container for quality laboratory testing. Individual containers shall be inspected for physical non-uniformity and also assessed by quality control testing. The results of testing of individual samples shall be evaluated for variability. A sampling and justification report shall be compiled to justify the sampling plan for the future batches supplied by the vendor. After the completion of the study, a change control shall be raised to initiate the change in sampling plan. A Sampling justification shall be at manufacturing unit for a given material and vendor source.

27.11 Definition

- *Batch:* A quantity of any drug produced during a given cycle of manufacture. If the manufacturing process is continuous, the batch originates in a defined period of time during which the manufacturing conditions are stable and have not been modified.
- *Sample:* A portion of a material collected according to a defined sampling procedure. The size of any sample should be sufficient to allow all anticipated test procedures to be carried out, including all repetitions and retention samples. If the quantity of material available is not sufficient for the intended analyses and for the retention samples, the inspector should record that the sampled material is the available sample (see Sampling record) and the evaluation of the results should take account of the limitations that arise from the insufficient sample size.
- *Random sample:* Sample in which the different fractions of the material have an equal probability of being represented. Representative sample obtained according to a sampling procedure designed to ensure that the different parts of a batch or the different properties of a non-uniform material are proportionately represented.
- *Retention sample:* Sample collected as part of the original sampling process and reserved for future testing. The size of a retention sample should be sufficient to allow for at least two confirmatory analyses. In some cases statutory regulations may require one or more retention samples, each of which should be separately identified, packaged and sealed.
- *Sampler:* Person responsible for performing the sampling operations.
- *Sampling method:* That part of the sampling procedure dealing with the method prescribed for withdrawing samples.
- *Sampling plan:* Description of the location, number of units and or quantity of material that should be collected and associated acceptance criteria.
- *Sampling procedure:* The complete sampling operations to be performed on a defined material for a specific purpose. A detailed written description of the sampling procedure is provided in the sampling protocol.

- ***Sampling record:*** Written record of the sampling operations carried out on a particular material for a defined purpose. The sampling record should contain the batch number, date and place of sampling, reference to the sampling protocol used, a description of the containers and of the materials sampled, notes on possible abnormalities, together with any other relevant observations, and the name and signature of the inspector.

- ***Sampling unit:*** Discrete part of a consignment such as an individual package, drum or container.

- ***Selected sample:*** Sample obtained according to a sampling procedure designed to select a fraction of the material that is likely to have special properties. A selected sample that is likely to contain deteriorated, contaminated, adulterated or otherwise unacceptable material is known as an extreme sample.

- ***Uniformity:*** A starting material may be considered uniform when samples drawn from different layers do not show significant differences in the quality control tests which would result in non-conformity with specifications. The following materials may be considered uniform unless there are signs to the contrary: organic and inorganic chemicals; purified natural products; various processed natural products such as fatty oils and essential oils; and plant extracts. The assumption of uniformity is strengthened by homogeneity, i.e. when the consignment is derived from a single batch.

- ***Attribute:*** Characteristics that are measured as either "acceptable" or "not acceptable", "Good or Bad", "Yes or No", thus have only discrete, binary, or integer values.

- ***Variable:*** Characteristics that are measured on a continuous scale in terms of percentage, numbers etc.

- ***KSM:*** An "API Starting Material" is a material used in the production of an API which is incorporated as a significant structural fragment into the structure of the API. An API Starting Material may be an article of commerce, a material purchased from one or more suppliers under contract or commercial agreement, or it may be produced in-house. API Starting Materials normally have defined chemical properties and structure.

- ***Good Receipt Note (GRN):*** Goods Receipt Note

- ***MOA:*** Method of Analysis provides the instructions for performing an analysis of a sample.

- ***Record of Analysis:*** Record of Analysis is used to record the details of the actual analysis performed for a particular batch of a material.

- ***COA:*** Certificate of Analysis is an authenticated document, issued by an appropriate authority that certifies the quality and purity of a product.

27.12 Quality Control Sample Management Flow

End to End Sample Management Process flow is as below

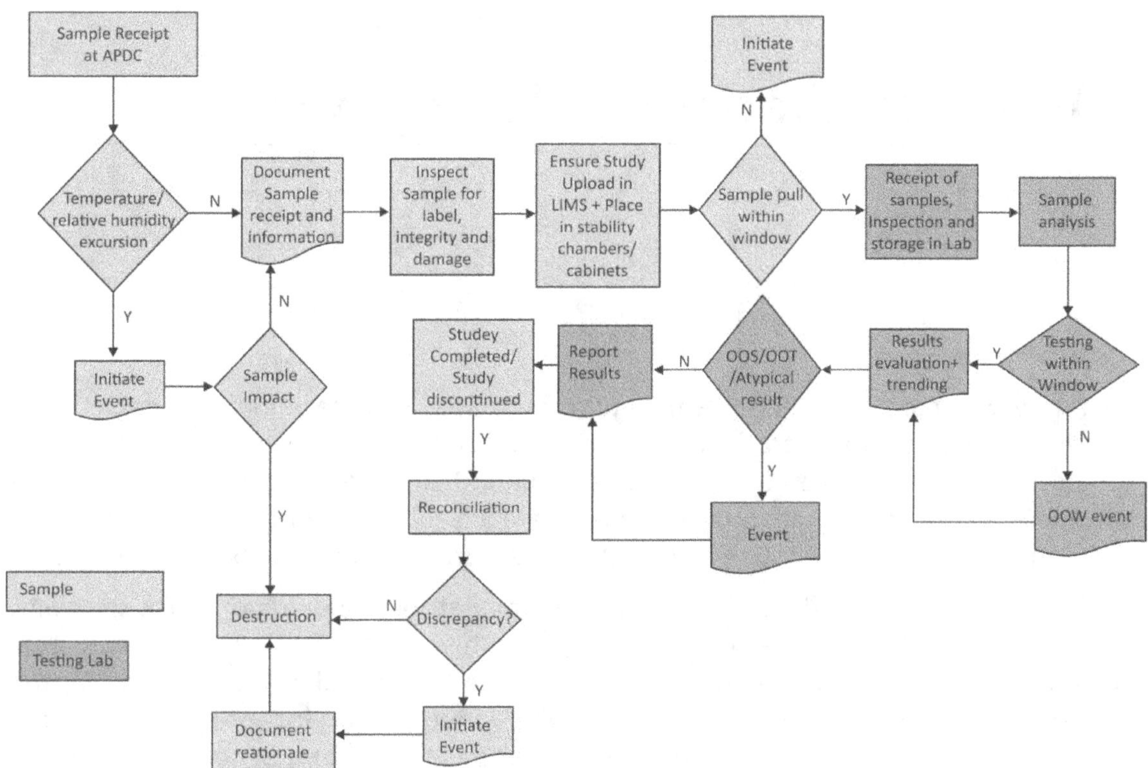

Chapter - 28

Laboratory Standard Management

Introduction

Laboratory Standards are an essential part of cGMP pharmaceutical manufacturing. Reference Standards are used for the assay, identification, purity test and /or analytical method validation and certified reference standard are used for the calibration of instrument or equipment. Laboratory standards are a highly purified sample of a particular compound that has been characterized so that an accurate content can be determined.

All laboratory standards (including certified laboratory standards or stock solutions, commercially supplied laboratory standards, or other material of documented purity) should be provided with a certificate of analysis (CoA) or equivalent documentation that includes, but is not limited to, the following information: lot or batch number, manufacturer, purity, identity, expiration or retest date, appropriate storage conditions, and clearly stated special handling requirements (e.g., light sensitivity).Laboratory standard are mainly used for Quantitative calibration ,Identity comparison ,Test of system suitability, Peak marker ,Fingerprinting and Visual comparison. Records of traceability of samples should be maintained and be available for review.

28.1 Source Material for Laboratory Standard

The purity requirements for a chemical substance depend upon its intended use. Laboratory standard chemical substances that are to be used in assays should possess a high degree of purity (purity of 99.5% or higher is desirable, calculated on the basis of the material in its anhydrous form or free of volatile substances).

- When source material to be used as a chemical reference sub stance is obtained from a supplier, the following should be supplied with the material:

- certificate of analysis with complete information on test method semployed, values found and number of replicates used, where applicable, and relevant spectra and/or chromatograms;
- results of any accelerated stability studies;
- information on optimal storage conditions required to ensure stability (temperature and humidity considerations);
- results of any hygroscopicity study and/or statement of the hygroscopicity of the source material;
- identification of impurities detected and/or specific information on the relative response factor as determined in compendia methods concerning the principal component, and/or the percentage mass of the impurity;
- updated material safety data sheet outlining any health hazards associated with the material

28.2 Traceability of Laboratory Standard

Laboratory standard name of material, manufacture, lot number shall be traceable. All information on the laboratory standard or laboratory material used for testing of the drug substance should be provided.

It is essential that a secondary reference substance is traceable to a primary reference substance, such as a pharmacopoeia or other officially recognized reference substance. In the cases of doubtful results or dispute when using secondary chemical reference substances, the test should be repeated using the primary standard

28.3 Reference Standards

Whenever official reference standards exist, these should preferably be used in analysis. All reference should be kept under the responsibility of a designated person in a secure area.

Secondary or working standards may be established by the application of appropriate tests and checks at regular intervals to ensure standardization. Reference standards should be properly labelled with at least the following information:

- name of the material;
- batch or lot number and control number;
- date of preparation;
- shelf-life;
- potency;
- storage conditions.

All in-house reference standards should be standardized against an official reference standard, when available, initially and at regular intervals thereafter. All reference standards should be stored and used in a manner that will not adversely affect their quality.

28.4 Steps Followed in Reference Standard Project Design

Following steps can be followed for reference project designs:

- definition of the reference material,
- design of a sampling procedure;
- design of a sample preparation procedure;
- selection of measurement methods appropriate for homogeneity and stability testing;
- design of the characterization of the reference material;
- sampling;
- sample preparation;
- choice of suitable methods for the characterization;
- homogeneity testing;
- stability testing;
- characterization of the reference material;
- combination of the results from homogeneity testing, stability testing and characterization, including a full
- evaluation of measurement uncertainty;
- design of a certificate and, if appropriate, a certification report

28.5 Receipt and Storage Standard Reference Standard

On the receiving a reference standard verify for current lot on validity, verify the certificate, quantity received .Verify the batch no/lot no of the vail against certificate of analysis. Verify the storage condition. Quantity received against the challan/invoice.

The labels on chemical reference substances should give the following information:

- The appropriate name of the substance: the international non proprietary name (INN) should be used wherever possible;
- The name of the issuing body;
- The approximate quantity of material in the container; and
- The batch or control number.

  Record the details in the reference standard log book with full details such as reference standard name, lot number, assay value, supplier, pharmaceutical status, date of receipt and validity if available.

Store the reference standard in original container as per the storage condition. Reference shall destroyed as per approved procedure. Record all disposal of the material. All reference standard shall be assign the unique number for the traceability. E.g: RS/XXXX/YYYY/001,: RS: Reference standard, XXXX: Name of the reference standard,YYYY:Year,0001: Serial number.

28.6 Working Standard

Working standard are also called as Secondary working standard and characterised against reference standard. Working standard reduce the cost and delay and delay in receiving the reference standard.

The traceability between the working standard and reference standard are very important.

Important steps to be followed in the working standard :

- Source material : the percentage purity is more than 95% (or 90% if for use in TLC).
- Reference Standard : Valid lot, authorized bodies ,Certificate
- Testing protocol and report with certificate of analysis
- Testing shall be performed as per pharmacopeia
- container and closure system,
- Stability data for working standard
- In used stability

Working substances should preferably be subdivided and presented as single-use units. However, if the working substance is kept in a multiuse container then re-testing will need to be more frequent because there is a greater risk of the uptake of moisture and/or decomposition of the reference substance. The testing methods should include the determination of water content and decomposition products. The maximum permitted variation from the assigned value should be predefined and if exceeded the batch should be re-established or replaced.

28.7 Receipt and Storage Standard Working Standard

On the receiving a working standard verify for current lot on validity, verify the certificate, quantity received. Verify the batch no/lot no of the vail against certificate of analysis. Verify the storage condition. Quantity received against the challan/invoice.

Record the details in the reference standard log book with full details such as working standard name, lot number, assay value, supplier, pharmaceutical status, date of receipt and validity if available.

Store the working standard in original container as per the storage condition. Working shall destroyed as per approved procedure. Record all disposal of the material. All reference standard shall be assign the unique number for the traceability. E.g: WS/XXXX/YYYY/001, RS: Working standard, XXXX: Name of the working standard, YYYY: Year, 0001: Serial number.

28.8 Chemical and Physical Methods used in Evaluating Substances

Standard should be full characterized

- Identification by IR absorption spectra
- Highly specific techniques, such as NMR spectroscopy, MS, or X-ray diffraction crystallography, may also be used for comparisons
- Ultraviolet spectrophones

  Analytical methods may be divided into three broad categories:

  - Those that require comparison with an external chemical reference substance (e.g. chromatographic or spectro photometric methods);
  - those that depend solely on an intrinsic dynamic property (e.g. phase solubility analysis and differential scanning calorimetry); and
  - other methods.

    A. Separation method : Chromatographic method : HPLC and GC
    B. Capillary electrophoresis.
    C. UV spectro photometric
    D. TLC method
    E. Methods based on intrinsic thermodynamic properties
       (i)   Differential scanning calorimetry
       (ii) Phase solubility analysis
    F. Other Methods
       (i)   Spectro photometric methods.
       (ii) Titrimetric methods.
       (iii) Optical rotation methods.
       (iv) Determination of water and organic volatiles.

28.9 Definition

- ***Chemical reference substance :*** The term chemical reference substance, as used in this text, refers to an authenticated, uniform material that is intended for use in specified chemical and physical tests, in which its properties are compared with those of the product under examination, and which possesses a degree of purity adequate for its intended use.

- ***Primary chemical reference substance:*** A designated primary chemical reference substance is one that is widely acknowledged to have the appropriate qualities within a specified context, and whose assigned content when used as an assay standard is accepted without requiring comparison with another chemical substance.

- ***Secondary chemical standard:*** A secondary chemical reference substance is a substance whose characteristics are assigned and/or calibrated by comparison with a primary chemical reference substance. The extent of characterization and testing of a secondary chemical reference substance may be less than for a primary chemical reference substance.

- ***International chemical reference substance:*** International Chemical Reference Substances (ICRS) are primary chemical reference substances established on the advice of the WHO Expert Committee on Specifications for Pharmaceutical Preparations. They are supplied primarily for use in physical and chemical tests and assays described in the specifications for quality control of drugs published in The International Pharmacopoeia or proposed in draft monographs. The ICRS may be used to calibrate secondary standards.

- ***Pharmacopoeial reference standards:*** "Pharmacopoeial standards and substances are established and distributed by pharmacopoeial authorities.

- ***Refence standard:*** reference standard, or reference material, is a substance prepared for use as the standard in an assay, identification, or purity test. It should have a quality appropriate to its use. It is often characterized and evaluated for its intended purpose by additional procedures other than those used in routine testing, For new drug substance reference standards intended for use in assays, the impurities should be adequately identified and/or controlled, and purity should be measured by a quantitative procedure.

- ***In-house primary reference material:*** An appropriately characterised material prepared by the manufacturer from a representative lot(s) for the purpose of biological assay and physicochemical testing of subsequent lots, and against which in-house working reference material is calibrated.

- ***In-house working reference material:*** A material prepared similarly to the primary reference material that is established solely to assess and control subsequent lots for the individual attribute in question. It is always calibrated against the in-house primary reference material.

- ***Impurities:*** Organic impurity levels can be measured by a variety of techniques, including those that compare an analytical response for an impurity to that of an appropriate reference standard or to the response of the new drug substance itself. In cases where the response factors of the drug substance and the relevant impurity are not close, this practice can still be appropriate, provided a correction factor is applied or the impurities are, in fact, being overestimated.

A. Annexure: Inventory Management of Laboratory standard

Inventory Management of Laboratory standard							
Sr. No	Laboratory standard	Date of receipt	Allocated number	Done By	Issuance date/Done By	Destruction date/Done by	Remark

Chapter - 29

Analytical Reagent, Indicators and Volumetric Solution Management

Introduction

In the laboratory chemical analysis, it is essential to use, reagent, indicators and volumetric solution should be adequate quality. The correct reagent, Indicators and Volumetric solution depends on how accurate result must be and the sensitivity limit of the method.

29.1 Requirement of Reagent, Indicators and Volumetric Solution

A. Labels according to regulation: All reagent, indicators and volumetric solution are labelled according to respective regulation and include the pictograms and the new hazard (H) and caution (P) phrases defined by this directive.

B. Material Safety Data Sheets (MSDS): MSDS are intended to give all available information enabling the safe management of chemicals and eliminating or minimizing adverse effects to humans and the environment

C. Certificate of Analysis: For each product batch, a certificate of analysis should be available. The certificate mentions the product name, quality, dates of manufacture and expiry.

29.2 Reagents Management

Reagents are mainly tested and approved by

- Reagents according to ACS (American Chemical Society),
- ISO (International Organization of Standardization) or both.
- Reagents that follow the requirements of the Pharmacopeias as analytical reagent.

Laboratories should verify the suitability of each batch of reagents critical for the test, initially and during its shelf-life

Laboratories should ensure that all reagents (including stock solutions) are adequately labelled to indicate, as appropriate, identity, concentration, storage conditions, preparation date, validated expiry date and/or recommended storage periods. They should be prepared and controlled in accordance with written procedures. The person responsible for preparation should be identifiable from records. The level of controls should be commensurate to their use and to the available stability data.

Where necessary, the date of receipt of any substance used for testing operations should be indicated on the container. Instructions for use and storage should be followed. In certain cases, it may be necessary to carry out an identification test and/or other testing of reagent materials upon receipt or before use.

Reagents are mainly used in

- Products of a suitable purity to be used as auxiliary reagent in chemical analysis.
- Laboratory glassware cleaning detergents.
- pH indicator papers.

   Reagent solution preparation:

Analytical Reagent, Indicators and Volumetric Solution Management

- Reagent solutions prepared by exact dilution.
- All bottles shall be labeled indicating the name and the strength of the solution, date prepared, signature of the person who prepared it and use before date.
- Records shall be maintained
- Reagent solutions should be prepared as per procedure in the individual monograph.
- Standardization notebook and checked by respective chemist.

   All indicator solutions should be used before expiry

29.3 Indicators Management

The end point of titration is usually found using an indicator .An indicator is a component or a mixture of components, which undergoes very visible changes, usually to the color, in response to specific chemical environments. Instrument as we well as indicators can also be used to reseals the equivalence point. These respond to characterise changes in the properties of solutions during titration. They include voltammeters, ammeters, ohmmeters, colorimeters, pH meters, temperature recorder and refractometer

Laboratories should ensure that all indicators (including stock solutions) are adequately labelled to indicate, as appropriate, identity, concentration, storage conditions, preparation date, validated expiry date and/or recommended storage periods. They should be prepared and controlled in accordance with written procedures. The person responsible for preparation should be identifiable from records. The level of controls should be commensurate to their use and to the available stability data.

Where necessary, the date of receipt of any substance used for testing operations should be indicated on the container. Instructions for use and storage should be followed. In certain cases, it may be necessary to carry out an identification test and/or other testing of indicator materials upon receipt or before use.

Indicator solution preparation:

- Indicator solutions prepared by exact dilution.
- All bottles shall be labeled indicating the name and the strength of the solution, date prepared, signature of the person who prepared it and use before date.
- Records shall be maintained
- Indicator solutions should be prepared as per procedure in the individual monograph.
- Standardization notebook and checked by respective chemist.
- All indicator solutions should be used before expiry

29.4 Volumetric Standard Management

Volumetric standard solution concentration is expressed in molarity or normality. Volumetric Analysis is to determine the volume of a solution of know concentration require to react quantitatively with a solution of substance to be analyzed. Volumetric analysis is also called as titrimetiric analysis.

- Titrate: Substance to be analyzed or to be determined
- Titrant: Reagent of know concentration
- Titration: The process of determining the volume
- Equivalence point: The point at which complete chemical reaction takes place and equivalent quantity of reagent is used
- End Point: The point at which the indicator changes its color
- Indicator: Auxiliary agents used to determine the end point of titration.
- Concentration: A defined quantity of substance in a defined volume of solution
- Normality: The number of gram equivalents of solute present in one liter of solution and represented by "N"
- Molarity: The number of moles of solute presents in one liter of solution and represented by "M"
- Molarity: The number of moles of solute per 1000gm of Solvent and represented by "m"

Importance of Volumetric analysis:

- High precision is obtained

- Simple apparatus is required

- Easy process and fast result

- Different methods for different types of substance

Laboratories should ensure that all volumetric solution (including stock solutions) are adequately labelled to indicate, as appropriate, identity, concentration, storage conditions, preparation date, validated expiry date and/or recommended storage periods. They should be prepared and controlled in accordance with written procedures .The person responsible for preparation should be identifiable from records.The level of controls should be commensurate to their use and to the available stability data.

Where necessary, the date of receipt of any substance used for testing operations should be indicated on the container. Instructions for use and storage should be followed. In certain cases it may be necessary to carry out an identification test and/or other testing of volumetric materials upon receipt or before use. In addition, for volumetric solutions, the last date of standardization and the last current factor should be indicated and storage bottle.

- Volumetric solutions shall be standardized by titration against a primary standard or by titration with a standard solution that has been recently standardized against a primary standard.

- Solutions of primary standard or highly purified salts that are of primary standard quality do not require standardization initially. Standard solutions may be prepared by accurately weighing a suitable quantity and dissolving it to produce a specific volume.

- Standard solutions prepared by exact dilution of a stronger standardized solution, do not require initial standardization but must be re-standardized periodically during use.

- All bottles shall be labeled indicating the name and the strength of the solution, date prepared, signature of the person who prepared it and use before date.

- Records shall be maintained for each solution starting with the value determined when the solution was prepared and continuing with the values determined throughout the shelf life of the solutions. This record shall be retained for at least one year after the solution is expended.

- Volumetric solutions should be prepared and standardized as per procedure in the individual monograph.

- Standardization notebook and checked by respective chemist.

- All volumetric solutions should be standardized or re-standardize before use or as per stability.

29.5 Batch Number System for Reagent, Indicators and Volumetric Solution

Batch numbering of reagent, indicators and volumetric solution:

A batch number shall be allotted to each reagent (RS), indicators (IS) and volumetric solution (VS) as follows:

VS-01/18/001

Where

VS= is the volumetric solution number and 01 is name of the volumetric solution

18= Preparation year

001= Serial number of the year 001,002,003-------

29.6 Label for Reagent, Indicators and Volumetric Solution

<table>
<tr><td colspan="2" align="center">Reagent /Indicators / Volumetric solution</td></tr>
<tr><td>Name of the solution and strength.
Reference No.
Date of Preparation
Date of Standardization.
Date of Re-standardization
Use before
Signature of Prepared by.
Signature of Checked by</td><td valign="top">Morality /Normality:</td></tr>
</table>

Chapter - 30

Statistical Tools for Pharmaceutical Industry

Introduction

Statistics are critical to the pharmaceutical industry, from clinical operations through manufacturing. However, clinical and manufacturing statistics represent entirely different worlds. According to current requirement of regulatory expectation and warning letter, extensive statistical knowledge is very important. Without extensive statistical knowledge, the requirements for GMP can be mysterious and intimidating. Qualitative or Quantitative statistical tool can be used for analysed of data. Statistical tool should not be only Data mining or data pooing. The usage of correct statistical tools at the right time is very important to understand quality level. This chapter consist of fundamental concepts in statistical analysis, concept of six sigma and lean manufacturing, statistical tools: measurement system analysis, control chart, capability, acceptance sampling, stability analysis, frequency, distributions, process capability, qualitative and quantitative analysis, patterns and trends.

Statistics is concerned with scientific methods for collecting, organizing, summarizing, presenting, and analysing data, as well as drawing valid conclusions and making reasonable decisions on the basis of such analysis.Statistics is a simple method of extracting information from what often seems at first glance to be a mass of random numbers.

Measures of central tendency and dispersion are the two most fundamental concepts in statistical analysis.

A. ***Measures of central tendency:*** Most frequency distributions exhibit a 'central tendency,' i.e., a shape such that the bulk of the observations pile up in the area between the two extremes. Central tendency is one of the most fundamental concepts in all statistical analysis. There are three principal measures of central tendency: **mean**-average of the total of the sample values, **median**-the middle value (midpoint) of a data, and **mode**-the value or number that occurs most frequently in a data set."

B. ***Dispersion:*** Dispersion is the variation in the spread of data about the mean. Dispersion is also referred to as variation, spread, and scatter. A measure of dispersion is the second of the two most fundamental measures of all statistical analyses. Data are always scattered around the zone of central tendency, and the extent of this scatter is called dispersion or variation. There are several measures of dispersion: **range**-difference between the maximumand minimum values in an observed data set, **standard deviation**-measures the extent of dispersionaround the zone of central tendency, and **coefficient of variation**-final measure of dispersion: the guaranteed existence of a difference between any two items or observations. Concept of variation states that no two observed items will ever be identical.

30.1 Concept of Six Sigma and Lean Manufacturing

- ***Six Sigma:*** A unique approach to improve the quality which was developed and implanted by Motorola.Six sigma is a statistical methodology that focuses on the reducing variation and defects, and mistake-proofing a process. Six sigma data-driven approach to process improvements is divided into five phase: ***Define, Measure, Analyse, Improve and control.***

- ***Lean Manufacturing:*** Lean manufacturing is a strategy for achieving the shortest possible cycle time by eliminating waste and unnecessary process steps, reducing inventory, reducing product development time and increasing customer responsiveness while providing high quality product. ***Lean manufacturing dependents on five basic principle:*** *Values-from customer perspective, Value steam-A series of activity, Flow-getting product to move without stopping, Pull-supplying for customer and Perfection-this is an iterative cycle of continuous improvement.*

30.2 Common Statistical Tools Pharmaceutical Industry can use to meet Regulatory Requirements

1. ***Measurement system analysis (MSA):*** an experimental and mathematical method of determining how much the variation within the measurement process contributes to overall process variability. ***There are five parameters to investigate in an MSA:*** *Accuracy, Repeatability, Reproducibility, Stability and Adequate Resolution.*

   DOE can be conducted by: Conduct an experiment where different people/ machines measure, plot the data, analyse the data. Use statistical techniques such as Analysis of Variance and improve the measurement process, if necessary.

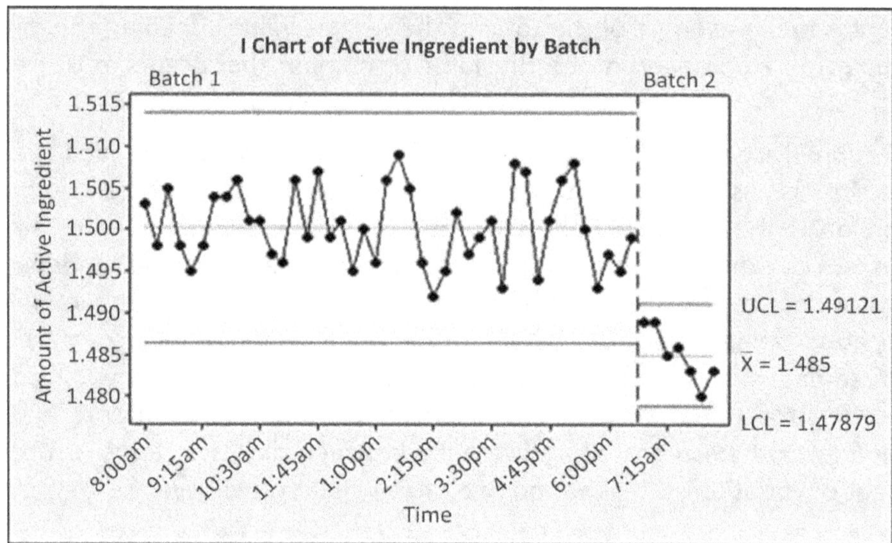

2. ***Control Chart:*** The control chart is a graph used to study how a process changes over time. Data are plotted in time order. A control chart always has a central line for the average, an upper line for the upper control limit and a lower line for the lower control limit. These lines are determined from historical data.

3. ***Capability:*** When your measurement system is adequate and your process is stable, you can accurately describe your quality level. There are four common measures of process quality level.(Brief description refer Process capability section).

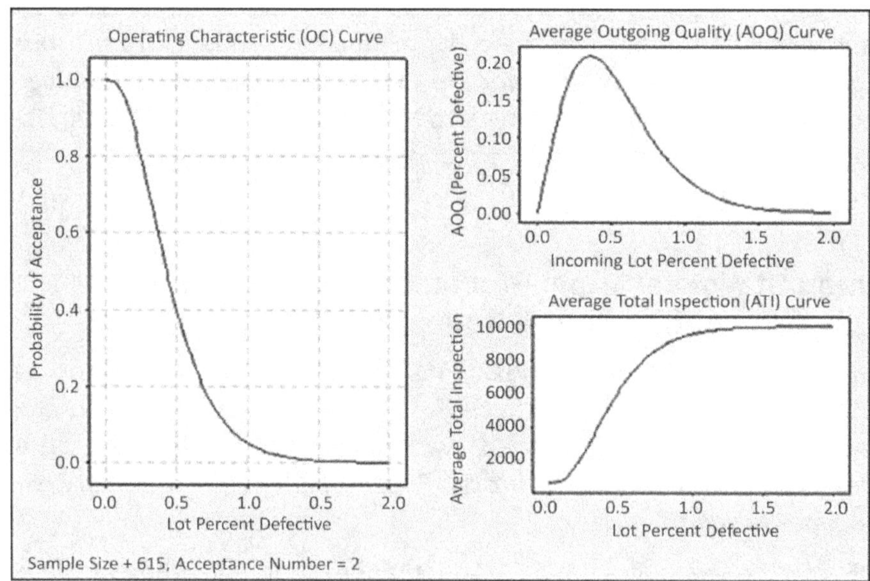

4. ***Acceptance sampling:*** Sampling is a complex subject. It can be divided into analytic: tries to predict what is going to happen and enumerative: tries to determine something about an existing population. *Two statistics acceptance sampling plans used in industry:* Average outgoing quality (AOQ): Represents the defect rate after implementing a sampling plan and Average total inspection (ATI): represents the average number of pieces that you inspect.

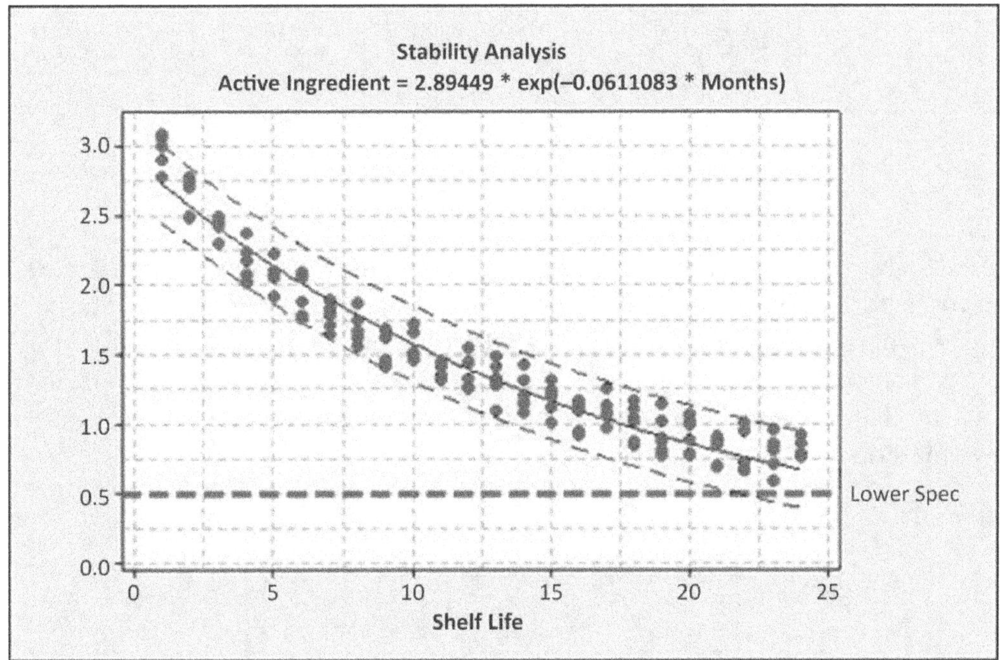

5. ***Stability Analysis:*** Stability analysis evaluates how your product degrades over time during shipment and storage. The data collection for stability involves taking samples from at least three batches, storing the samples at the manufacturing facility, and measuring samples from each batch at regular time intervals.

30.3 Other Statistical Tools used in Pharmaceutical Industry

Frequency Distributions.......

A frequency distribution is a tool for presenting data in a form that clearly demonstrates the relative frequency of the occurrence of values as well as the central tendency and dispersion of the data. Raw data are divided into classes to determine the number of values in a class or class frequency. The data are arranged by classes, with the corresponding frequencies in a table called a frequency distribution. When organized in this manner, the data are referred to as grouped data, as in the above table. The data in this table appear to be normally distributed. Even without constructing a histogram or calculating the average, the values appear to be centered around the value18. In fact, the arithmetic average of these values is 18.02.

Class Boundaries	Mid-point	Frequency	Cumulative frequency
14.5-15.4	15	2	2
15.5-16.4	16	9	11
16.5-17.4	17	24	35
17.5-18.4	**18**	29	64
18.5-19.4	19	25	89
19.5-20.4	20	8	97
20.5-21.4	21	3	100

The histogram provides a graphic illustration of the dispersion of the data.Histogram may be used to compare the distribution of the data to specification limits in order to determine where the process is centered in relation to the specification tolerances. Frequency distributions are useful for evaluating process performance and presenting the evidence of their analysis. Not only is a histogram a simple tool to use, it is also an effective method of illustrating process results.

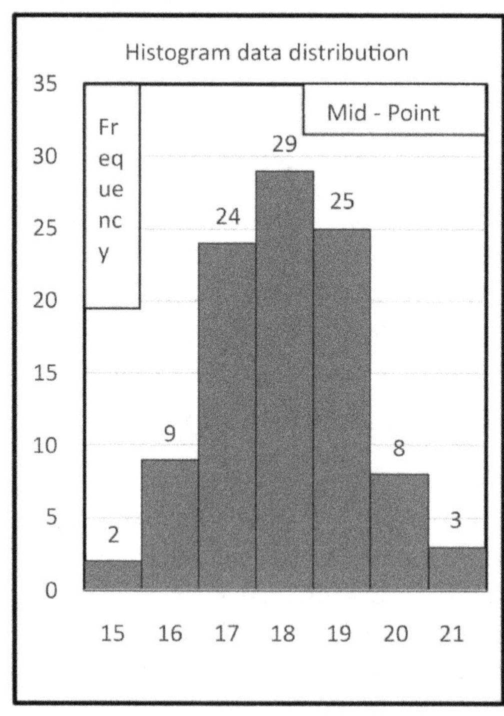

Process Capability (PC)..........

Process capability provides a quantified prediction of the adequacy of a process. It also measures the capability of a process to produce products that meet specifications by measuring the inherent uniformity of the process.

PC indices numerically express the relationship between the distribution and the specification limits. A process is in a state of statistical control when the plotted data points are within the calculated UCL and LCL almost all the time (such as 99.73 percent) and the data do not display a particular form that would indicate an out-of-control condition or instability. *A process is capable when the distribution of individual piece data is within or equal to the upper specification limit (USL) and the lower specification limit (LSL).*

A process capability index expresses a numerical relationship between the process variability and the specification limits. **The indices currently used are C_r, C_p, C_{pk} and C_{pm}.**

$$C_p = \frac{USL - LSL}{6\sigma}$$

is the reciprocal of Cr

$$C_p = \frac{6\sigma}{USL - LSL}$$

- **C_r and C_p are ratios of the specification tolerance (USL – LSL) to the process spread (6σ). The difference between them is that Cp.**

  If the UCL and the LCL are superimposed on the USL and the LSL, the Cp would equal 1. This assumes that the mean of the process is exactly centered within the specification limits. If the control limits fall outside the specification limits, the value of Cp would be less than 1.

$$C_{pk} = Min\frac{USL - X, X - LSL}{3\sigma 3\sigma}$$

- **Cpk is a method of measuring process capability when the mean of the process is not centered with regard to the specification limits.**

  A C_{pk} of 1 or greater means that the process is capable at a level of at least

  99.73 percent conforming, which is the limit associated with an X– and an R chart. It is extremely important when using these capability indices that the process be in control because the theory is based entirely on a normal distribution. If the distribution is not normal, the indices are meaningless

$$C_{pm} = Min\frac{USL - X, X - LSL}{6\sigma^2 + (X - \mu^2)}$$

- *C_{pm} takes into account variation between the process average and a target value.*

  If the process average and the target are the same value, C_{pm} will be the same as Cpk. If the average drifts from the target value, C_{pm} will be less than the Cpk value.

  Calculations of process capability are based on the assumption that the data are taken from a normal distribution. Doers should be aware of the relationship of the two most commonly used measures, Cp and Cpk, which are considered to be measures of long-term capability. Together, these two measures provide information regarding the process variation and where the process is centered in relation to the specification limits. If the values are the same, the process is centered. The process is less centered as the difference between these two values increases. The less centered the process, the lower the Cpk will be and the greater the probability of measures outside the specification limits.

30.4 Qualitative and Quantitative Analysis

Quantitative data means either that measurements were taken or that a count was made, such as counting the number of defective pieces removed (inspected out), the number of customer complaints, or the number of cycles of a molding press observed during a time period. In short, the data are expressed as a measurement or an amount.

In contrast, **qualitative data refers** to the nature, kind, or attribute of an observation Qualitative data must be unbiased and traceable. Quantitative information help in a direct comparison between the information and the requirements.

30.5 Patterns and Trends

- **Pattern analysis** involves the collection of data in a way that readily reveals any kind of clustering that may occur.

  No specific single tool exists to determine patterns and trends. Patterns and trends can used to help determine whether a problem is a systemic issue or not.

- *Line/Trend graphs* connect points that represent pairs of numeric data, to show how one variable of the pair is a function of the other. As a matter of convention, independent variables are plotted on the horizontal axis, and dependent variables are plotted on the vertical axis. *Line graphs are used to show changes in data over time.*

  A trend is indicated when a series of points heads up or down. Nonrandom patterns indicate a trend or tendency.

- *Bar graphs* also portray the relationship or comparison between pairs of variables, but one of the variables need not be numeric. Each bar in a bar graph represents a separate or discrete, value. Bar graphs can be used to identify differences between sets of data.

- ***Pie charts*** are used to depict proportions of data or information in order to understand how they make up the whole. The entire circle, or "pie," represents 100 percent of the data. The circle is divided into "slices," with each segment being proportional to the numeric quantity in each class or category.

- ***Matrices are two-dimensional tables*** showing the relationship between two sets of information. They can be used to show the logical connecting points between performance criteria and implementing actions, or between required actions and personnel responsible for those actions. In this way, matrices are used for determining what actions and/or personnel have the greatest impact on an organization's mission.

30.6 Summary

When you use the correct statistical tools at the right time, your manufacturing processes become more understandable, they achieve a higher quality level, and you become better prepared to explain how your processes work to others. For regulatory inspections, you'll be well-prepared to demonstrate the correct use of statistical quality tools in your response. More important, ensuring that your processes are producing the quality your customers need means that you won't receive a warning letter in the first place.

Failure Investigation: To Prevent Reoccurrence

Introduction

Failure refers to the state or condition of not meeting a desirable or intended objective, and may be viewed as the opposite of compliance or in other words is a Non-conformity of Product failure (Release testing/In-process testing/stability testing), Failure of Utility, Quality System failure, Market complaints, Deviation or incident. Failure investigation is the process of collecting and analyzing data to determine the cause of a failure. Failure Investigation is a detailed investigation, rather than a quick fix, to prevent the re-occurrence of a similar failure.

31.1 Failure Investigation Regulatory Expectation

Must have the following qualities:

- *Objectivity:* In order to provide reliable results, the investigation must be objectively based on analysis of data about the failure. The results should be independently verifiable and subjected to the investigator's familiarity with the system.

- *Comprehensiveness:* The investigation must be comprehensive to cater for all possible causes of the failure, which must all be tested.

- *Reproducibility:* A different investigator should obtain the same results by following the same procedure as the initial investigators. This best achieved through a standard investigation process, which eliminates the risks of subjectivity.

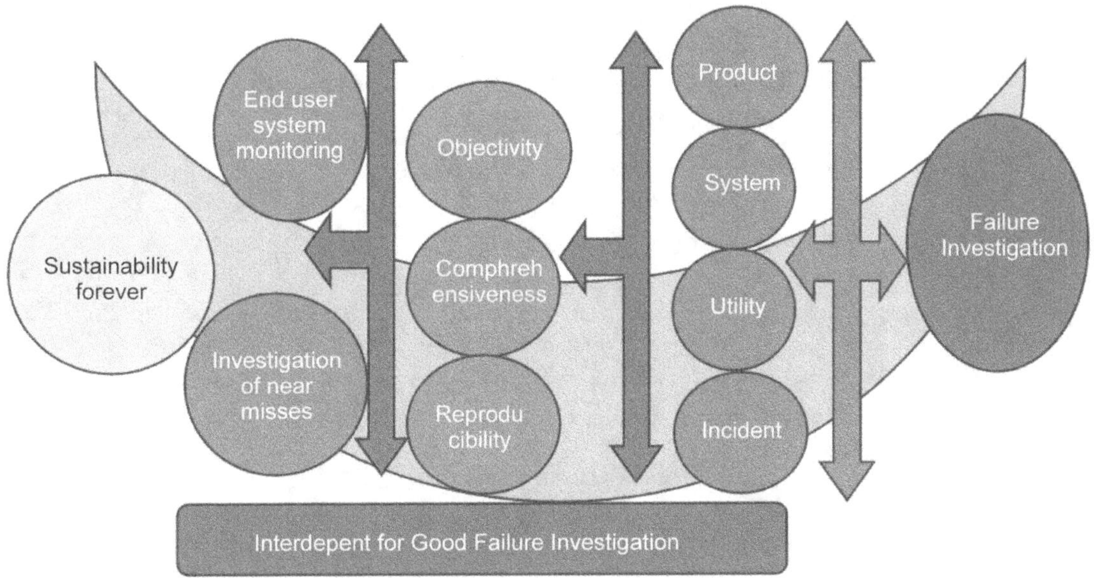

31.2 Failure Investigation Approach

The failure investigation approach must also ensure the following:

- *End user system monitoring:* Monitoring of a systems operation and behaviour is required to avoid accidental or fortuitous discovery or failure and to allow for the real-time detection of errors and risk conditions that are conducive to failure.

- *Investigation of near misses:* Identifying and addressing the root causes of near misses, even though they are harmless, can prevent a more serious accident from developing and can also provide valuable information for conducting the root-cause analysis of the subsequent failure.

31.3 Why do we need Investigation?

"To satisfy the regulatory agencies?"

"Aim should be to highlight the negative impact of failures and to identify the limitation"

Recent Era of Regulatory Outlook on Failure investigation:

Investigation failure is an important tools in the Pharmaceutical industry which identify the reason of failure, to correct and prevent reoccurrence .Recently all regulators are bomming on the depth and effectiveness of investigation failure and co-relate the product quality and effectiveness of GMP for each and every element. Following are the top regulatory observations, which give overall umbrella that regulators have eyed on each and every item of QMS which lead to the deficiencies, companies' reputation, question the training and controls on the QMS effectiveness.

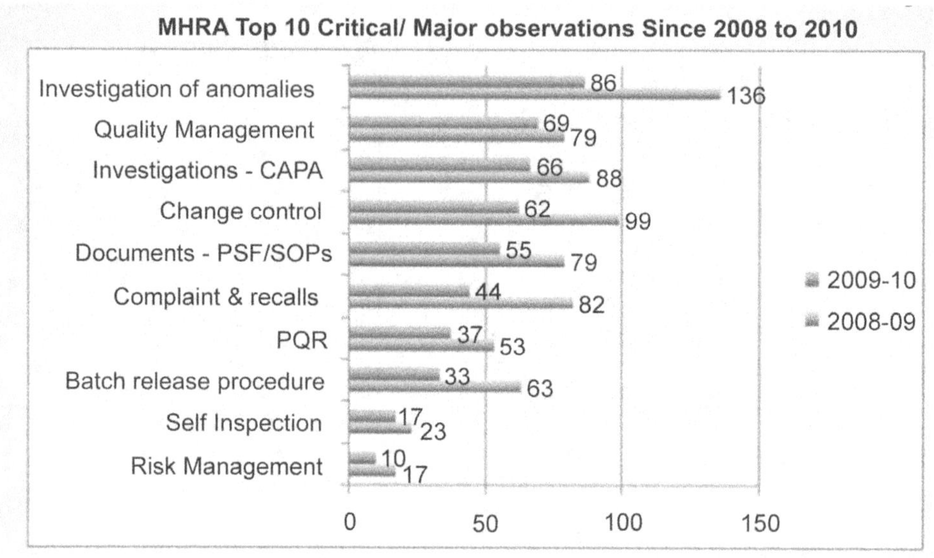

MHRA Top 10 Critical/ Major observations Since 2008 to 2010
Investigation of anomalies
Quality Management
Investigations - CAPA
Change control
Documents - PSF/SOPs
Complaint & recalls
PQR
Batch release procedure
Self Inspection
Risk Management
86
136
69
79
66
88
62
99
55
79
44
82
37
53
33
63
17
23
10
17
2009-10
2008-09
0
50
100
150

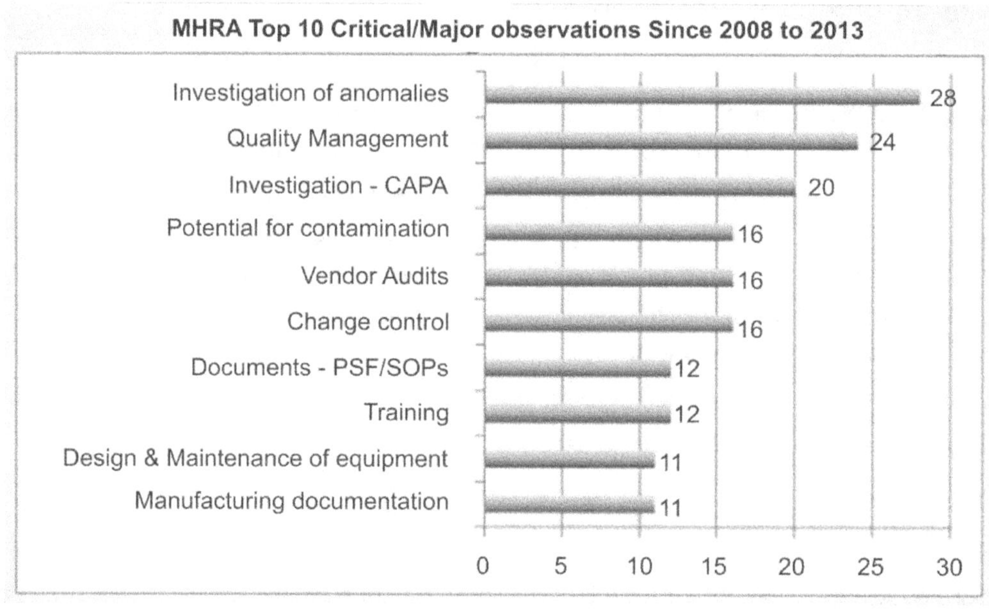

MHRA Top 10 Critical/Major observations Since 2008 to 2013
Investigation of anomalies
Quality Management
Investigation - CAPA
Potential for contamination
Vendor Audits
Change control
Documents - PSF/SOPs
Training
Design & Maintenance of equipment
Manufacturing documentation
28
24
20
16
16
16
12
12
11
11
0
5
10
15
20
25
30

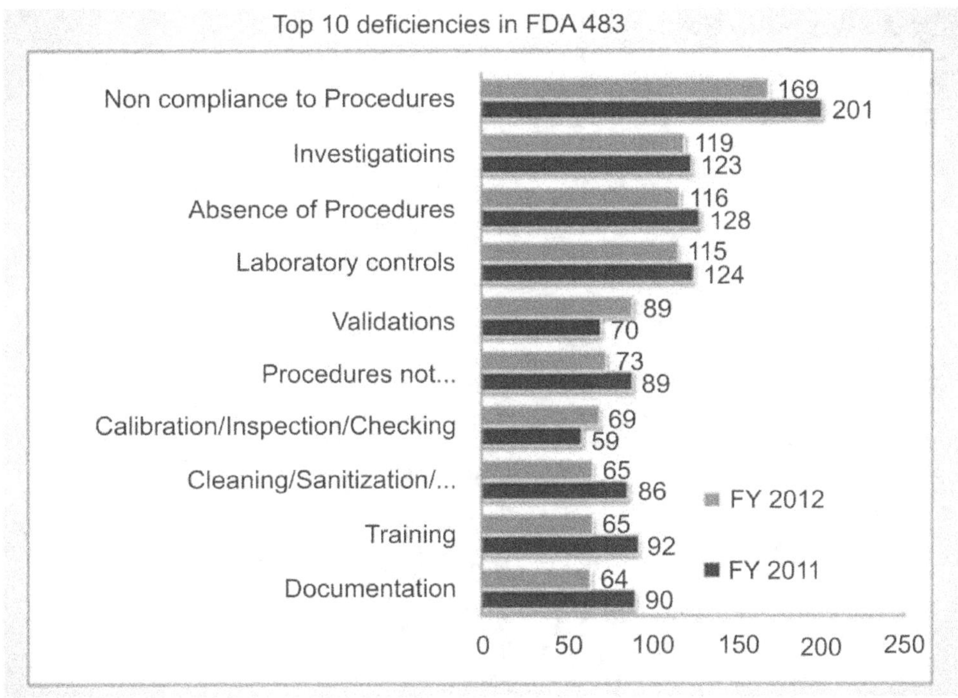

Review of FDA Warning Letter

Quality Systems, lab control systems and production system contribute to ~85% of all FDA observations globally over last 5 years

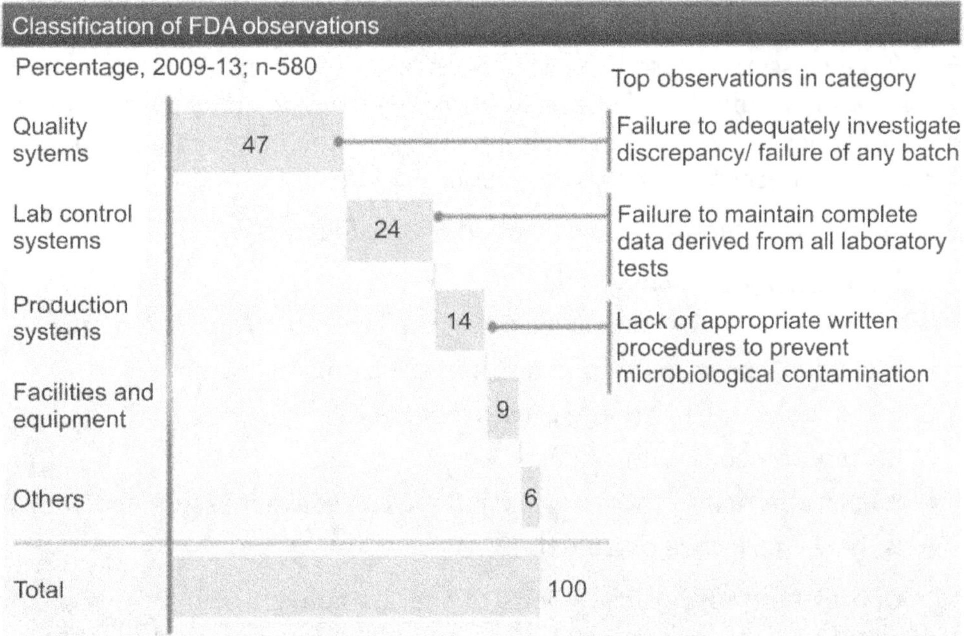

31.4 Why do we have Failures?

Even though we had good System, Equipment, Facility, Trained Manpower & fully validated processes, then also industry had deficiency in failure investigation.

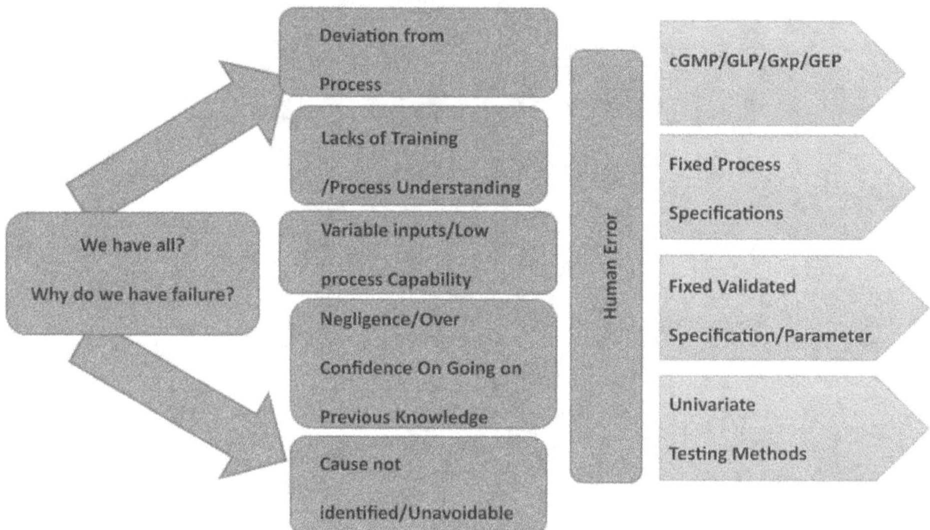

31.5 Failure Investigation Methodology:
A Key to Success in a World Failure

The question is not how many failures happen, but how to avoid the same failures in the future. One of the key success procedures is the investigation process in a pharmaceutical company. Effective science-based investigations lead to error prevention, operational excellence and patient safety. And it is an industry challenge to apply appropriate scientific rigor to investigations. As a GMP fundamental approach the investigation reports must be well documented. This seems to be an easy task.

First, we need to decide what a investigation Methodology is.

A. Right People on Discussion Table

- Are the right people and expertise at the evaluation/discussion table?
- Is the appropriate data complete and is it being reviewed?
- Is the root cause or are the causes identified?
- Is there someone with a holistic view?
- Is each aberrant event reviewed and are corrective measure(s) made?
- Is there a collective perspective?

  One of the most crucial points of the list above - also from my industry & consulting experience - is the first one. Just imagine that the right people are not at the table. How would it be possible to support that all of the appropriate data is reviewed? Which in turn questions the validity of the assessment and identification of the causes. And finally, are you sure to implement and execute the appropriate corrective measures?

B. Develop and brief investigation plan

- Use 6Ms (Six Sigma) approach
- Identify information to understand event scope
- Check if impacted products are on drug shortage list
- Draft investigation timetable
- Develop plan to determine root cause

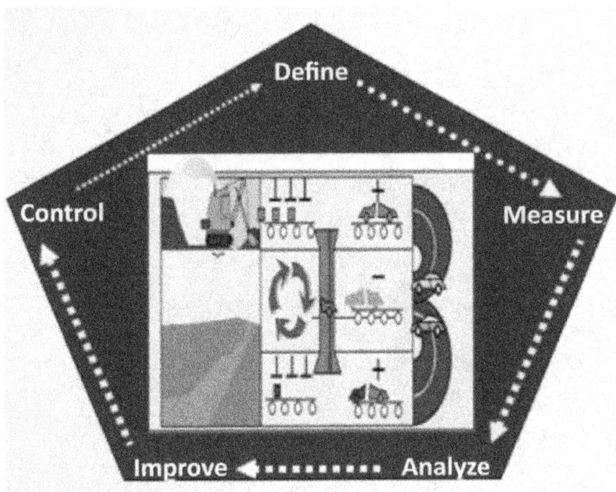

Define the problem and customer requirements

Measure defect rates and document the process in its current incarnation.

Analyze process data and determine the capability of the process.

Improve the process and remove defect causes.

Control process performance and ensure that defects do not recur.

C. Investigation Skills

- Product & process understanding
- Thorough equipment understanding
- Risk assessment
  - Identification of sources of variability and their impact on product quality
  - Data to be collection based on Risk assessment
- Statistical tools
- Analytical ability for data analysis & meaningful conclusions
- Team approach
- Presentation of findings

D. Statistical Quality Control (SQC) Tools for investigation

- Brain storming
- Five Whys?

- Fault tree analysis
- Process flow diagrams
- Cause and Effect Diagram
- FMEA
- Histogram
- Pareto Chart
- Scatter Diagram
- Regression analysis
- Design of Experiments
- Control Charts

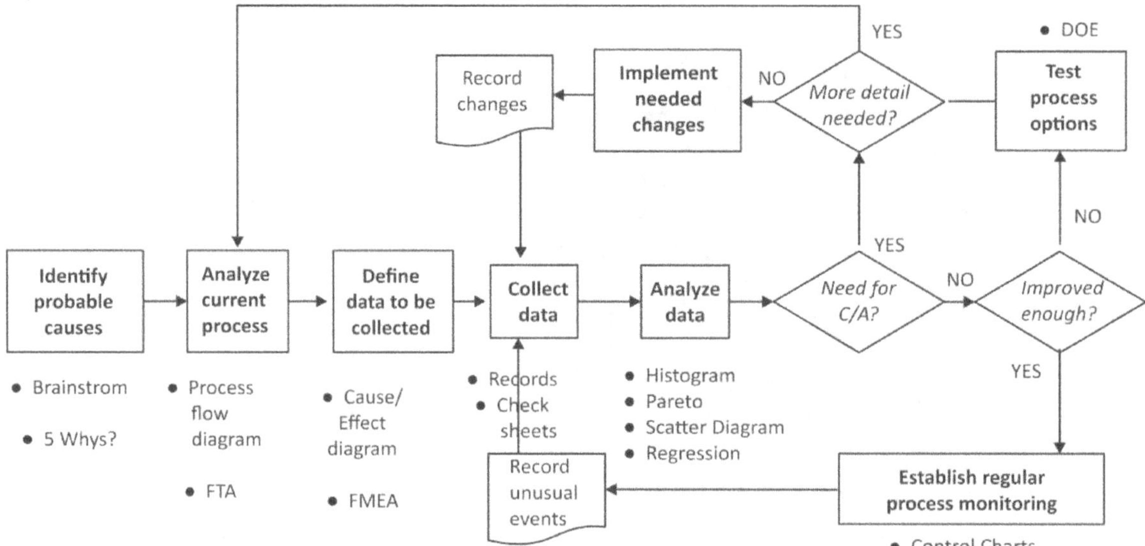

E. Brain storming

- Generation of ideas
- Involvement of people
- Allow people to speak without interruption
- Random / sequential
- Note down all the ideas without elimination

F. Five Whys(5W)?

(a) The 5 Whys is a questions-asking method used to explore the cause/effect relationships underlying a particular problem.

(b) Ultimately, the goal of applying the 5 Whys method is to determine a root cause of a defect or problem.

(c) Keep asking Why till you reach root cause

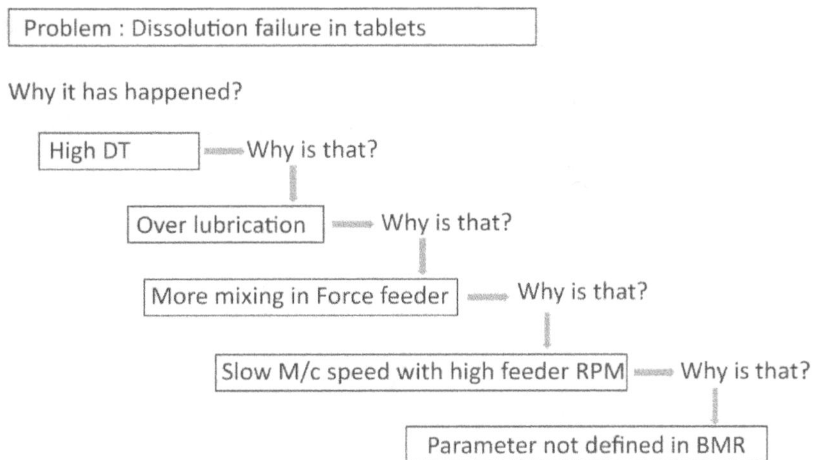

G. Process flow diagram

(a) Helps in understanding the process

(b) Identifies critical steps, critical parameters, control measures

(c) Helps in identifying the data to be collected

(d) Helps in identifying probable causes

H. Fault Tree Analysis (FTA)

(a) Graphical representation of the major faults, the causes for the faults, and potential counter measures.

(b) Helps identify areas of concern for new product design or for improvement of existing products.

(c) Helps identify root cause of failure

(d) Helps identify corrective actions.

(e) Can be useful both in designing or investigation.

I. Cause & Effect Diagram

(a) Developed by Kaoru Ishikawa

(b) Used to explore potential causes (6 M's) that can result in single undesirable effect (UDE).

(c) Potential causes arranged according to hierarchy

(d) Helps in identifying potential problem areas under each of 6 M's

J. Factors to be explored for 6 M's

- **Man Power**
  - Sill, Knowledge, competency and attitude
  - Adequacy of supervision & support
  - Clarity about job role
  - Experience, training
  - Shift in which the activity was done
  - Conductive work environment?
  - Availability of tools/ equipment

- **Machine**
  - Age of euipment or machine
  - Maintenance history
  - Clarity about job role
  - Experience, training
  - Shift in which the activity was done
  - Conductive work environment?
  - Availability of tools/ equipment

- **Material**
  - Change in Source of material
  - Change in process
  - Age of material v/s stability
  - Test results at incoming stage/re-test
  - Material packing
  - Storage condition
  - Correctness of Quantity
  - Quality trends

- **Method**
  - Is the process well defined
  - Critical control points
  - Adequacy of Control parameters
  - Robustness of the process
  - Process capbility
  - Recent changes if any
  - Dviations in execution
  - Trend analysis of Process parameters
  - Safety mechanisms & challenges

- Milieu (Environment)
  - Control of Environmental conditions (Temp/RH)
  - Impact of environmental conditions on the processes
  - Impact of environmental condition on the meterials

- **Measurement**
  - Method validation Specificity & robustness
  - Analyst training
  - Equipment calibration
  - Standards used
  - Frequency of Inspection
  - Other analysis along with the failing batch
  - Execution of methodology

K. Failure mode effect analysis

(a) Scientific Risk assessment of process steps

(b) Analyzes each step with respect to its impact on the quality, probability of occurrence & detectability

(c) Helps in filtering of probable causes

(d) Helps in identifying the data to be collected

L. Tools for data analysis

(a) *Histogram*: Understanding the process centre& spread and Estimation of process capability

(b) *Pareto analysis:* Prioritizing defects & causes

(c) *Scatter plots:* Studying correlation between Xs & Ys

(d) *Regression analysis:* Establishing correlation between different variables

M. Control charts
   (a) Process monitoring
   (b) Trend analysis of past data
   (c) Monitoring effectiveness of CAPA

31.6 Failure Investigation Data Review and Evaluation

Data review and evaluation is a solid ground work for the failure investigation usually involves the review of hard copy and electronic documents. The purpose of the data review is to verify that data was recorded, and that the data quality is good, judging by whether it is reasonable.

Data to be reviewed in case of Different types of failures:

A. Assay failures
   (a) Correctness of Formulation : Label claim / Salt / Water molecule / Batch size
   (b) Correctness of quantities : Weights / reconciliation of leftover / yields
   (c) Working standard : Validity / Response comparison / WS changes / Other batches
   (d) Loss / gain of material : Yield / water content / physical verification
   (e) Segregation
   (f) Sampling technique
   (g) Sample solution preparation

B. Failure in Impurities
   (a) Development report : Pathway of formation / Structure / Control measures
   (b) Forced degradation study
   (c) Manufacturing conditions
   (d) Solution stability : Testing conditions
   (e) Past trend
   (f) Stability data (exhibit / commercial)
   (g) Possibility of contamination: Testing / manufacturing

C. Drug Release failures
   (a) Verification of dissolution medium
   (b) pH / Degassing
   (c) Critical component quality
   (d) Release polymers, binders, disintegrants
   (e) Critical process parameters

(f) Granulation (Power / Current, PSD, BD / TD)

(g) Drying (Inlet temperature, outlet temperature, drying time, LOD)

(h) Mixing (Over lubrication)

(i) Compression (Speed, force feeder, compaction force)

(j) Coating (Air flow, inlet temperature, bed temperature, exhaust temperature, RPM, spray rate, distance of spray gun, left over solution, weight gain)

(k) Capsule filling (machine setting, machine principle)

31.7 What is Corrective & Preventive Actions?

According to Regulatory Authority "

A. ISO 9000:2005, ICH Q-10

- **Corrective Action**
    - Action to eliminate the cause of a detected non-conformity or other undesirable situation
    - Corrective action is taken to prevent recurrence of non-conformity

- **Preventive Action**
    - **Action** to eliminate the cause of a potential non-conformity or other undesirable potential situation.
    - **Preventive** action is taken to prevent occurrence of potential non-conformity.

B. Various Actions in (21CFR 820.100)

- **Correct** ("correction") nonconforming product and other quality problems
- **Prevent** recurrence ("corrective action") of nonconforming product and other quality problems
- **Eliminate** the cause of potential ("preventive action") nonconforming product and other quality problems

C. Case Study of CAPA

- **Problem** : Twinning of tablets observed
- **Correction:** Sorting of twin tablets by inspection
- **Corrective action:** Standardization of coating parameters and fixing recipes to eliminate the root cause of improper rolling of tablets during initial spray.
- Modification of tablet shape to eliminate flat surfaces
- **Preventive action:** Review of shapes of other products to identify possibility of twinning.

31.8 Failure Investigation Report

Good investigation writing gives the gleam of person's knowledge and writing skill. Quality of investigation reports will influence your professional success. They depict your competence: how well you think, how well you gather, assemble and analyse data, how well you draw conclusions and recommendations from data, how well you support your assertions, and how well you create messages that meet the needs of your readers. Your credibility is on the line every time you prepare a report.

Inverted Pyramid Style is used mostly used in writing good investigation failure. This is mainly used in journalism. This style is called an 'inverted' pyramid simply because it is an upside-down pyramid with the most important information at the top

Inverted Pyramid Structure

Most Important Information

Least Important Information

A. What should be included in good failure investigation report?

- Problem definition
- Discussion on analytical results / observations
- Summary of data reviewed
- Conclusions based on data review
- Experimental plan
- Summary of results of experiments
- Discussion & Conclusion of experiment results
- Final conclusion about root cause
- Corrective & Preventive actions
- Batch disposition

B. Problem definition in good failure investigation report

- Use correct technical terminology
- Describe observed results along with the limits
- Include extent of the problem

C. Discussion of analytical results and observations in good failure investigation report

- Original Test results
- Repeat analysis results
- Results of experimental testing
- Results of related samples (control samples, other batches analyzed together etc)
- Observation of the complaint samples
- Discussions and conclusions based on analysis of results

D. Summary of data review in good failure investigation report

- Past history
- Trend data of quality parameters
- Stability data
- Trend of Manufacturing parameters
- Process timings and hold times
- Equipment usage
- Cleaning history / line clearances
- People deployed in operations
- Materials used
- Environmental conditions

E. Conclusions of data review in good failure investigation report

- Provide data analysis
- Scientific justification
- Include both positive and negative conclusions
- Collect more data if data collected earlier is not resulting definite conclusion or suggest review of some other parameters not seen earlier.

F. Experimental Plan and Results in good failure investigation report

- Scientific rationale & purpose of experiment
- Description of experiments & Measurement methods
- Summary of results
- Conclusions based on the evaluation of the results and scientific rationale
- Need for additional experimentation if conclusion is not reached
- Conclusion about root cause or probable causes

G. Investigation CAPA

- Identification of corrective actions to eliminate identified root cause or probable causes
- Identification of Preventive actions to eliminate other potential non-conformities
- Target dates of completion
- Evaluation of effectiveness of CAPA

H. Actions on the batches

- Impact assessment
- Decision about the involved batches
- Actions to salvage if any
- Actions for safety of the consumers

I. Investigation Summary

- Investigation of failures is critical in Pharmaceutical both from improvement and regulatory perspective.
- Investigation skills are crucial for Pharma personnel
- Key skills
  - Domain knowledge
  - Analytical ability
  - Systematic approach (DMAIC)
  - Statistical tools
  - Report / Presentation

Chapter - 32

Laboratory Failure Investigation Management Report Writing

Introduction

In the current Era or 21 century, Quality and investigation management are the top most deficiency area identified by regulators (USFDA and MHRA) during the inspection. These are mainly due to improper investigation, not identify the right root cause with effective CAPA or logical investigation, which was interlinked with product quality.

Top 10 US-FDA Inspection Observation (2015 & 2016)

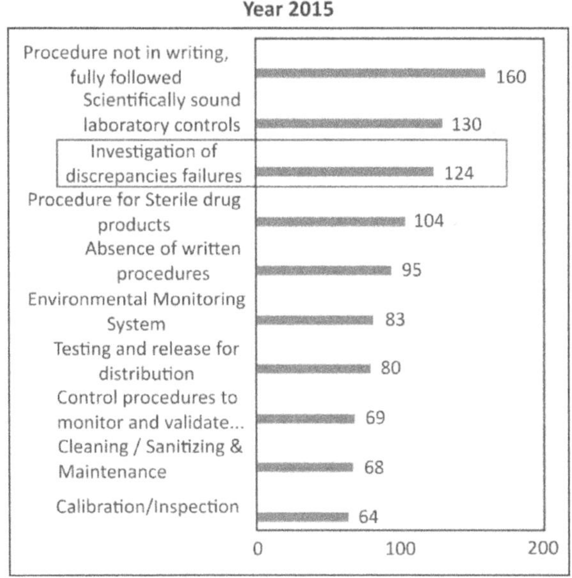

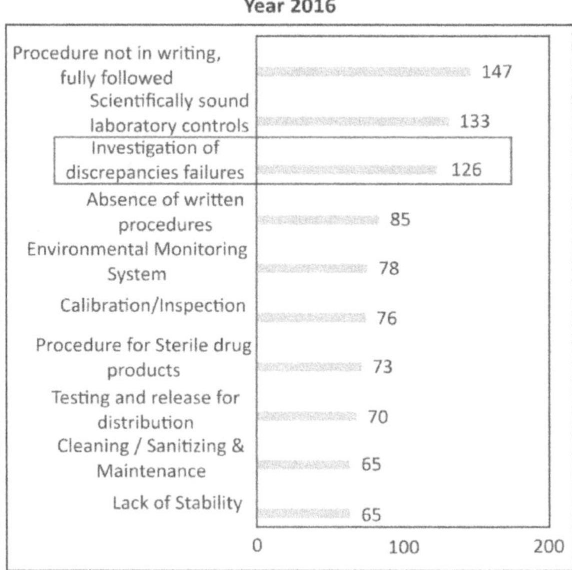

Top 10 MHRA Inspection Observation (2015 & 2016)

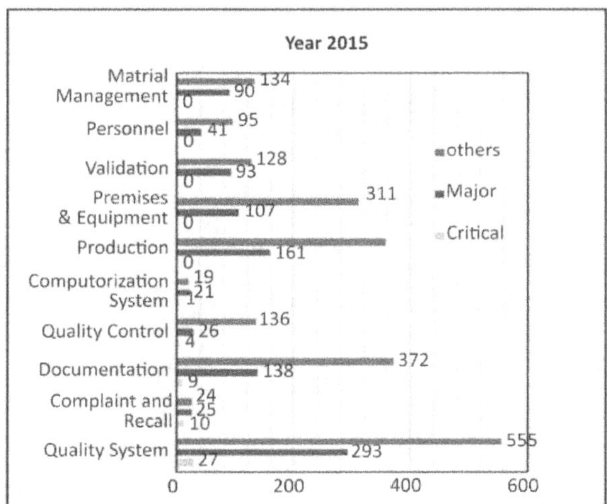

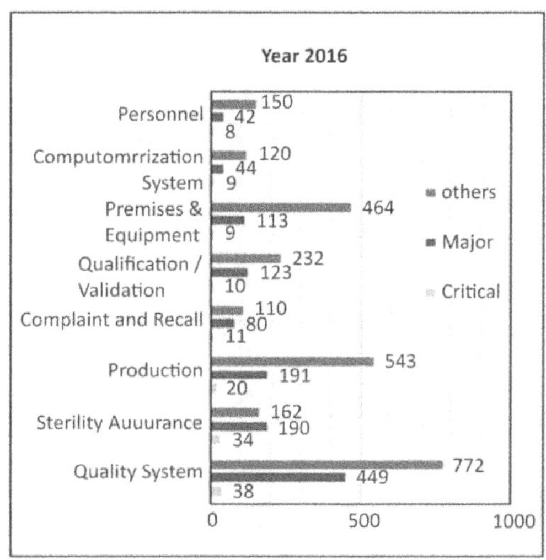

Regulators requires that all conformance (Invalid OOS, Non Substantial Complaint) and non-conformance results must be investigated by clear understanding of regulatory expectations and in sequence order. Lack of consistency around investigation and root-cause analysis processes will lead to error and expensive pharmaceutical activities. First we needs to clarify or to understood what are the regulatory expectation.

Deficiency	Regulatory Expectation
Investigation of Anomalies	Deviation system must requires all deviations to be classified or ranked based on the risk to patient. Risk assessment must be documented. For those considered more serious, an effective impact assessment and root cause analysis must be performed and documented. Appropriate Correction should be implemented and corrective action should be taken to ensure the problem does not reoccur. Internal audits should look for evidence of recurrence and deviation should be trended for the same purpose.
CAPA	All elements of the quality system can give rise to opportunities for improvement. CAPAs are by defining improvements, provided that they are effective. Any failure, complaints, deviations and audit findings should be prioritized based on risk and fed into the CAPA system. CAPA must be implemented in a defined time frame and the effectiveness of the CAPA systems must be monitored by various systems, e.g internal audits, Product Quality review system, trending of deviation parameter or other parameter, changes, CAPA, etc.

32.1 Investigation Management

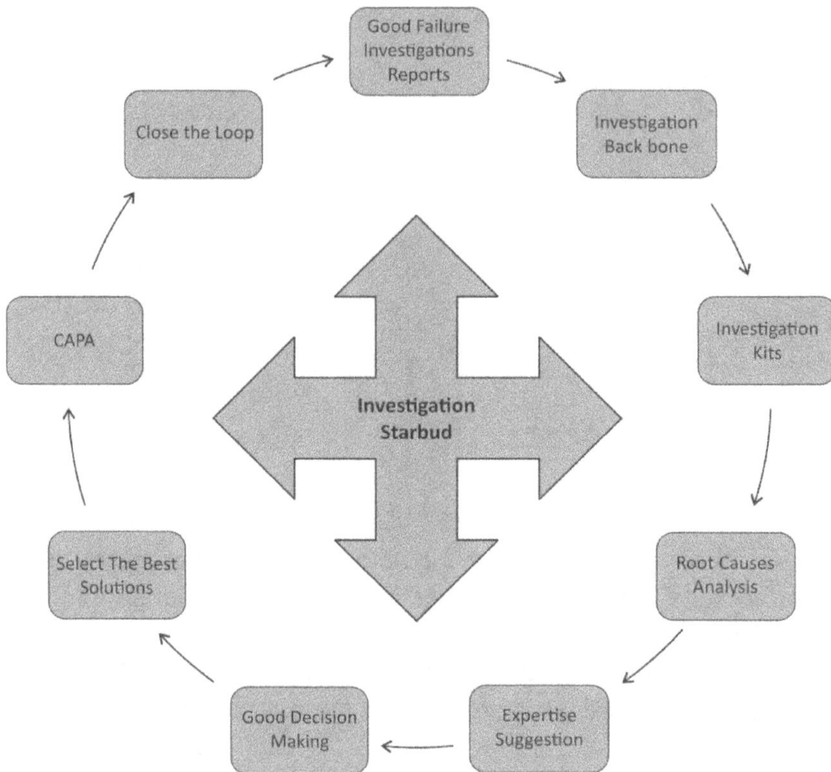

An investigation is an information gathering activity conduct for many different reasons to determine and document facts concerning a particular issue. Experience gained on the job play help to determine the proper course of action, root cause analysis in the investigation.

A good thumb rule of investigation is to first summarize what you did, why or give the reason for the investigation and briefly state the findings. After this, you can go into detail about how you conducted the investigation and what you found. Reporting the course of your investigation and your findings chronologically works in many situations. For long narratives, using headings will make it easier for the reader to follow your reporting. Some types of investigations have forms that need to be completed in addition to the narrative.

32.2 Investigation Star Bugs

A. Writing and ensuring Good Failure Investigations Reports

Good investigation writing gives the gleam of person knowledge and writing skill. Quality of investigation reports will influence your professional success. They depict your competence: how well you think, how well you gather, assemble, and analyze data, how well you draw conclusions and recommendations from data, how well you support your assertions, and how well you create messages that meet the needs of your readers. Your credibility is on the line every time you prepare a report.

Inverted Pyramid Style is used mostly used in writing good investigation failure. This is mainly used in journalism. This style is called an 'inverted' pyramid simply because it is an upside-down pyramid with the most important information at the top.

Inverted Pyramid Structure

Most Important Information

Least Important Information

(a) Benefits of 'inverted' pyramid

- Readers can quickly assess whether they want to read your entire investigation.
- Readers can stop reading at any point and still come away with the main point of your investigation.
- By starting with your conclusion, the first few sentences on your investigation will contain most of your relevant keywords, boosting your investigation.
- By front-loading each paragraph, you allow your readers to skim through the first sentences of every paragraph to get a quick overview of your entire investigation.

(b) Steps involves in Writing and ensuring Good Failure Investigations Reports

I. *Front-load Investigation:* Write a brief summary or overview of investigation in a few sentences. Include an executive summary at front of reports – with all critical information and clear recommendations. Include most important keywords in the summary and put it at the beginning of investigation. This allows users to quickly assess what investigation is about, and helps search engines to identify your most important keywords.

II. Front-load every paragraph

   (a) Limit each paragraph to one main point or idea.

   (b) Start each paragraph with the main point or conclusion in the first one or two sentences.

   (c) Then go on to explain your point

   (d) By doing this, you allow your readers to skim your entire investigation and get an overview of it by simply reading the first sentence or two of each paragraph. They can also scan your content, looking for points of interest, without having to read every paragraph to the end.

III. *Front-load headlines:* Start your headings with significant words. The first words then communicate the subject matter, and catch the eye of your reader. Well-composed headings at the beginning of each section will help your readers to skim and scan the entire article looking for points of interest.

B. Investigation Backbone

Content of investigation backbone depends on 5'W' (Which, When, When, Which and What), 'How', 'Are' and 'IS 'question.

Investigation Backbone		
What are the issue and the conclusion?	Which words or phrases are ambiguous?	What are the value conflicts and assumptions (descriptive or otherwise)?
What are the reasons? What reasonable conclusions are possible?	Are there any fallacies in the reasoning?	How good is the evidence?
Are there rival causes?	Are the statistics deceptive?	What significant information is omitted?

C. Investigation Kits

   (a) Notify your supervisor or QA promptly (within24 hours).

   (b) Draft a clear, complete problem statement and Assess patient safety and product impact.

   (c) Use your company's forms/SOPs, and fill forms out completely.

   (d) Describe immediate corrective action(s) taken.

(e) Investigation Plan Investigation Team and Subject Master File)

(f) List all possible causes. Create a timeline or chronology. Look for any prior occurrences.

(g) Look for any changes or differences or "ripple effect."

(h) Brainstorm root causes.

(i) Carry out the investigation; most should be done within 30 days.

(j) Ask good questions: who, what, when, where, why and how. Ask "why" 5 times.

(k) Analyze the information. Brainstorm and select best possible solutions; apply your criteria.

(l) Create an action plan. Follow up. Measure CAPA effectiveness

(m) Document well; use the inverted pyramid writing style.

(n) Get the investigation approved.

(o) Communicate the results to everyone involved, including other sites.

D. Root Cause Analysis

- Working with a group of peers, brainstorm possible root causes for major or more critical problems
- Always look for at least two causes in form of pre-existing conditions and actions (catalysts)
- Effort should extend beyond effects of problem to discover its most fundamental cause(s)
- Prepare investigation plan.

E. Expertise Suggestion

- Seek different opinions.
- Challenge experts.
- Ensure everyone has a voice, not only technical experts or high-ranking individuals.
- Become comfortable with nuance, uncertainty, doubt.
- Work with experts to help figure things out, but be aware of their limitations and our own

F. Good Decision Making

- Think about own thinking process.
- Take more control of decisions and how to make them.
- Ask for help when needed.
- Choose individuals who have the skill set(s) you need.

- Look into your own psyche; identify ways you may be sabotaging your own decision making.

G. Select The Best Solutions

- Working from confirmed causes, brainstorm possible solutions. Try not to go for quick fix (retrain operator).
- Apply criteria to select best solution:
  - (a) Solution that prevents recurrence of problem, including at other locations
  - (b) Solution that does not cause unacceptable problems
  - (c) Solution is within your organization' control
  - (d) Solution that provides good value for its cost

H. Corrective and Preventive Action (CAPA)

CAPA procedures: Implementing an effective corrective or preventive action capable of satisfying quality assurance and regulatory documentation requirements is accomplished in six basic steps:

- ***Identification:*** The initial step in the process is to clearly define the problem. It is important to accurately and completely describe the situation as it exists now. This should include the source of the information, a detailed explanation of the problem, the available evidence that a problem exists.

- ***Evaluation:*** The situation that has been described and documented in the "Identification" section should now be evaluated to determine first, the need for action and then the level of action required. The potential impact of the problem and the actual risks to the company and/or customers must be determined. Essentially, the reasons that this problem is a concern must be documented.

- ***Investigation:*** In this step of the process a procedure is written for conducting an investigation into the problem. A written plan helps assure that the investigation is complete and nothing is missed. The procedure should include: an objective for the actions that will be taken, the procedure to be followed, the personnel that will be responsible, and any other anticipated resources needed.

- ***Analysis:*** The investigation procedure that was created is now used to investigate the cause of the problem. The goal of this analysis is primarily to determine the root cause of the problem described, but any contributing causes are also identified. This process involves collecting relevant data, investigating all possible causes, and using the information available to determine the cause of the problem. It is very important to distinguish between the observed symptoms of a problem and the fundamental (root) cause of the problem.

- **Action Plan:** By using the results from the Analysis, the optimum method for correcting the situation (or preventing a future occurrence) is determined and an action plan developed. The plan should include, as appropriate: the items to be completed, document changes, any process, procedure, or system changes required, employee training, and any monitors or controls necessary to prevent the problem or a recurrence of the problem. The action plan should also identify the person or persons responsible for completing each task.

- **Follow Up:** One of the most fundamental steps in the CAPA process is an evaluation of the actions that were taken.

  Several key questions must be answered:

  (a) Have all of the objectives of this CAPA been met? (Did the actions correct or prevent the problem and are there assurances that the same situation will not happen again?)

  (b) Have all recommended changes been completed and verified?

  (c) Has appropriate communications and training been implemented to assure that all relevant employees understand the situation and the changes that have been made?

  (d) Is there any chance that the actions taken may have had any additional adverse effect on the product or service?

I. Close the Investigation Loop

- Management with executive responsibility must be aware of and review CAPA results. Organization must take prompt action when violative products or situations are discovered.

- A surveillance system is only as good as the information it receives; we must be able to rapidly pick up on important signals.

- Follow up is crucial. Close loop by providing input into design control/R&D requirements/study design.

- Nonconforming product identified and corrected

32.3 Challenges in Investigation

The main pitfall of investigation strategy is that their solution is completely dependent upon and limited by their expertise. If the root cause happens to lie outside the scope of their expertise levels, they are not likely to find it. Therefore, they must design a series of closely monitored experiments to test their hypotheses and to determine if they are on the right track in locating the root cause. The problem with designing and implementing these experiments is that, they are invasive requiring personnel, equipment laboratory resources, down time and funding. A unique and simple tool is used to synthesize the collected data into information that tells the root cause story.

Chapter - 33

Deviation Management in Good Laboratory Practices

Introduction

Deviation Management is a part of QMS providing efficient support for controlling deviation or incidents in implementing corrective measures, helping avoid their recurrence, and for taking a proactive approach to continuous quality improvement. Deviation is applicable to all deviation observed during any stage of receipt, handling, storage, transfer, testing, documentation and other activities from approved procedure.

"Any departure from the approved process or procedure is called deviation" Type of deviations are:

- **Planned Deviations:** Planned and know before its occurs
- **Unplanned Deviations:** Occur without intimation.

33.1 Deviation Categorization

- **Major Deviations:** When the deviation affects a quality attribute, a critical process parameter, an equipment or instrument critical for process or control, of which the impact to patients (or personnel/environment) is unlikely, the deviation is categorized as Major requiring immediate action, investigation, and documented.
- **Critical Deviations:** When the deviation affects a quality attribute, a critical process parameter, an equipment or instrument critical for process or control, of which the impact to patients (or personnel or environment) is highly probable, including life threatening situation, the deviation is categorized as Critical requiring immediate action, investigated, and documented.

- ***Minor Deviation:*** When the deviation does not affect any quality attribute, a critical process parameter, or an equipment or instrument critical for process or control, it would be categorized as Minor

33.2 Management of Deviation

Any deviation noted during the analysis or any laboratory activities against approved procedure should be reported to the department head and Quality Assurance. Quality Assurance should issue the deviation form after assigning the deviation number in Deviation log. Quality Assurance should enter the details of deviation" Planned/Un-Planned "in Deviation log. Deviation number consist year and chronological serial number. A new serial number should be start at every year for deviation. Deviation form is one time use.

Example: CC/Year/**Serial** number (Serial number should be chronological order =CC/2017/001

Brief description and reason for deviation should be recorded in deviation form.

Investigation team should be investigate the reason for devisaiton. Investigation of the deviation should be performed in sequence manner with chronological order which leads to deviation occurrence. Root cause of deviation should be investigated and corrective and preventive action shall be taken to prevent recurrence.

Risk analysis shall be performed to confirm deviation are critical, major or minor or critical. Based on the evaluation and closure out date should be mentioned by Quality Assurance in the deviation form.

After impact assessment the deviation should be assign for cross function team to evaluate the changes to their respective end and gives their comments on the deviation form and submit back to Quality Assurance for further action plan. The comments shall be written by concerned or relevant department head or his designee .After receiving the comments from all cross function team Quality Assurance should approve the deviation. Time line should be assign for the closure of change control not exceed than 30 days from date of occurrence. Deviation should be closed after completion of the corrective and preventive action. If deviation is not closed within the time line, two extension should be taken on scientific justification and approved by Quality Assurance. Trend analysis should performed periodically to know the repeat deviation and systematic correction .In case of batch release deviation must be closed before batch dispatched .Deviation number shall be recorded in test analysis report or/and batch record. A review of all Critical, major and minor deviation should be done during Product Annual Quality Review.

Quality Assurance should be monitor the status of deviation periodically at least 15 days. The records can be archived in paper form or electronically form.

33.3 Steps Involved in Deviation Investigation

- Login of Deviations and description
- Event Detection and investigation
- Risk Analysis and Impact Assessment
- Deviation Categorization
- Root cause analysis
- CAPA
- Efficacy of corrective action and conclusion
- Verification and closure

33.4 Most Common Deviation in the Laboratory, but not Limited to

- Sample breakdown during analysis
- Wrong entry of method parameters, sequences in soft ware
- Wrong dilution or weight
- System suitability failure
- Sample spillage
- Mistake calculation
- Contamination of diluent, sample or standard
- OOT and border line results
- Baseline drift

33.5 Deviation Trend Analysis

Periodical trend analysis shall be performed for identification of repeat deviation and verification of corrective and preventive action. Protocol and report shall be prepared for trend analysis.

33.6 Deviation Case Study

A. Deviation Form

Deviation No: 201600023		
Deviation Raised on Date: 28.09.2016	**Deviation Occurred on:** 28.09.2016	**Deviation Reported on:** 28.09.2016
Department : Laboratory Management	**Location of Deviation :** (Room/Area, Line etc.)	
Deviation related to (Facility/Equipment/Material/ Product/ Process/ Analytical/ Document/ Others –Specify the details	Personnel	

Table *Contd...*

Deviation No: 201600023	
Product Name	ABC Tablets 600mg
Batch No.	TP17029.
Material code	212327801
Reference Procedure / BMR / BPR / Specification / Test Procedure / others	NA
Equipment/instrument ID No.	NA

Details of Incident:

Description of the Deviation:

The ABC Tablets 600mg for Batch No: TP17029. Identification by NIR analysis performed by using Uncalibarted instrument (Inst ID: QC347) the analysis was executed on 20.09.2016.

Immediate actions / Corrections done :

1. Verified the instrument status and found that subjected NIR instrument was under calibration from 14-09-2016..
2. Intimated to supervisor and the event was addressed through quality notification.: Yes
3. Whether manufacturing was continued? No
4. If yes mention the justification and the interim controls applied :NA

Deviation Form Completed and Reported By /date(Initiator / Observer / Identifier / Supervisor)

Name :Ravi Malhotra

Date : 28.09.2016

Supporting Documents Attached::

(a) Calibration report #QC1234

(b) SOP#ABS-01

Impact Assessment by User

Initial impact assessment was done as per the report and categorized under category-2. Further investigation shall be performed for identifying the probable root cause.

Name: HariKesav

Date: 30.09.2016.

Intimation and review by HOD/Designee

Categorization is Appropriate: Yes

Initial Impact assessment is performed: Yes

Review Comments & Conclusion: The investigation is appropriate.

Recommendations /Corrective actions: After calibration of the instrument, reanalysis shall be done for the subjected sample.

Name: Trupti

Date : 30.09.2016

Table *Contd...*

Deviation No: 201600023
Intimation & review by QA Head/Designee Categorization is Appropriate: Yes Initial impact assessment is appropriate and approved by QA: Yes Recommendations: Investigation to be carried out to identify the assignable cause. Further Investigation Requirement: Yes Name: Sita Patel Date: 01.10.2016
Immediate actions / Corrections done As an immediate action rechecked the calibration and preventive maintenance (PM) status label on the instrument and confirmed that instrument was due for calibration. Analyst had defaced the calibration label and the PM label immediately when the instrument had crossed the calibration due date. However the same was not ensured by the analyst before use of the instrument for the analysis. As an action plan the preventive maintenance and the calibration was performed and data found satisfactory. Hence in the above event noticed no discrepancy was noticed with respect to the instrument functioning. Hence the instrument shall be released for routine operations. Name: Siddi Patel Date: 07.10.2016
Investigation: Identification test parameter of Sevelamer Carbonate tablets 800 mg (Batch No: TP17029 and AR No: 29081530) was performed by the analyst by FTNIR Spectrophotometer (Inst ID: QC347) using approved STP (STP No: ANR-12567-01), as per analysis plan made on 20-09-2016. After completion of the analysis, the analyst to enter in to the INSTRUMENT LOG BOOK of the subjected instrument found that the log book was not in the designated place due to which the analyst did not make relevant entries whereas online entries was not made by the analyst at the starting time of the instrument usage. Verified the instrument status and found that the Status labels of the instrument were defaced by the lab support team and had removed the particular instrument log book from their designated place to avoid the usage after the calibration due date as the subjected instrument was due for the calibration. The subjected instrument was due for preventive maintenance and calibration, hence status labels of the FTNIR were defaced and not valid beyond 14-09-2016 Further the preventive maintenance was carried out on 29-09-2016 for the subjected Instrument (Inst ID: QC347) by external engineer and found satisfactory. Further calibration was performed with AR No: JMS162101 as per SOP CTSPQC146 and found satisfactory. Verified both the system data as well as instrument log book (Register No: TAPQC0017/F02-0029) of the subjected instrument (Inst ID: QC347) and no analysis was found from due date of calibration i.e. 14-09-2016 to current calibration date i.e. 29-09-2016 except the above subjected analysis. Hence from the above investigation it was

Table Contd...

Deviation No: 201600023

concluded that the event was occurred due to analyst error as he had not ensured the status labels before starting the analysis.

Name: Karuna

Date : 20.10.2016

Categorization of Incident :Major

Investigation review by HOD/Designee

Reviewed complete investigation and concluded that deviation was occurred due to inadvertently analyst used the under calibration instrument. However analysis was invalidated and repeated the analysis after completion of calibration. Hence no impact on quality and reviewed proposed CAPA found satisfactory. (For more details refer attached investigation report).

Name: Mahi

Date : 20.10.2016

Review& Quality Head/Designee

Reviewed investigation and conclusion, found that the usage of un-calibrated instrument for routine IR testing is due inadvertent error by the analyst. Invalidation of initial impacted analysis and repetition analysis after successful completion of calibration of subjected instrument was found satisfactory. Successful completion of FT NIR instrument preventive maintenance and calibration were found satisfactory. There is no impact on any of the data due to this incident. Proposed actions and of implementation of under calibration board display and awareness to team as an additional controls were found satisfactory. Proposed actions shall be tracked through CAPA and therefore investigation, conclusion and proposed actions were approved.

Name: Aarya

Date: 20.10.2016

Final review of Deviation& closure by QA

1. Was there any change in the categorization of incident? No
2. If yes, mention the category of the incident: NA
3. Investigation is completed and investigation report is approved: Yes
4. CAPA addresses the root cause: Yes
5. Target date/time allotted for implementation or each CAPA is adequate: Yes
6. Extends to other related processes, sites, products, equipment procedures: No
7. CAPA effectiveness plan is adequate and time line is justified: Yes
8. CAPA Notification is raised: No
9. CAPA Notification No.: NA
10. Decision on product disposition is documented: Yes
11. Documentation is clear, succinct, logical, and written as per good documentation practices: Yes
12. All supporting documents are available and uploaded / attached to the Deviation report: Yes

Table *Contd...*

Deviation No: 201600023
Comments / Remarks: Based on the investigation report, it was concluded that the Deviation was occurred due to analyst error i.e analyst inadvertently used uncalibrated instrument for the IR testing. Invalidated the initial data and repeated the analysis after successful completion of calibration of FTNIR instrument, obtained results found satisfactory and considered for reporting. No impact on the functionality of the instrument and the quality of the product. Name: Radha Date : 20.10.2016
Closed By : Radha **Date :** 20.10.2016

B. Investigation Form : Investigation Report for Deviation No: 201600023:

Investigation Report		
Deviation No: 201600023		
Deviation Raised on Date: 28.09.2016	**Deviation Occurred on:** 28.09.2016	**Deviation Reported on:** 28.09.2016
Department : Laboratory Management		
Deviation related to (Facility/Equipment/ Material/ Product/ Process/ Analytical/ Document/ Others –Specify the details	Personnel	
Product Name	ABC Tablets 600mg	
Batch No.	TP17029.	
Material code	212327801	
Reference Procedure / BMR / BPR / Specification / Test Procedure / others	NA	
Equipment/instrument ID No.	NA	
Description of the Deviation: The ABC Tablets 600mg for Batch No: TP17029. Identification by NIR analysis performed by using Uncalibarted instrument (Inst ID: QC347) the analysis was executed on 20.09.2016.		
Summary of Investigation : • As an immediate action rechecked the calibration and preventive maintenance status labels on the instrument and confirmed that instrument was due for calibration. • The lab support team analyst had already defaced the calibration label and the preventive maintenance label on the day of due date: 14.09.2016. • On interaction with the Lab support team upon the delay for the calibration of her subjected instrument and understood the subjected FTNIR preventive maintenance shall be carried out by the external engineer (vendor) and delay in attending for preventive maintenance by external engineer caused for delay in completion of calibration against schedule. • However the same was not ensured by the analyst who was performing the subjected analysis before use of the subjected instrument.		

Table *Contd...*

Investigation Report

Deviation No: 201600023

- The affected analytical data was invalidated and subjected batch kept under hold for further release.
- Preventive maintenance followed by calibration was performed for the subjected FTNIR instrument and no discrepancy was noticed with respect to the instrument functionality, hence release for routine analytical applications. The discrepancies observed with respect to the usage for routine analytical application. The discrepancy observed with respect the usage of instrument after calibration due date was happened unintentionally by the analyst without ensuring defaced status labels of preventive maintenance and calibration.

Investigation Team :

Sr. No	Name	Department	SME in
1	Nisha	Quality Control	Head of Department
2	Kishana	Quality Control	Calibration Member
3	Venu	Quality Assurance	Approver

Investigation Findings:

- Identification test [parameter of ABC Tablets 600mg for Batch No: TP17029 was performed by analyst by FTNIR spectrophotometer (Inst ID: QC347) using approves STP (STP No: ANR-12567-01), as per analysis plan made on 20-019-2016
- After completion of the analysis, the analyst searched for the instrument log book (Register No: TAPQC0017/F02-0029) of the subjected instrument to enter analysis details and log book was not found in the designated place due to which the analyst dose not make relevant entries, whereas online entries was not made by the analyst at the starting time of the instrument.
- As a part of regular practice during delay in calibration of any instrument due to certain reason, lab support team shall deface the respective status labels and remove the particular instrument log book from the designed place.
- Immediately the analyst verified the preventive maintenance and calibration status label of the FTNIR and found the status labels were defaced and not valid beyond 14-09-2016.
- The subjected discrepancies was intimated to the group leader and the event was address through quality notification.
- Further the preventive maintenance was carried out on 29-09-2016 for the subjected instrument # by external engineer and found satisfactory. Further calibration was performed with Analytical reference number (A.R.No.) #TAY201918 as per SOP CTSPQC146 and found satisfactory.
- Verified there subjected analyst (Mr. Ravi) training records and the Qualification and found that analyst was trained on the subjected SOP and qualified for the technique on 25-05-2016.

Table *Contd…*

Investigation Report
Deviation No: 201600023

- Based on interaction with the analyst it was understood that, he had missed to ensure status labels before stating the analysis.
- Future reviewed the previous history of the subjected analyst and no such instance on usage of instrument beyond its calibration validity were noticed.

Conclusion:

- Based on the investigation it was concluded that the subjected deviation occurred due to analytical error however the analyst inadvertently failed to verify the instrument status of the subjected instrument before staring the analyst as it was due for calibration. As the schedule calibration of FTNIR Spectra photometer performed on 29-09-2016 was found satisfactory as per SOP, hence there is no impact on the functional aspects of instrument as the observed discrepancy was attributed to the analytical error specific to this particular analysis and not related to the instrument functioning. It was concluded by depicting that there was no discrepancy due to instrument functionality and considering the incident as an isolated case for the subjected FTNIR instrument.

Impact Assessment and Batch Disposition:

- The calibration results of the instrument performed on 29-09-2016 was found satisfactory, evidencing that there is no discrepancies in the instrument functionality.
- Based on the impact assessment it was noticed that the incident was an isolated case and restricted to the particular analysis.
- Initial analysis data and repeat analysis data for identification test were compared and found comparable. Hence there is no impact on the quality of the product.

Recommended Correction, Corrective and Preventive Action:

A. Correction: As a correction initial analysis data performed on the subjected instrument after calibration due data was invalidated.

B. Corrective Action: Repeat analysis was performed using the same instrument after calibrating the instrument and data found satisfactory.

Training session shall be conducted for concerned personnel on "Good Laboratory Practice" and "Good Documentation Practices "

C. Preventive Action: As an additional control, Under Calibration" status label shall be display during calibration stage.

List of Attachment:

- Training report
- Calibration report
- Analysis certificate

Prepared By : Somesh	**Checked By:** HariDas	**Approved By:** Nikitha
Date : 20.10.2016	**Date :** 20.10.2016	**Date:** 20.10.2016

Handling of Out of Trend (OOT) Result in Laboratory Management

Introduction

An out of trend result is the result that does not follow the expected trend, either in comparison with other batches or with respect to previous results collected. Trend limits is to be generated on sufficient analytical data by averaging the value. Variable values must not be average as average value may hide variability and may result into in appropriate limits.

The greater trend limits must not be exceeding the release limits. The limit can be by finalized on complied date from all the batches by approaches such as 3 sigma or 5 sigma or 6 sigma. The back update must be attached for trend analysis. Trend must be reviewed periodically. OOT investigation can be performed as per OOT investigation. Avoiding potential issues with marketed product, as well as avoid potential regulatory issues apply of OOT control in the analysis is a best practice in the industry. In summary, the issue of OOT is an important topic both from a regulatory and business point of view. Despite this, little has been discussed in the scientific literature or in regulatory guidance on this topic. This article will introduce some approaches that might be used to identify OOT data and discuss some issues that companies will likely need to address before implementation and during use of an OOT identification procedure.

34.1 Reporting and Login of the OOT

Upon observation of any OOT result, analyst shall immediately report the same to supervisor within 24 hours. Login of OOT may be an manual or electronic. A unique number shall be allocated for each type of OOT. In case, if multiple OOT are reported within a single analytical reference number, the OOT number assigned shall be only once, however the investigation formats shall be issued to individual test specific, cross-referencing the same OOT number with proper remarks.

34.2 Laboratory Investigation-Phase-I

The investigation shall be initiated at the earliest time after the OOT is reported and purpose of investigation is to determine, whether any laboratory error has resulted for reported OOT result. The designated personnel of lab shall issue a copy of "Phase I-Laboratory investigation" format to the investigator for initiating lab investigation. The investigator shall enter the relevant OOT information in the copy of format issued. If any additional portions for relevant sections is warranted, the same shall be annexed with additional pages/sheets & cross-referenced.

I. ***Interference of Preliminary assessment:*** Conducted the primary assessment with help of investigator with a focus to assess the accuracy of reported OOT result

- As part of preliminary assessment shall examine the raw data such as instrument printouts/chromatogram/ diffractogram /spectra /histogram, etc. The examination shall also extend to instrument used & associated equipment such as chambers, drying oven, etc including any breakdown, calibration, review of previous data.

- The investigation team shall also visit the work bench/actual place of testing and examine the glassware's used, homogeneity of actual preparations and solutions available, in initial analysis at the earliest possible time from the reporting of OOT result.

- If obvious error (Calculation error, wrong weight / glassware, equipment failure, wrong/processing integration parameter etc) has been identified, the same shall be documented with supportive evidence, where appropriate and necessary corrections shall be recommended. The initial reported result shall be 'Invalidated' & impact assessment (such as during calculation error in automated template, etc) shall be conducted, NO FURTHER INVESTIGATION IS REQUIRED.

- If the assessment does not tell the presence of any obvious error, further detailed investigation should be initiated to identify the assignable/probable cause for OOT result.

- The detailed investigation and outcomes shall be documented. The Remeasurement plan shall be reviewed & authorized person, prior to initiating the re-measurement activity

II. ***Interference of Hypothesis analysis:*** Details of proposed hypothesis studies shall be documented and reviewed prior to initiating the hypothesis studies. The proposed details shall have following information at minimum; Suspected Probable causes, Proposed plan for hypothesis studies, Type of samples (i.e.) Batch which is under investigation and/or approved batch to be used and their execution of testing, Number of sample replicates, and how the data generated will be evaluated. On investigation appropriate CAPA shall be assigned with all impact assessment document.

III. **Interference of Re sampling** shall be allowed only in the following scenarios;

Findings/Observations from detailed investigation *and/or* discussion with analyst *and/or* Re measurement indicates the;-

   (a) *Sample Integrity*: Sample has been compromised due to improper packing/ handling including the storage of sample prior to analysis, (ie) retention at warehouse for outsourced materials/at manufacturing area for products processed at In-house *or*

   (b) *Sampling error:* Not representative of homogeneous batch

   - Insufficient amount of sample for subsequent investigation process.

     Justification for Re sampling (supported with appropriate evidence) shall be detailed in relevant section of "Phase I-Laboratory investigation" format as per annexure GQA035/A02, by investigator, followed by review & authorization of QA head/designee, prior to initiating resampling activity

IV. *Interference of Retesting:* Shall be performed by analyst 2 on justification of number of sample. The approach for 'Resampling' should be same (ie) qualified/standard method that was used during initial sampling. If assignable cause is not identified or when probable cause is not confirmed, RETESTING SHOULD NOT BE PERFORMED, the batch shall be rejected & proceed for Phase II investigation to identify the manufacturing aberration. The outcomes of retest including any other observations encountered during laboratory investigation process shall be evaluated by investigator for its suitability to batch disposition with respect to Completeness & variation between individual results, wherever applicable, Closeness of the reportable value to the target or with earlier trend, Past history of the product including the stability data, wherever applicable and Clarity of observations/findings documented.

   - The decision shall be based on collective information available from ; Evaluation of initial results, all retest outcomes, Conclusion of laboratory investigation made by investigation team ,Observations cited during document review and/or retesting, if any, and Appropriateness of proposed CAPA /corrective measures, if any.

   - The approval of phase I investigation summary report shall be prepare and reviewed by authorized person before the investigation is concluded as 'Invalid OOT'

   - If the Phase I, laboratory investigation OOT is unable to ascertain any lab error, a full scale investigation shall be initiated.

34.3 Phase II: Full Scale Investigation

- When the initial assessment does not determine the laboratory, error caused the OOT result and testing results appear to be accurate, a full –scale investigation using predefined procedure must be conducted. This investigation may consist of a production review and/or laboratory work. The objective of such an investigation must be to identify the root cause analysis of the OOT result and take appropriate corrective and preventive action. A full-scale investigation must include a review of production and sampling procedures and will often include additional laboratory testing. Such investigation must be give high priority.

- A full-scale OOT investigation must consist of a timely, through and well documented review. A written record of the review must include

  (a) ***Interference of Production Review***: The investigation must be conducted by Quality assurance along with other department including manufacturing, process developments maintenance and engineering. Potential problem must be identified and investigated. The records and documentation of the manufacturing process must be fully reviewed to determine the possible cause of the OOT result.

    - A clear statement of the reason for the investigation.

    - A summary of the aspects of the manufacturing process that may have caused the problem.

    - The results of documentation review, with the assignment of actual or probable cause.

    - The results of a review made to determine if the problem has occurred previously,

    - A description action taken.

    - If this part of the OOT investigation confirms the OOT results and is successful in identifying the root cause, the OOT investigation may be terminated and the product rejected. However a failure investigation must be completed with predefined time. If any material is reprocessed after additional testing, the investigation must include comments and the signatures of appropriate production and quality control and quality assurance personnel.

  (b) ***Interference of Additional Laboratory testing***: **A** full-scale OOT investigation may include additional laboratory testing. A number of practices are used during the laboratory phase of an investigation. These include retesting a portion of the original sample and re-sampling.

  (c) ***Interference Retesting***: The sample used for retesting must be taken from same homogeneous material that originally collected from the lot, tested and yielded the OOT results .For a liquid , it may be from the original unit liquid product or composite of the liquid product and for solid , it may be from the same sample composite prepared for the original sample.

(d) *Interference Decision* of retest must be based on the objectives of testing and scientific judgment. It is often important for the predefined retesting plan to include retests performed by an analyst other than the one who performed the original test. A second analyst performing a retest must be at least as experienced and qualified in the method as the original analyst. The maximum number of retests to be performed on sample must be specified in advance in a written standard operating procedure. The number of retest must not be adjusted depending on the results obtained.

- If **no** laboratory or calculation errors are identified in the first test, there is no scientific basis for invalidating initial OOT results in favor of passing results. All test results, both passing and suspect must be reported and considered in batch release decisions.

(e) *Interference Resampling*: After Quality Assurance approval for resamples the original sample from an batch where OOT is occurred. Resampling must be performed by the same qualified, validated methods that are used for the initial sample. However, if the investigation determines that initial sampling methods was inherently inadequate; a nee accurate sampling method must be developed, documented, and reviewed and approved by the Quality Assurance.

(f) *Interference of Reporting Testing* **Results:** Averaging the results of the original test that prompted during an OOT investigation and additional retest or results obtained during the OOT investigation is not appropriate because it hides variability among the individual sample.

(g) *Conclusion* **the** *investigation*: To conclude the investigation, the results must be investigated and the findings of the investigation, including retest results, must be interpreted to evaluate the batch and reach a decision regarding release or rejection. A confirmed OOT result indicates that the batch does not meet established standards or specification must be rejected and disposed with proper documentation. Impact on the other batches and product must be studied. For inconclusive investigations-in case where an investigation I(1) does not reveal a cause for the OOT test results and (2) does not confirm the OOT results must be give full consideration in batch or lot disposal decision.

34.4 OOT Trend Analysis

Trending of OOT test results shall be done periodically. The trend analysis shall include classification of OOT results based on causes (analyst error, instrument error, analytical method error or specification related, equipment error, process error, operator error, input material, etc). Any emerging trend (example, increase in instrument error) shall be evaluated for taking necessary CAPA and to avoid recurrence. Trends shall be reviewed & evaluated during management review meetings. The OOT trend shall be evaluated to monitor the effectiveness of CAPA implemented for earlier OOT of past 2 year and to prevent recurrence of the same in future.

34.5 OOT Methodology Calculation

- A minimum of 15-30 batches data shall be compiled for fixing the Trend range.
- Results that shall be obtained from the 20 batches tabulated, average value, minimum and maximum values are noted.
- Standard deviation will be calculated for these 20 batches. Validated Excel spread sheet shall be used for Standard deviation calculation.
- Standard deviation will be multiplied by 3or 6 to get the 3 sigma or 6 sigma value.
- Trend range for upper and lower limits.
- Any value that shall be out of this range will be considered as Out of Trend
- Wherever specification has only Not more than, then only Maximum limit for trend can be considered. Minimum limit should be excluded.
- Wherever specification has range then both the Maximum and Minimum limits for trend should be considered.

34.6 Challenges in Implementation of OOT

- What statistical approaches are used to determine OOT criterion? What data are used to determine OOT limits?
- What are the minimum data requirements? What evaluation is performed if the minimum data requirement is not met?
- What data should be used to update limits?8 The investigation requirements (i.e., who is responsible, what is the timeline, how is it documented, who should be notified must be clearly defined.
- Who is responsible for comparing the result with the OOT criterion?
- How is an OOT result confirmed? What additional analytical testing or statistical analyses are appropriate?
- What actions should be taken if an OOT result is confirmed as an unusual result?
- How are OOT investigations incorporated into the annual product review?

34.7 Definition

- *Assignable causes:* The underlying special cause, identified (either during lab /manufacturing investigation) based on scientific evidence, which is responsible for OOT.

- *Atypical result:* A laboratory test result that is within its specification, but which is markedly different from values obtained from series of test results, which is considered to be aberrant/abnormal/irregular and need to be investigated.

- *Attribute data:* Discrete categorical data that can be counted or classified in some meaningful way but which cannot be measured. These 'count' data may be expressed as 'Pass/Fail' 'Yes/No' 'Presence of defect/Absence of defect'

- *Control chart:* A Graphical method of recording results (obtained by measuring a continuous variable), in order to distinguish between random causes and assignable causes of variation in a process. It presents a graphic display of process stability or instability over time.

- *Control limit:* Numerical value derived (through control chart) for a process such that any result obtained beyond that value indicates some change has occurred to the process and that action should be taken to investigate and/or correct for the change. These limits are assigned either side of the center line and are designated as Upper control limit (Ucl) and Lower control limit (Lcl).

- *Out of Trend:* In case of release testing of commercial batches, a laboratory test result which is atypical and indicates that manufacturing process could be out of control. (Outside the normal analytical /sampling variation & normal change over the time). In case of stability evaluation, a laboratory test result that is significantly different and not in agreement with the results obtained from previous stations. An OOT in stability sample may provide early indication of a potential OOT result, which may also necessitate corrective and preventive action.

- *Random causes:* These are the common causes of variation that cannot be precisely identified and are part of normal variability of the process or testing. Example. Machine vibration, Temperature fluctuations, etc.

- *Variable data:* Data which is obtained by measurement (for a specific process:- manufacturing/testing) and has a definite numerical value. Within the context of this procedure, variables refers to test results of "Assay, Chromatographic purity, Related substances, LOD, Water content, Dissolution, Acceptance value for Content Uniformity, Preservative content, Residual solvents, Microbial limit test which are either included as part of release and/or shelf life specification.

34.8 Annexes

A. Log book for OOT

| OOT log Book for Year _________ | | | | | | | | | | | |
Date of reporting	OOT. No	Product/ Stage	Batch No	Test Parameter	Spec. Limit	Initial Result	Entered by sign & date	Final Result	Root cause	Status of the batch	Entered by sign & date

B. Format for Phase –I Laboratory Investigation

Format for Phase –I: OOT Investigation			
Notification/Ref. No		Issued on	
Issued by		Due date	
Issued to			

OOT/ information			
Name of material /product		Stage	
Batch. No		AR. No.	
Test		Ref. Specification. No	
Observed results		Limit	
Reported on		Lab Book/Ro A	
Date of initiation		Name of the analyst_1	

Section A- Preliminary assessment
Observations:
Corrections:
Sign & date of the investigator

Section B- Checklist for Detailed investigation					
Category	Ref. #.	Checkpoints	Yes	No	NA
Calculation		Whether any calculation error has occurred?			
		Are there any reasons to suspect or evidence that any un authorized or un validated changes have been made to programmed calculation methods, if used?			

Table *Contd…*

Section B- Checklist for Detailed investigation					
Category	**Ref. #.**	**Checkpoints**	**Yes**	**No**	**NA**
		Any transcription error occurred in calculations used to convert raw data values in to a final result?			
Training		Is the analyst-1, NOT qualified/ evaluated for the specific technique, through which testing has been performed. If 'NO' date of qualification for the specific technique______			
Sample		Is there any evidence to suspect the sampling approach followed for the material/product			
		Are there any reasons to suspect or evidence that the original sample is invalid due to improper packing, storage, handling etc?			
		Are there reasons to suspect or evidence that sample integrity not maintained? (wrong container sampled, chain of custody)			
		Are there reasons to suspect or evidence that sample tested is incorrect? B.No as indicated in the label_________			
		Are there reasons to suspect or evidence that sample contamination has occurred during testing? (sample left open to air or unattended)			
		Are there reasons to suspect or evidence that original sample is not representative of the batch?			
Methods		Is there any evidence that correct method not followed? Approved procedure &its version # which was examined as part of assessment.________			
Observation		Are there any abnormalities noticed specific to the Method or Operation which may or may not be captured in the method of anlysis or approved SOP?			
Standards / Samples/ Chemicals		Is there any evidence that sample is impacted due to nature of the material/product (ie hygroscopic, light sensitive, etc...)			
		Is there an evidence that valid analytical standards is NOT used?			

Table *Contd...*

Section B- Checklist for Detailed investigation					
Category	**Ref. #.**	**Checkpoints**	**Yes**	**No**	**NA**
		Name of analytical standard: Code.No: Validity:			
		Is there any evidence that analytical standards were not stored properly?			
		Is there an evidence of degradation of standards/sample/chemicals used for testing (before or after preparation)? Time of re-examination of sample preparation/solution:			
		Is there a problem identified in weights, preparation and dilution of. Standards/Samples/Chemicals?			
		Is there a problem identified in labelling of Standards/ Samples/Chemicals?			
		Is there any evidence that Standards/Samples/Chemicals and their solutions are not used within their expiration or use before date?			
		Is there an evidence that chemicals used for testing is NOT having the desired grade, make , purity/concentration , etc. as specified in STP			
Sample / Standard / Preparation / Solution		Are there reasons to suspect or evidence of contamination during the sample/standard preparation or transfer?			
		Is there any evidence that correct reagents, solvents standards glassware, or other auxiliary equipments (E.g. balance) were not used?			
		Is there any evidence of error in weighing, dilution and sonication etc...			
		Is there any abnormality noticed like color change etc. identified before, during or after the analysis?			
Reagents / Solutions		Is there any evidence that Solutions/Reagents are not prepared according to the procedure?			
		Is there any evidence that Solutions/Reagents are not within the valid date for use?			
		Whether the visual examination performed on Solutions/ Reagents retained, reveal any abnormal appearance?			

Table *Contd...*

Section B- Checklist for Detailed investigation					
Category	**Ref. #.**	**Checkpoints**	**Yes**	**No**	**NA**
Glass wares		Is there any evidence that incorrect glassware have been used?			
		Is there evidence that glassware used is contaminated?			
Instruments		Is there evidence for malfunctioning of any component of the instrument? Ex. leakage at injector, pump, column connection, flow cell (before or during the analysis likely to impact the testing)			
		Is there any evidence that the instruments involved in testing are not calibrated within the period and also not calibrated properly? Name of instrument :_________________ Code No:_____________________________ Valid up to:__________________________			
Chromatography		Do the consecutive blank pattern are NOT comparable (as an evidence for inadequate column equilibration)			
		Do the System suitability results NOT comply with the criteria? (those before analysis and during analysis)			
		Is there any abnormality noticed in standard responses?			
		Is there any abnormality noticed in pressure channel/data?			
		Is there any abnormality observed in the injection vials ?			
		Is there any evidence that correct method not used?			
		Is there any evidence that right columns were not used?			
		Is there any evidence that correct mobile phase not used?			
		Whether the chromatograms reveal any anomalous or suspect information like noisy resolution, poor peak resolution, poor injection			

Table Contd...

Section B- Checklist for Detailed investigation					
Category	**Ref. #.**	**Checkpoints**	**Yes**	**No**	**NA**
		reproducibility and improper integration/ processing methods, etc ?			
		Are there reasons to suspect or evidence that potentially interfering or incompatible testing activities occurring at the time of test?			
		Are there reasons to suspect or evidence that vials used for sample or standards for injections is improperly crimped or improper septa , etc?			
Stability testing		Were pinholes observed in the sample packing?			
		Is there an evidence of improper sealing of samples or any abnormalities in packing condition?			
		Is there any abnormal trend (product behavior w.r.t time) observed during review of stability data of other batches?			
		Any abnormalities in the sample pre-test storage			
		Is there any impact of temperature & humidity due to stability chambers breakdown.			
History		Is there a history of OOT result for the same parameter that has occurred previously for the same product &stage?. (Review of OOT data for at least past 6 months.) If 'YES', also review the outcome of the earlier investigation in light of the current failure for existence any recurring error. If 'NO' in case of stability OOT, review of stability data obtained from Exhibit, Process validation batch indicates any abnormal trend, with the batch which has obtained OOT?			
		Are there reasons to suspect lab error based on the history of product, type of test being performed and any other test results obtained at earlier stages?			
		Is there any evidence available, for the noticed abnormality? Attach photographic evidence, if relevant.			

Table *Contd...*

Section B- Checklist for Detailed investigation					
Category	**Ref. #.**	**Checkpoints**	**Yes**	**No**	**NA**
If answer to any of the reference questions is 'YES' the investigator shall elaborate, the details of same appropriately. Sign & date of the investigator					
Observations/Findings based on document review and/or discussion with analyst Sign & date of the investigator					
Recommendations: Sign & date of the investigator					

Section C- Remeasurement	
Details of proposed Remeasurement and plan:	
Proposed by investigator	
Review comments by QA designee /designee for remeasurement	
Remeasurement Authorization by QA head/designee	
Assigned by team leader	
Result of re-measurement(s)	
Re-measured by	
Observations and/or Recommendations from Remeasurement Sign & date of the investigator :	

Authorisation for Re-analysis (strike out if not required)	
Review comments by QA head/designee for reanalysis	*Applicable only in case either if an Obvious error or Assignable cause has been identified in Phase I*
Reanalysis Authorization by QA	

Section D- Hypothesis studies
Suspected probable causes *Proposed hypothesis studies to prove the Suspected probable causes as root cause for failure.* *Type of samples [Batch which is under investigation and /or approved batch] and their execution of testing :*

Table *Contd...*

Section D- Hypothesis studies	
Number of sample: *Evaluation approach:*	
Requested by	
Remarks/comments by QA	
Authorization for Hypothesis by QA	
Observation from Hypothesis studies *Root cause for laboratory failure:* *Immediate corrections, if any or CAPA* Sign & date of the investigator :	
Impact Assessment: *of other batches analysed in the same run/sequence and/or Consideration of other test results, if any on the same batch under investigation and its relevance:* Sign & date of the investigator :	

AUTHORISATION FOR RESAMPLING(strike out if not required)	
Justification for Resampling: *Insufficient sample/Evidence of sample integrity/Observation of sampling error* Quantity:	
Requested by	
Review comments by QA head/designee for resampling	*Opinion to proceed further (based on evidences/facts supported) either re-analysis or Hypothesis*
Resampling Authorization by QA	

Justification for timeline extension of phase I investigation (Strike out if not required)	
Requested by :	**Authorized by :**

Section E- RETESTING	
Batches requested for retesting /number of samples (5) retest plan	
Requested by	
Remarks/comments by QA	

Table Contd...

Section E- RETESTING			
Authorization by QA for retest			
Individual results of retest by Analyst 2			
Outcomes of retest	Whether all outcomes meetsspecification? **Y E S / N O**		
Retested by		Lab Book/RoA	

Section F- Evaluation of retest outcomes & conclusion of lab investigation
Evaluation of retest outcomes :   Sign & date of the investigator :
Conclusion of Laboratory investigation :   *Reasons for failure results*   Sign & date of the QC designee

Section G- Batch disposition and approval of investigation
QA evaluation & approval of investigation :   Sign & date of the QA :

C. Format for Phase –II: Full Scale OOT Investigation :

Format for Phase –II: Full Scale OOT Investigation			
Notification/Ref. No		Issued on	
Issued by		Due date	
Issued to			

Failure information			
Name of /product		Stage	
Batch. No		AR. No.	
Test		Ref. Specification. No	
Observed results		Limit	
Date of initiation			

Section A-Document review				
Ref. #.	**Checklist**	**Yes**	**No**	**NA**
1.	Is current version of Batch Production/Manufacturing record is NOT used for manufacturing? Version of BPR/BMR_________________			
2.	Is there any evidence that operating instructions as per production record/processed are NOT followed during execution of batch? Counter check through questioning			
3.	Whether the operator and supervisor are UNABLE to interpret the basic process flow, product nature and other systemic requirements? Confirm by questioning.			
4.	Is there any evidence of inconsistent/ illogical sequence steps followed during manufacturing/packing?			
5.	Any abnormal observations (such as equipment malfunction, extended or frequent stoppage, etc) during manufacturing/packing are noted/recorded? Verify with operators and supervisors?			
6.	Any in-process checks are missed /not performed during manufacturing?			
7.	Whether the sampling collected is NOT as per sampling procedure/plan ? Confirm with applicable checklist, if available.			
8.	Is there any evidence for extended time/ hold time for any stage/unit operation? If yes, whether it has exceeded the established hold time. Established hold time____________			
9.	Any significant difference or variation in yield, when compared against previous /next batch? Yield of atleast 2 previous & 2 next batch, if available.			
10.	Whether environmental conditions (such as humidity, temperature) as mentioned in production record is NOT within the limit during batch processing?			
11.	Any significant difference or variation in process parameters			
12.	Whether any stages of manufacturing process is NOT validated? If yes which stage is NOT validated____________			
13.	Is there similar history (atleast 1 year) noted from any earlier investigation.			
14.	Are there any new personnel involved in manufacturing, Review the training document and its scope.			

Table *Contd...*

Section A-Document review				
Ref. #.	**Checklist**	**Yes**	**No**	**NA**
15.	Whether the particular stage/unit operation has extended more than one shift of operation?			
16.	Is there evidence of operator change within a shift? Countercheck with other supportive data/information?			
17.	Any measuring devices used to measure process parameters are NOT calibrated?			
18.	Are the balances used in dispensing, manufacturing including in process area are NOT calibrated?			
19.	Any stoppage or failure of utilities supply during any particular stage or unit operation?			
20	Any changes in manufacturing equipment train without authorization?			
If answer to any of the reference questions is 'YES' the investigator shall elaborate, the details of same appropriately. Sign & date of the investigator				
Observations/Findings based on document review and /or discussion with operator and/or supervisor Sign & date of the investigator :				

Section B- Hypothesis studies	
Suspected probable causes and rationale for each probable cause *Proposed/Related hypothesis studies in order to establish/evidence the identified probable causes as root cause for failure.* *Type of samples [Batch which is under investigation and /or approved batch] and their execution of testing :* *Number of sample:* *Evaluation approach:*	
Requested b	
Remarks/comments by QA	
Authorized by QA	
Observation from Hypothesis study: *Root cause for Manufacturing aberration:* *Immediate corrections, if any or CAPA* Sign & date of the investigator :	
Impact Assessment: *Other batches of same product and/or Different product with validated processes similar to that which has resulted failure* Sign & date of the investigator :	

Justification for timeline extension of phase II investigation (Strike out if not required)	
Requested by :	Authorized by :

Evaluation by QA, in case Error remains Unclear or Hypothesis is inconclusive (strike out if not required)	
Comments by QA designee as part of Independent evaluation	
Evaluated by	*Sign & Date of QA designee*

Section C- RETESTING		
Batches requested for retesting /number of samples (5) retest plan		
Requested by		
Remarks/comments by QA designee		
Authorization by QA designee for retest		
Individual results of retest by Analyst 2		
Outcomes of retest	Whether all outcomes meets specification? Y E S / N O	
Retested by		Lab Book/RoA	

Section D- Conclusion of full scale OOT investigation
Evaluation of outcomes :   Sign & date of the investigator :
Conclusion of full scale OOT investigation :   *Reasons for failure results*   *CAPA proposal with TCD*   Sign & date of the Operations designee

Section E- Batch disposition and approval of full scale OOT investigation
QA evaluation & approval of full scale investigation :   Sign & date of the QA designee:

Chapter - 35

Handling of Out of Specification Laboratory Results

Introduction

The term OOS results include all test results that fall out of the specifications or acceptance criteria established in item or production must be thorough, timely, unbiased, well documented and scientifically sound. OOS investigation consist of two phase:

- Phase I : Laboratory investigation
- Phase II : Full-Scale OOS investigation.

35.1 Reporting and Login of the OOS

Upon observation of any OOS result, analyst shall immediately report the same to supervisor within 24 hours. Login of OOS may be an manual or electronic. A unique number shall be allocated for each type of OOS. In case, if multiple OOS are reported within a single analytical reference number, the OOS number assigned shall be only once, however the investigation formats shall be issued to individual test specific, cross-referencing the same OOS number with proper remarks.

35.2 Phase I: Laboratory Investigation

In Phase I Analyst report the OOS to laboratory supervisor. Supervisor log OOS. Initial investigation is carried out by involving analyst by interviewing him / her for correct procedure, data, calculation chemicals, standards, glassware used. Checklist can be prepared for the same. This phase also includes verification by reinjection the initial preparation for hypotheses regarding what might have happen such as dilution error, instrument error. Original sample must not be destroy till the investigation is completed, if the test solution stability is less, then justification for same must be documented with back up data.

- If clear laboratory error is identified, the firm must determine the source of that error and take corrective and preventive action. All the laboratory assessment must be documented. Initial investigation must be completed in defined time with scientifically sound conclusion. After conclusion In-valid the previous data.

- If clear laboratory error is not identified or un-clear a full scale investigation must be carried out .Both the initial laboratory assessment and following OOS investigation must be documented fully. In this case initial data must not be done in- valid.

Steps followed in Laboratory Investigation-Phase-I

I.   *Laboratory investigation-Phase-I:* The investigation shall be initiated at the earliest time after the OOS is reported and purpose of investigation is to determine, whether any laboratory error has resulted for reported OOS result. The designated personnel of lab shall issue a copy of "Phase I-Laboratory investigation" format to the investigator for initiating lab investigation. The investigator shall enter the relevant OOS information in the copy of format issued. If any additional portions for relevant sections is warranted, the same shall be annexed with additional pages/sheets & cross-referenced.

II. *Preliminary assessment:* Conducted the primary assessment with help of investigator with a focus to assess the accuracy of reported OOS result

- As part of preliminary assessment shall examine the raw data such as instrument printouts/chromatogram/ diffract gram /spectra /histogram, etc. The examination shall also extend to instrument used & associated equipment such as chambers, drying oven, etc including any breakdown, calibration, review of previous data.

- The investigation team shall also visit the work bench/actual place of testing and examine the glassware's used, homogeneity of actual preparations and solutions available, in initial analysis at the earliest possible time from the reporting of OOS result.

- If obvious error (Calculation error, wrong weight / glassware, equipment failure, wrong/processing integration parameter etc) has been identified, the same shall be documented with supportive evidence, where appropriate and necessary corrections shall be recommended. The initial reported result shall be 'Invalidated' & impact assessment (such as during calculation error in automated template, etc) shall be conducted, NO FURTHER INVESTIGATION IS REQUIRED.

- If the assessment does not tells the presence of any obvious error, further detailed investigation should be initiated to identify the assignable/probable cause for OOS result.

- The detailed investigation and outcomes shall be documented. The Remeasurement plan shall be reviewed & authorized person, prior to initiating the re-measurement activity

III. *Hypothesis analysis:* Details of proposed hypothesis studies shall be documented and reviewed pi or to initiating the hypothesis studies. The proposed details shall have following information at minimum; Suspected Probable causes , Proposed plan for hypothesis studies, Type of samples (ie) Batch which is under investigation and/or approved batch to be used and their execution of testing ,Number of sample replicates ,and how the data generated will be evaluated. On investigation appropriate CAPA shall be assigned with all impact assessment document.

IV. *Resampling:* shall be allowed only in the following scenarios ;

- Findings/Observations from detailed investigation *and/or* discussion with analyst *and/or* Remeasurement indicates the
    - **(a)** *Sample Integrity:* Sample has been compromised due to improper packing/ handling including the storage of sample prior to analysis, (ie) retention at warehouse for outsourced materials/at manufacturing area for products processed at In-house *or*
    - **(b)** *Sampling error:* Not representative of homogeneous batch
        - Insufficient amount of sample for subsequent investigation process.
        - Justification for Resampling (supported with appropriate evidence) shall be detailed in relevant section of "Phase I-Laboratory investigation" format as per annexure GQA035/A02, by investigator, followed by review & authorization of QA head/designee, prior to initiating resampling activity

V. **Retesting** shall be performed by analyst 2 on justification of number of sample. The approach for 'Resampling' should be same (ie) qualified/standard method that was used during initial sampling. If assignable cause is not identified or when probable cause is not confirmed, RETESTING SHOULD NOT BE PERFORMED, the batch shall be rejected & proceed for Phase II investigation to identify the manufacturing aberration. The outcomes of retest including any other observations encountered during laboratory investigation process shall be evaluated by investigator for its suitability to batch disposition with respect to Completeness & variation between individual results, wherever applicable, Closeness of the reportable value to the target or with earlier trend, Past history of the product including the stability data, wherever applicable and Clarity of observations/findings documented.

- The decision shall be based on collective information available from ; Evaluation of initial results, all retest outcomes, Conclusion of laboratory investigation made by investigation team, Observations cited during document review and/or retesting, if any, and Appropriateness of proposed CAPA /corrective measures, if any.

- The approval of phase I investigation summary report shall be prepare and reviewed by authorized person before the investigation is concluded as 'Invalid OOS' If the Phase I, laboratory investigation is unable to ascertain any lab error, a full scale investigation shall be initiated.

35.3 Phase II : Full Scale Investigation

- When the initial assessment does not determine the laboratory, error caused the OOS result and testing results appear to be accurate, a full –scale investigation using predefined procedure must be conducted. This investigation may consist of a production review and/or laboratory work. The objective of such an investigation must be to identify the root cause analysis of the OOS result and take appropriate corrective and preventive action. A full-scale investigation must include a review of production and sampling procedures and will often include additional laboratory testing. Such investigation must be give high priority.

- A full-scale OOS investigation must consist of a timely, through and well documented review. A written record of the review must include

    (a) ***Production Review***: The investigation must be conducted by Quality assurance along with other department including manufacturing, process developments maintenance and engineering. Potential problem must be identified and investigated. The records and documentation of the manufacturing process must be fully reviewed to determine the possible cause of the OOS result.

    - A clear statement of the reason for the investigation.

    - A summary of the aspects of the manufacturing process that may have caused the problem.

    - The results of documentation review, with the assignment of actual or probable cause.

    - The results of a review made to determine if the problem has occurred previously,

    - A description action taken.

    - If this part of the OOS investigation confirms the OOS results and is successful in identifying the root cause, the OOS investigation may be terminated and the product rejected. However a failure investigation must be completed with predefined time. If any material is reprocessed after additional testing, the investigation must include comments and the signatures of appropriate production and quality control and quality assurance personnel.

(b) ***Additional Laboratory testing***: A full-scale OOS investigation may include additional laboratory testing. A number of practices are used during the laboratory phase of an investigation. These include retesting a portion of the original sample and re-sampling.

(c) ***Retesting***: The sample used for retesting must be taken from same homogeneous material that originally collected from the lot, tested and yielded the OOS results. For a liquid, it may be from the original unit liquid product or composite of the liquid product and for solid, it may be from the same sample composite prepared for the original sample.

(d) ***Decision*** of retest must be based on the objectives of testing and scientific judgment. It is often important for the predefined retesting plan to include retests performed by an analyst other than the one who performed the original test. A second analyst performing a retest must be at least as experienced and qualified in the method as the original analyst. The maximum number of retests to be performed on sample must be specified in advance in a written standard operating procedure. The number of retest must not be adjusted depending on the results obtained.

- If no laboratory or calculation errors are identified in the first test, there is no scientific basis for invalidating initial OOS results in favor of passing results. All test results, both passing and suspect must be reported and considered in batch release decisions.

(e) ***Resampling***: After Quality Assurance approval for resamples the original sample from an batch where OOS is occurred. Resampling must be performed by the same qualified, validated methods that are used for the initial sample. However, if the investigation determines that initial sampling methods was inherently inadequate; a nee accurate sampling method must be developed, documented, and reviewed and approved by the Quality Assurance.

(f) ***Reporting Testing*** **Results**: Averaging the results of the original test that prompted during an OOS investigation and additional retest or results obtained during the OOS investigation is not appropriate because it hides variability among the individual sample.

(g) ***Conclusion*** **the** ***investigation***: To conclude the investigation, the results must be investigated and the findings of the investigation, including retest results, must be interpreted to evaluate the batch and reach a decision regarding release or rejection. A confirmed OOS result indicates that the batch does not meet established standards or specification must be rejected and disposed with proper documentation. Impact on the other batches and product must be studied. For inconclusive investigations-in case where an investigation I(1) does not reveal a cause for the OOS test results and (2) does not confirm the OOS results must be give full consideration in batch or lot disposal decision.

35.4 OOS Trend Analysis

Trending of OOS test results shall be done periodically. The trend analysis shall include classification of OOS results based on causes (analyst error, instrument error, analytical method error or specification related, equipment error, process error, operator error, input material, etc). Any emerging trend (example, increase in instrument error) shall be evaluated for taking necessary CAPA and to avoid recurrence. Trends shall be reviewed & evaluated during management review meetings. The OOS trend shall be evaluated to monitor the effectiveness of CAPA implemented for earlier OOS of past 2 year and to prevent recurrence of the same in future.

35.5 Field Alter Report (FAR)

Field Alter report is submitted to USFDA regulation within 3 working days for those product that are subjected of approved full abbreviated new drug applications, which are not meeting approved specification. OOS test results on these products are considered to be one kind of "information concerns any failure". Unless the OOS results on the distributed batch is found to be invalid within 3 days, an initial FAR must be submitted follow up FAR must be submitted when the OOS investigation is completed.

35.6 Definition

- ***Acceptance criteria:*** Numerical limits, ranges, or other suitable measures for acceptance of test results.
- ***Assignable cause:*** A scientifically justified explanation of the reason for an OOS test result identified and documented during investigation.
- ***Analyst-1:*** Analyst who has performed the analysis of initial reported OOS results.
- ***Analyst-2:*** A 2nd analyst involved in performing additional testing / retest as per requirement in the OOS investigation, who is atleast as experienced and qualified in the method as the analyst-1.
- ***Batch disposition:*** Action to be taken on the batch in question based on conclusion of investigation. Actions may include reject /reprocess/ rework/ release/ recall (in case, if OOS is confirmed on stability samples of commercially distributed batches)
- ***Experimental study:*** Studies conducted with an objective to identify any probable causes that could have resulted for failures. Such studies should be conducted only after prior approval from QA. The discrete outcome of experimental activity should NOT be used for taking decisions on batch disposition or reporting the final results.
- ***Full scale OOS investigation (Phase II) Detailed*** investigation into manufacturing of the product and process to identify the root cause for batch failure. This

includes review of various records and process simulation & /or experimentation where required and may also include retesting of same sample in statistically significant replicates, if necessitated.

- *Hypothesis studies:* Activities conducted to prove or disprove the probable cause(s) from the assumptions made i.e., what might have happened (such as during sample filtration /centrifugation /holding/ processing/ packaging, etc…) that can be Studied? .Such activities should be conducted only after prior authorization from QA, by recreating the analysis condition or by simulation/repetition of a process, which is identified as a probable cause. Multiple hypothesis can be explored. The individual outcome of hypothesis activity should NOT be used for taking decisions on batch disposition.

- *Invalidated test:* A test is considered invalid, when the investigation has determined the laboratory error, which has resulted for OOS result.

- *Investigation:* General process of information or data gathering, analysis and checking possible underlying reasons to identify the root cause. The investigation shall be thorough, timely, unbiased, well documented and scientifically justifiable.

- *Laboratory error:* An error associated with the performance of a test procedure or due to laboratory equipment malfunction or failure.

- *Laboratory investigation (Phase I):* Investigation carried out to assess the accuracy of laboratory testing and recording in order to identify whether there was any laboratory error have resulted OOS result for the batch in question. This investigation includes review of sample handling, storage of sample prior to analysis, examination of standard & sample preparation and testing in the laboratory against the documented procedures and also the practices associated with testing which may not be covered in test procedure.

- *Non routine samples:* Samples which are other than routine samples, such as Vendor evaluation sample, Innovator or other market sample, samples employed for method transfer, Outputs of Lab scale batches wherein an evaluation is done by supervisor (if required through phase I investigation format) only to ascertain any laboratory error.

- *Obvious error:* Errors which are clearly evident such as wrong glassware used, wrong weight taken, incorrect instrument settings, inappropriate number of dosage units, power outage, calculation error , equipment failure , spillage of sample solution, incomplete transfer of sample composite which are identified by the analyst and/or investigator shall be referred as 'Obvious error'. These type of error should be substantiated with appropriate supportive evidence (such as Photograph, etc) , wherever applicable.

- *Out of specification:* An observation of a test result(s) or a measurement that does not comply with the pre-determined specifications or internal acceptance criteria filed in drug applications, drug master files, approved marketing authorization or official compendia or criteria determined by company's in-house documentation.

- ***OOS result:*** A result which is NOT confirmed and reported as failure by the analyst-1 through first analysis, which is considered to be doubtful and need to be investigated. This also takes account of a discrete test result (specifically when the testing is performed in replicates) irrespective of whether the average is within the pre-determined /established specifications or acceptance criteria or not.

- ***Probable cause:*** Identifiable factor(s), which are most likely to have resulted for batch failure, but has not been scientifically proven and which needs to be further established through hypothesis activities.

- ***Random error:*** Generally arise from variety of uncontrolled sources causing small arbitrary variations which affects the measured value to differ from true/accepted value, when the measurement is repeated number of times. Known sources of indeterminate errors include variables such as temperature, pressure, humidity, vibrations, noise, electrical properties such as current, voltage, resistance, frequency which cannot be clearly identified (ie) represents experimental uncertainty and can be only minimized.

- ***Raw material:*** In case of drug substance manufacturing refers to starting materials, reagents, solvents, intermediate or another active substance, which has an approved specification and employed in production of an Active pharmaceutical ingredient. In case of formulations refers to API's, excipients, including hard/soft gelatin capsules, which has an approved specification and used in manufacturing of dosage forms.

- ***Re-measurement:*** Replicating the analysis by repeating one or more steps with an objective to identify any assignable cause(s) and/or to ascertain any errors due to instrument malfunction / dilution errors, which will guide to investigate further. Specifically, with respect to chromatography, it refers to reinjection of the original aliquot solutions, provided the solutions are stable. The values obtained through re-measurement should not be used for taking decision on the batch disposition or reporting the final result.

- ***Re-analysis:*** Repeat Testing performed using the retained solution/preparations, which were used during initial analysis, provided an obvious error/ assignable cause has been identified during Phase I and which was responsible for initial reported OOS result.

- The number of samples (including any replicates, etc for re-analysis shall be as similar to that mentioned in approved procedure. If retained solution / preparation are not valid then "fresh portion" of sample (from which OOS results are reported) available in lab shall be used.

- ***Resampling:*** Activity of collecting additional (as a 2nd instance) units/ quantity of sample, from a lot of material/ from the specific batch product.
- ***Retesting:*** Analyzing of the original homogeneous composite sample [if it has not been compromised and/or is still available] that yielded the OOS results either;
  - After root cause for lab error is established, during phase I or
  - Performed as part of phase II, after the discretion by QA
- ***Root cause:*** The underlying cause, identified based on scientific evidence, which is responsible for lab error/batch failure.
- ***Root cause analysis:*** A structured documented investigation that aims to identify the true cause(s) of failure with assumption that "best way to solve problems is by eliminating their underlying causes" and with belief that "symptoms serves as indicators for underlying causes"
- ***Specification:*** A list of tests, references to analytical procedures, and appropriate acceptance criteria that are numerical limits, ranges, or other criteria for the test described. It establishes the set of criteria to which a material should conform to be considered acceptable for its intended use. "Conformance to specification" means that the material, when tested according to the listed analytical procedures, will meet the listed acceptance criteria. Specifications are critical quality standards that are proposed and justified by the manufacturer and approved by regulatory authorities as conditions of approval.
- ***Systemic/determinate errors:*** Generally, arise from identifiable sources causing the measured value to differ from a true or accepted value, which can be eliminated by careful observation and record keeping, equipment maintenance and adequate training. Three basic sources of determinate errors are analyst, equipment/instrument including its operating environment and the method/procedure based on which the personnel have operated the equipment/instrument.

35.7 Annexes

A. Log book for OOS

OOS log Book for Year __________											
Date of reporting	OOS. No	Product/ Stage	Batch No	Test Parameter	Spec. Limit	Initial Result	Entered by sign & date	Final Result	Root cause	Status of the batch	Entered by sign & date

A. Format for Phase –I Laboratory Investigation

Format for Phase –I: OOS Investigation			
Notification/Ref. No		Issued on	
Issued by		Due date	
Issued to			

OOS/ information			
Name of material / product		Stage	
Batch. No		AR. No.	
Test		Ref. Specification. No	
Observed results		Limit	
Reported on		Lab Book/Ro A	
Date of initiation		Name of the analyst_1	

Section A- Preliminary assessment
Observations:
Corrections:
Sign & date of the investigator

Section B- Checklist for Detailed investigation					
Category	Ref. #.	Checkpoints	Yes	No	NA
Calculation		Whether any calculation error has occurred?			
		Are there any reasons to suspect or evidence that any un authorized or un validated changes have been made to programmed calculation methods, if used?			
		Any transcription error occurred in calculations used to convert raw data values in to a final result?			
Training		Is the analyst-1, NOT qualified/ evaluated for the specific technique, through which testing has been performed. If 'NO' date of qualification for the specific technique______			
Sample		Is there any evidence to suspect the sampling approach followed for the material/product			
		Are there any reasons to suspect or evidence that the original sample is invalid due to improper packing, storage, handling etc?			

Table *Contd...*

Section B- Checklist for Detailed investigation					
Category	**Ref. #.**	**Checkpoints**	**Yes**	**No**	**NA**
		Are there reasons to suspect or evidence that sample integrity not maintained? (wrong container sampled, chain of custody)			
		Are there reasons to suspect or evidence that sample tested is incorrect? B.No as indicated in the label__________			
		Are there reasons to suspect or evidence that sample contamination has occurred during testing? (sample left open to air or unattended)			
		Are there reasons to suspect or evidence that original sample is not representative of the batch?			
Methods		Is there any evidence that correct method not followed? Approved procedure & its version # which was examined as part of assessment.________			
Observation		Are there any abnormalities noticed specific to the Method or Operation which may or may not be captured in the method of analysis or approved SOP?			
Standards / Samples/ Chemicals		Is there any evidence that sample is impacted due to nature of the material/product (ie hygroscopic, light sensitive, etc...)			
		Is there an evidence that valid analytical standards is NOT used? Name of analytical standard: Code.No: Validity:			
		Is there any evidence that analytical standards were not stored properly?			
		Is there an evidence of degradation of standards/ sample/chemicals used for testing (before or after preparation)? Time of re-examination of sample preparation/ solution:			
		Is there a problem identified in weights, preparation and dilution of. Standards/Samples/ Chemicals?			
		Is there a problem identified in labelling of Standards/ Samples/Chemicals?			

Table Contd...

Section B- Checklist for Detailed investigation					
Category	Ref. #.	Checkpoints	Yes	No	NA
		Is there any evidence that Standards/Samples/Chemicals and their solutions are not used within their expiration or use before date?			
		Is there an evidence that chemicals used for testing is NOT having the desired grade, make, purity/concentration, etc. as specified in STP			
Sample / Standard / Preparation/ Solution		Are there reasons to suspect or evidence of contamination during the sample/standard preparation or transfer?			
		Is there any evidence that correct reagents, solvents standards glassware, or other auxiliary equipments (E.g. balance) were not used?			
		Is there any evidence of error in weighing, dilution and sonication etc...			
		Is there any abnormality noticed like color change etc. identified before, during or after the analysis?			
Reagents / Solutions		Is there any evidence that Solutions/Reagents are not prepared according to the procedure?			
		Is there any evidence that Solutions/Reagents are not within the valid date for use?			
		Whether the visual examination performed on Solutions/ Reagents retained, reveal any abnormal appearance?			
Glasswares		Is there any evidence that incorrect glassware have been used?			
		Is there evidence that glassware used is contaminated?			
Instruments		Is there evidence for malfunctioning of any component of the instrument? Ex. leakage at injector, pump, column connection, flow cell (before or during the analysis likely to impact the testing)			
		Is there any evidence that the instruments involved in testing are not calibrated within the period and also not calibrated properly? Name of instrument :_________________ Code No:_____________________ Valid upto:___________________			

Table *Contd...*

Section B- Checklist for Detailed investigation					
Category	**Ref. #.**	**Checkpoints**	**Yes**	**No**	**NA**
Chromato-graphy		Do the consecutive blank pattern are NOT comparable (as an evidence for inadequate column equilibration)			
		Do the System suitability results NOT comply with the criteria? (those before analysis and during analysis)			
		Is there any abnormality noticed in standard responses?			
		Is there any abnormality noticed in pressure channel/data?			
		Is there any abnormality observed in the injection vials ?			
		Is there any evidence that correct method not used?			
		Is there any evidence that right columns were not used?			
		Is there any evidence that correct mobile phase not used?			
		Whether the chromatograms reveal any anomalous or suspect information like noisy resolution, poor peak resolution, poor injection reproducibility and improper integration/processing methods, etc ?			
		Are there reasons to suspect or evidence that potentially interfering or incompatible testing activities occurring at the time of test?			
		Are there reasons to suspect or evidence that vials used for sample or standards for injections is improperly crimped or improper septa, etc?			
Stability testing		Were pinholes observed in the sample packing?			
		Is there an evidence of improper sealing of samples or any abnormalities in packing condition?			
		Is there any abnormal trend (product behavior w.r.t time) observed during review of stability data of other batches?			
		Any abnormalities in the sample pre-test storage			
		Is there any impact of temperature & humidity due to stability chambers breakdown.			

Table *Contd...*

Section B- Checklist for Detailed investigation					
Category	**Ref. #.**	**Checkpoints**	**Yes**	**No**	**NA**
History		Is there a history of oos result for the same parameter that has occurred previously for the same product & stage?. (Review of oos data for atleast past 6 months.) If 'YES', also review the outcome of the earlier investigation in light of the current failure for existence any recurring error. If 'NO' in case of stability ooS, review of stability data obtained from Exhibit, Process validation batch indicates any abnormal trend, with the batch which has obtained oos?			
		Are there reasons to suspect lab error based on the history of product, type of test being performed and any other test results obtained at earlier stages?			
		Is there any evidence available, for the noticed abnormality? Attach photographic evidence, if relevant.			

If answer to any of the reference questions is 'YES' the investigator shall elaborate, the details of same appropriately.
Sign & date of the investigator

Observations/Findings based on document review and/or discussion with analyst
Sign & date of the investigator

Recommendations:
Sign & date of the investigator

Section C- Remeasurement	
Details of proposed Remeasurement and plan:	
Proposed by investigator	
Review comments by QA designee /designee for remeasurement	
Remeasurement Authorization by QA head/designee	
Assigned by team leader	
Result of re-measurement(s)	
Re-measured by	

Observations and/or Recommendations from Remeasurement
Sign & date of the investigator :

Authorisation for Re-analysis (strike out if not required)	
Review comments by QA head/designee for reanalysis	*Applicable only in case either if an Obvious error or Assignable cause has been identified in Phase I*
Reanalysis Authorization by QA	

Section D- Hypothesis studies	
Suspected probable causes *Proposed hypothesis studies to prove the Suspected probable causes as root cause for failure.* *Type of samples [Batch which is under investigation and /or approved batch] and their execution of testing :* *Number of sample:* *Evaluation approach:*	
Requested by	
Remarks/comments by QA	
Authorization for Hypothesis by QA	
Observation from Hypothesis studies *Root cause for laboratory failure:* *Immediate corrections, if any or CAPA* Sign & date of the investigator :	
Impact Assessment: *of other batches analysed in the same run/sequence and/or Consideration of other test results, if any on the same batch under investigation and its relevance:* Sign & date of the investigator :	
authorisation for Resampling (strike out if not required)	
Justification for Resampling: Insufficient sample / Evidence of sample integrity / Observation of sampling error Quantity:	
Requested by	
Review comments by QA head/designee for resampling	Opinion to proceed further (based on evidences/facts supported) either re-analysis or Hypothesis
Resampling Authorization by QA	

Justification for timeline extension of phase I investigation (Strike out if not required)	
Requested by :	Authorized by :

Section E- RETESTING	
Batches requested for retesting /number of samples (5) retest plan	
Requested by	
Remarks/comments by QA	
Authorization by QA for retest	
Individual results of retest by Analyst 2	
Outcomes of retest	Whether all outcomes meets specification? Y E S / N O

Retested by		Lab Book/RoA	

Section F- Evaluation of retest outcomes & conclusion of lab investigation
Evaluation of retest outcomes : Sign & date of the investigator :
Conclusion of Laboratory investigation : *Reasons for failure results* Sign & date of the QC designee

Section G- Batch disposition and approval of investigation
QA evaluation & approval of investigation : Sign & date of the QA :

B. Format for Phase –II: Full Scale OOS Investigation

Format for Phase –II: Full Scale OOS Investigation			
Notification/Ref. No		Issued on	
Issued by		Due date	
Issued to			

Failure information			
Name of /product		Stage	
Batch. No		AR .No.	
Test		Ref. Specification. No	
Observed results		Limit	
Date of initiation			

Section A-Document review			
Ref. #.	**Checklist**	**Yes**	**No**	**NA**
1.	Is current version of Batch Production/Manufacturing record is NOT used for manufacturing? Version of BPR/BMR________________			
2.	Is there any evidence that operating instructions as per production record/processed are NOT followed during execution of batch? Counter check through questioning			
3.	Whether the operator and supervisor are UNABLE to interpret the basic process flow, product nature and other systemic requirements? Confirm by questioning.			
4.	Is there any evidence of inconsistent/ illogical sequence steps followed during manufacturing/packing?			
5.	Any abnormal observations (such as equipment malfunction, extended or frequent stoppage, etc) during manufacturing/packing are noted/recorded? Verify with operators and supervisors?			
6.	Any in-process checks are missed /not performed during manufacturing?			
7.	Whether the sampling collected is NOT as per sampling procedure/plan? Confirm with applicable checklist, if available.			
8.	Is there any evidence for extended time/ hold time for any stage/unit operation? If yes, whether it has exceeded the established hold time. Established hold time____________			
9.	Any significant difference or variation in yield, when compared against previous /next batch? Yield of at least 2 previous & 2 next batch, if available.			
10.	Whether environmental conditions (such as humidity, temperature) as mentioned in production record is NOT within the limit during batch processing?			
11.	Any significant difference or variation in process parameters			
12.	Whether any stages of manufacturing process is NOT validated? If yes which stage is NOT validated____________			
13.	Is there similar history (atleast 1 year) noted from any earlier investigation.			
14.	Are there any new personnel involved in manufacturing, Review the training document and its scope.			

Table Contd…

	Section A-Document review			
Ref. #.	**Checklist**	**Yes**	**No**	**NA**
15.	Whether the particular stage/unit operation has extended more than one shift of operation?			
16.	Is there evidence of operator change within a shift? Countercheck with other supportive data/information?			
17.	Any measuring devices used to measure process parameters are NOT calibrated?			
18.	Are the balances used in dispensing, manufacturing including in process area are NOT calibrated?			
19.	Any stoppage or failure of utilities supply during any particular stage or unit operation?			
20.	Any changes in manufacturing equipment train without authorization?			

If answer to any of the reference questions is 'YES' the investigator shall elaborate, the details of same appropriately.

Sign & date of the investigator

Observations/Findings based on document review and /or discussion with operator and/or supervisor

Sign & date of the investigator :

	Section B- Hypothesis studies

Suspected probable causes and rationale for each probable cause

Proposed/Related hypothesis studies in order to establish/evidence the identified probable causes as root cause for failure.

Type of samples [Batch which is under investigation and /or approved batch] and their execution of testing :

Number of sample:

Evaluation approach:

Requested by	
Remarks/comments by QA	
Authorized by QA	

Observation from Hypothesis study:

Root cause for Manufacturing aberration:

Immediate corrections, if any or CAPA

Sign & date of the investigator :

Impact Assessment: Other batches of same product and/or Different product with validated processes similar to that which has resulted failure

Sign & date of the investigator :

Justification for timeline extension of phase II investigation (Strike out if not required)	
Requested by :	Authorized by :

Evaluation by QA, in case Error remains Unclear or Hypothesis is inconclusive (strike out if not required)	
Comments by QA designee as part of Independent evaluation	
Evaluated by	Sign & Date of QA designee

Section C- Retesting			
Batches requested for retesting /number of samples (5) retest plan			
Requested by			
Remarks/comments by QA designee			
Authorization by QA designee for retest			
Individual results of retest by Analyst 2			
Outcomes of retest	Whether all outcomes meets specification? YES/NO		
Retested by		Lab Book/RoA	

Section D- Conclusion of full scale OOS investigation
Evaluation of outcomes : Sign & date of the investigator :
Conclusion of full scale OOS investigation : Reasons for failure results CAPA proposal with TCD Sign & date of the Operations designee

Section E- Batch disposition and approval of full scale OOS investigation
QA evaluation & approval of full scale investigation : Sign & date of the QA designee:

Chapter - 36

Pharmaceutical Change Control Management

Introduction

Change control is important an element of pharmaceutical Quality Management system and is closely interwoven with regulatory compliance.The change control system provides transparency and a structured approach towards end- to end changes.Change control system provides checks and balances in the quality system by initiating, reviewing, approving, distribution, tracking and storing change historythroughout the life cycle. It also links to other quality and regulatory processes such as corrective and preventive action (CAPA) and product registration tracking.

The scope of the change control program must be border than change control, which typically applied to one change at a time. The change control management should have a broad set of possibilities view on changes to product formulation or design, upgrades to facilities, utilities, equipment and computer systems, manufacturing instructions, SOPs, test methods and specifications, any new raw materials as well as any changes in policy.

Benefits of Change control System:

(a) Structured and consistent approach towards managing change

(b) Documenting the details of change

(c) Routing of change requests to appropriate individuals/team for approvals

(d) Documentation of change approvals and implementation

(e) Maintenance of change history and easy retrieval of information

(f) Tracking changes effectively and providing an audit trail

(g) Demonstrate compliance to regulations

(h) The change management system should include the following, as appropriate for the stage of the lifecycle:

36.1 Regulatory Expectation in Change Control Managements

21 CFR 211.160 states: The establishment of any specifications, standards, sampling plans, test procedures,or other laboratory control mechanisms required by this subpart, including anychange in such specifications, standards, sampling plans, test procedures, or other laboratorycontrol mechanisms, shall be drafted by the appropriate organizational unit andreviewed and approved by the quality control unit.

Approved and written procedure should be available for change control handling procedures should describe the actions to be taken if a change is proposedto a product component, process equipment, process environment (or site), method ofproduction or testing or any other change that may affect product quality or support systemoperation and should covers the following points.:

(a) What types of changes does change control take into account; for which areas does this operating instruction apply?

(b) Who can suggest/initiate changes?

(c) How changes are requested (forms, methods of communication)?

(d) How changes are graded, who is responsible for the grading?

(e) How are the measures necessary for carrying out the change determined; who compiles the directions required?

(f) Who is responsible for the execution and monitoring of all necessary measures?

(g) How is the change controls committee assembled?

(h) How the change is documented (format, content, storage)?

(i) Who is responsible for authorizing changes?

(j) What are the special regulations for urgent changes?

36.2 Steps Involved in Change Control System

Change control is not department-specific, rather the task of the whole company. Change control consist of five main Section:

(a) Change control request

(b) Impact Analysis and Risk assessment

(c) Implementation of changes and obsoleting the previous documents (If required)

(d) Effectiveness check of changes made

(e) Closure of Change control

Respective department should identify changes and Quality assurance should issue the change control form after assigning the change control number in change control log.

Change control number consist year and chronological serial number.A new serial a change control should be start at start of the year.

Example: CC/Year/Serial number (Serial number should be chronological order =CC/2017/001

Brief description and reason forchanges should be recorded in change control form. Impaction changes on system, facility, process, product parameter validation, qualification, documentation, etc shall be verified. Risk analysis shall be performed to confirm changes are major or minor.Based on the evaluation and impact assessment, tentative closure out date should be mentioned by Quality Assurance in the change control form.

After impact assessment the change control should be assign for cross function team to evaluate the changesto their respective end and gives their comments (Acceptable or not Acceptable) on the change control form and submit back to Quality Assurance for further action plan. The comments shall be written by concerned or relevant department head or his designee .After receiving the comments from all cross function team Quality Assurance should approve or reject the change control on the sound scientific justification.If the changes are not acceptable or cannot be implemented at any stage, change control should be rejected with justification and reason in change control form.

Implementation of changes should after existing document are withdrawn and stamped obsolete,training implemented on revised documentsand evaluation of change for its effectiveness check after implementation.After implementation, an evaluation of the change should be undertaken to confirm the change objectives were achieved and that there was no deleterious impact on product quality. Quality assurance should review the change control to ensure the changes are implanted before change control closure.Quality Assurance should be monitor the status of change control periodicallyat least 15 days.The records can be archived in paper form or electronically form.

Steps involved in Change control Management
 (a) Login of Change Control and description
 (b) Description of Changes: Reason for change, Proposed changes, justification for change
 (c) Type of Change Control : Facility /System/ Documentation / Product
 (d) Assigning the task to Cross function teams
 (e) Risk Analysis and Impact Assessment
 (f) Change control Categorization Major / Minor Change Control
 (g) Approval of Change control :
 (h) Implementation of Change Control
 (i) Results/ Effectiveness of the Actions
 (j) Verification and closure

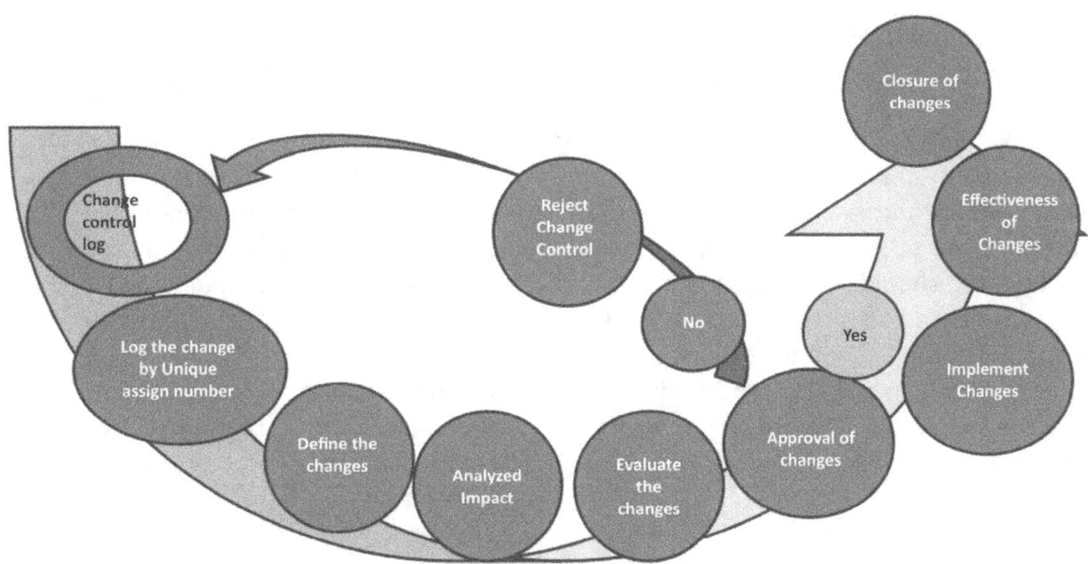

36.3 ICH Q8, Q9 and Q10 Change Management and Application of Changes

Change Management: A systematic approach to proposing, evaluating, approving, and implementing and reviewing changes. (ICH Q10)

The change management system should include the following:

(a) Quality risk management should be utilized to evaluate proposed changes;

(b) The level of effort and formality of the evaluation should be commensurate with the level of risk;

(c) Proposed changes should be evaluated relative to the marketing authorization, including current product and process understanding and/or design space, where established;

(d) Expert teams, with appropriate expertise and knowledge, should evaluate proposed changes;

(e) An evaluation of the change should be undertaken after implementation to confirm the change objectives were achieved.

(f) Request for changes

Application of changes:

Pharmaceutical development	Technology transfer	Commercial manufacturing	Product discontinuation
Change is an inherent part of the development process and should be documented; the formality of the change management process should be consistent with the stage of pharmaceutical development.	The change management system should provide management and documentation of adjustments made to the process during technology transfer activities.	A formal change management system should be in place for commercial manufacturing. Oversight by the quality unit should provide assurance of appropriate science and risk based assessments	Any changes after product discontinuation should go through an appropriate change management system.

36.4 E-Change Control System

Electronic change control management helps to managing all change control processes in a centralized and harmonized manner.It helps in proper coordination across stakeholders through automated workflows and alerts if any timelines are crossed. Itincreases operational transparency through automated alerts, summary , dashboard functionality, and extensive reporting capabilities. It also increases accountability though assignments, process step sign-offs, and automated audit trails. Electronic change control processes increase compliance, enable companies to reduce costs, liability and patient risk.

36.5 Format Used in Change Control

    A. Change Control Log

    B. Change control Form

A. Format for Change control log

Change control log for Year 2017											
Sr. no	Change control No	Date	Department	Product/ Facility/ Documentation/Equipment	Description of changes	Major /Minor	Tentative closing date	Sign/ date	Closed out date	Sign/ Date	Remark

B. Format for Change control Form

Change Control Form			
Date	Change control No.	Initiation department	Applicable to :
Initiated By		Details of change: New/ Additional/ Deletion of changes	
1. Description of change : Sign and Date :			
2. Justification of Change : Sign and Date :			
3. Impact of Changes: Impacted document /SOPs/Process/Facility/Equipment/System/Critical parameters etc.			
4. Risk Assessment :			
5. Assessment of Impact changes :			
6. Proposal Methodology of implementation :			
7. Categorization of Changes :			
8. Tentative closure Date :			
9. Assigning task for Cross Function team with respective department comments and justification:			
10. Approval /Rejection of Changes by Quality Assurance with sound justification :			
11. Implementation of changes : • Exiting document made withdraw or obsoleted • Revised document number (If Applicable) • Training imparted • Effective batch No.(If applicable) • Effective date			
12. Evaluation of Changes :			
13. Verified By :Quality Assurance (Date/Sign) :			
14. Closure of Changes :			
15. Done By(Sign/Date) by Quality Assurance :			

36.6 Definition

1. ***Change:*** Any modification to documents, policies, documents, facility, equipment's, product, process etc. called as the change.

2. ***Major Changes:*** Changes have the direct effect on the quality purity and efficacy of the product is called as major changes.

3. ***Minor Changes:*** Changes have the does not have direct effect on the quality purity and efficacy of the product is called as major changes.

4. ***Change control:*** "A formal system by which qualified representatives of appropriate disciplines reviewproposed or actual changes that might affect the validated status of facilities, systems, equipment or processes. The intent is to determine the need for action that would ensureand document that the system is maintained in a validated state."

Data Integrity in Pharmaceutical Quality Control Laboratories

Data integrity problems in pharmaceutical quality control laboratories are driving more regulatory action than ever before. What has changed to drive all this activity? While plenty of information is available, much of it seems to confuse rather than clarify. This article will dispel common myths by looking at facts, based on a study of available resources of U.S. Food and Drug **Administration** (FDA) staff and EU. It will discuss the following:

- How to put the current enforcement environment into historical context.
- How to apply critical thinking skills to various myths regarding data integrity.
- How to evaluate current laboratory software and associated processes against new expectations.
- How vendors are redesigning laboratory software to help respond to new realities.

37.1 Procedural versus Technical Controls

Let's consider the statement: "This software is Part 11 compliant." There are a few problems with this statement. First of all it is a logical impossibility. There are components of Part 11 that are not meant to be satisfied by technical controls within a computerized system. For example, CFR Part 11.10(j) requires policies for the use of electronic signatures. This is a requirement that a chromatography data system is not going to satisfy. It is an element of the regulation, but it is not something that is expected to be a technical control.

The software, in fact, does not comply with regulations. The software itself is inert; software contains the technical controls to support compliance with the applicable regulations. In addition to technical controls, procedural controls must also be in place. A discussion about procedural controls versus technical controls is often seen in FDA warning letters, particularly when gaps in a system's ability to support technical controls required by various regulations have been exploited.

A standard operating procedure (SOP), used as a procedural control, can substitute for a technical control as long as:

- People are trained on that SOP
- The SOP is followed
- Adherence to the SOP is confirmed by quality oversight and/or compliance auditing

Often, however, even if SOPs exist, they are not followed, and adherence isn't properly verified. Consequently, the FDA will demand system remediation to prevent a recurrence of the behavior.

Audit trails within computerized systems are an example of technical controls. The software must be able to generate audit trails that contain all the components the regulations require, and then those controls must be enabled.

Audit trail records automatically (21 CFR Part 11.10(e)) show that the system is working and it's doing its job properly, and can be consulted in an audit or an inspection without any extra work by the humans.

37.2 Certificates of Software Validation or Capability: Do they Provide Value?

Many software vendors provide a Certificate of Software Validation, or they may issue a certificate along the lines of 21 CFR Part 11 Readiness Claims. Such a certificate has limited value, because the FDA expects the software to be validated for its intended use by the users in the environment where it will be used. While vendors should engage customers to build and design systems according to customer needs, and spend considerable time testing that software before they deliver it, the development and testing work does not (and cannot) substitute for the customer declaring their intended use and then validating their system according to that intended use.

Data Integrity in Pharmaceutical Quality Control Laboratories

37.3 Your Responsibilities: Audit, Assess, and Validate

Data integrity compliance responsibilities include vendor audits, a computer system assessment, and software validation. Assessment components include:

- Procedures
- Training of personnel
- Software development activities

- Testing activities
- Quality management systems
- Infrastructure

37.4 Assessing the Software's (and the Vendor's) Compliance Support

Assessing a software's ability to support compliance requires attention to all the regional regulations where a regulated company does business or may intend to do business. Some regulated companies, if they are solely doing business within the United States, or solely within Europe, may choose to pay attention only to Part 11, or only Annex 11 requirements. Part 11 and Annex 11 share commonalities. However, any software evaluation for compliance should be based on evidence rather than hearsay (that shiny Part 11 Certificate).

Vendors, receives numerous customer checklists, and provides straightforward responses with product answers. However, it is important that the evaluation of the responses be based on evidence, rather than being strictly limited to what the vendor has said that the system will do. Areas for review should include data integrity issues, access controls, audit trails, device checks, etc., as per applicable regulations. This review is valuable during a software assessment process because observed gaps indicate where procedural controls or customization may be required. Feedback to the vendor regarding any gaps observed is therefore important so that the procedural controls or custom solutions do not become permanent.

37.5 Laboratory Audit Trail

The Audit Trail Is an Integral Part of the Application: The audit trail must be designed into the software application, not added on as an afterthought. A database is the only way to ensure a secure and robust audit trail that cannot be tampered with the U.S. or the EU, the audit trail requirements are similar good laboratory practices include tracking the identity of the person making the change, the time of the change and the ability to review the audit trail to determine what was deleted.

37.6 Audit Trails and Design

An audit **trail** (also called **audit log**) is a security-relevant chronological record, set of records, and/or destination and source of records that provide documentary evidence of the sequence of activities that have affected at any time a specific operation, procedure, or event.

There are two main design philosophies for audit trails in a regulated application:

- One large audit trail covering all entries from log on or off to data modification. This approach requires very effective search routines to be able to pull all entries related to a specific analysis and then mark them as reviewed.

- Two audit trails. One at the application level dealing with system-level activities and one at the work package level so that each analysis has the relevant entries directly and close with the analytical work. Search routines will still be required, by the way.

37.7 Audit Trail Beginning

(a) ***For data records:*** If the data is recorded directly to electronic storage by a person, the audit trail begins the instant the data hits the durable media. It should be noted, that the audit trail does not need to capture every keystroke that is made before the data is committed to permanent storage. This can be illustrated in the following example involving a system that manages information related to the manufacturing of active pharmaceutical ingredients. If during the process, an operator makes an error while typing the lot number of an ingredient, the audit trail does not need record every time the operator may have pressed the backspace key or the subsequent keystrokes to correct the typing error prior to pressing the "return key" (where pressing the return key would cause the information to be saved to a disk file). However, any subsequent "saved" corrections made after the data is committed to permanent storage, must be part of the audit trail.

(b) ***For document records:*** If the document is subject to review and approval, the audit trail begins upon approval and issuing the document. A document record undergoing routine modifications, must be version controlled and be managed via a controlled change process. However, the interim changes which are performed in a controlled manner, i.e. during drafting or review comments collection do not need to be audit trailed. Once the new version of a document record is issued, it will supersede all previous versions.

37.8 System Audit Trail Features

The main purpose of the audit trail is to provide assurance with regards the integrity of the electronic record.

The audit trail must be

- ***Automated:*** The audit trail entries must be automatically captured by the computer system whenever an electronic record is created, modified or deleted.

- ***Secure:*** Audit trail data must be stored in a secure manner and must not be editable by any user.

- **Contemporaneous:** Each audit trail entry must be time stamped according to a controlled clock which cannot be altered. The time should either be based on central server time or a local time, so long as it is clear in which time zone the entry was performed.

- **Traceable:** Each audit trail entry must be attributable to the individual responsible for the direct data input. Updates made to data records must not obscure previous values and where required by regulation the reason for changing the data must also be recorded.

- **Archived:** The audit trail must be retained as long as the electronic record is required to be stored.

- **Available:** The audit trail must be available for agency review and copying.

37.9 Audit Trail Content

For each audit trail entry, the following information should be recorded.

- **Identification of the User making the entry:** This is needed to ensure traceability. This could be a user's unique ID, however there should be a way of correlating this ID to the person

- **Date and Time Stamp:** This is a critical element in documenting a sequence of events and vital to establishing an electronic record's trustworthiness and reliability. It can also be effective deterrent to records falsification.

- **Link to Record:** This is needed to ensure traceability. This could be the record's unique ID

- **Original/New Value:** This is needed in order to be able to have a complete history and to be able reconstruct the sequence of events

- **Reason for Change:** This is only required if stipulated by the regulations pertaining to the audit trailed record.

37.10 Audit Strategies

There are two fundamentally differing strategies to achieving a compliant audit trail, version control and audit trail control

(a) **Version control:** The distinction between the two strategies can be made clear by examples from paper-based systems. A paper-based example of version control is updating a standard operating procedure (SOP). Here, the 21 CFR Part 58 predicate rule states: "A historical file of standard operating procedures, and all revisions thereof, including the dates of such revisions, shall be maintained." (Part 58.81) Therefore, when an SOP is updated, an entire copy of the superseded version must be retained

(b) ***Audit trail control:*** For audit trail control of an analytical record, the same document states: "Any change in entries shall be made so as not to obscure the original entry, shall indicate the reason for such change, and shall be dated and signed or identified at the time of the change." (Part 58.130) Therefore, in contrast to modifying an SOP, it is not necessary to retain an entire copy of the unmodified record and reissue a completely new amended copy when modifying a paper analytical record; it is acceptable to amend and annotate the existing paper copy.

37.11 Regulatory Main Elements of Data Integrity and Audit Trail Reviews

- Classification of data - which are critical data?
- Classification of systems - which systems are relevant?
- What Audit Trails are important?
- What shall I do with legacy systems without audit trail?
- Who shall review Audit Trails?
- How is the review documented?
- Process and documentation in case of deviations?

37.12 Audit Trail Review

Review of audit trail entries by exception also needs to be balanced by the need to consider the second person review and the risk of data falsification:

- Has the analysis been carried out correctly, with, for example, consistent date and time stamps on records?
- Have the data been interpreted correctly according to sound science and applicable procedures?
- For critical applications, instead of review by exception, perhaps a sampling process of audit trail entries could be adopted.

A periodic review of audit trail reports should be performed on each computer system generating GMP data that could impact product safety, purity and efficacy. Focusing the review on the more critical computer systems based on a documented risk assessment of the data produced, is a way to minimize the workload of the task. A monthly routine audit trail review should be implemented for critical GMP systems and functions, or within days if required to investigate an error, security breech or other deviation. Some of the more complex computer systems generate volumes of audit trails, so it is beneficial to apply a documented risk assessment to the more critical functions of the computer, to identify the highest risk audit trail reports to review. Consideration at highest risk compliance should be given to systems that require an electronic signature to meet predicate rule requirements (electronic batch records, etc.) and critical functions of system that might include product disposition (release, reject). The review should be detailed in a written procedure and signed/dated by the reviewer. The audit trail may also be supported by and should be checked against cross-referenced information in a lab notebook and/or documented lab events/investigations.

37.13 Beyond the Audit Trail

As stated above, the review of an audit trail report may not in all cases be used independently to conclude electronic record integrity. Assuming the audit trail of data collected from equipment indicates no data manipulation, security or sequential issues are noted, the electronic data files that are collected from laboratory equipment, for example, may be stored locally on that computer, and could be unprotected from manipulation. It's therefore important to identify system connections to other computers and storage devices.

Unless the files are locked when they are stored, data can be manipulated after it is collected, and resaved under the same file name. Real-time storage of data to secured servers is the only way to assure data is protected once it's stored. Uploading of data to other software, such as Excel spreadsheets, for example, may generate additional electronic records which also require limited access and protection from changes. Audit trails are not generated in the typical use of Excel spreadsheets, so data should be protected through security and locked cells and files.

In short, security access, data integrity, sequencing of events and activity detail can readily be reviewed in an electronic audit trail report which, compliant with 21 CFR Part 11, details every action that occurred from the point of system logon to logoff. This allows for the identification of unauthorized changes or deletion of data, alteration or deletion of files, and process/procedures not being followed, with the intent to ensure electronic records are trustworthy and reliable

37.14 Data Integrity from a Regulators' Point of View
- ALCOA principle
- Audit trail
- Data review
- File format
- Storage media
- Encryption
- User management (access control)
- Review of the data life cycle
- Handling of raw data
- Changes to test parameters
- Changes to data processing parameters

- Data deletion
- Data modifications
- Analyst actions
- Data manipulation
- Excessive integration of chromatography peaks
- Security breaches related to data
- Data changes
- Data deletions
- Unauthorized access or transactions

These points have always been part of the inspection's agenda

(a) ***Audit Trail of Standalone system:*** An audit is a system that traces the detailed transactions relating to a database or file.

Prepared and review the list of standalone instrument.

During the audit trail review verify the generated data for re-naming of file, wrong naming of file, Duplicate measurements, deletion of file, data saved in correct folder/ path and record the observation in the format.

A. List of Standalone Instrument/Equipment

List of Standalone Instrument/Equipment			
Sr. No	Instrument Name	Software	Remark

B. Checklist for Standalone Equipment/Instrument Review

Checklist for Standalone Equipment/Instrument Review			
Sr. No	Observations to be Checked	Observation	Performed by
	Re-Naming of file		
	Wronging Naming of file		
	Duplicate Measurements		
	Deletions of Ile		
	Data saved in correct folder/ path		
	Instrument/Equipment time matching with satellite time		
	Change in Previlage		
	Any other observations		

Chapter - 38

Alarm Management in Good Laboratory Practices

Introduction

Alarm management principles, together with the ISPE, GAMP, ICH Q9, Part 11 and FDA inspection guidelines for risk management, provide a good foundation for the implementation of alarm systems in pharmaceutical manufacturing processes. This chapter gives an overview of the key principles of Alarm Management and how these can be used by the pharmaceutical industry to implement a risk based approach to alarm management to enhance productivity and implement exception reporting. A brief overview of alarm management literature and explanation of the alarm concept are provided with the aim of furthering insight and understanding.

Alarm is typical computer system or of built-in alarms to alert personnel to some out-of-limits situation or malfunction." System of alarm may be lights, buzzers, whistles, etc.

"Alarm: Audible and/or Visible means of indicating to the operator an equipment malfunction, process deviation, or abnormal condition requiring a timely response" (Figure 1).

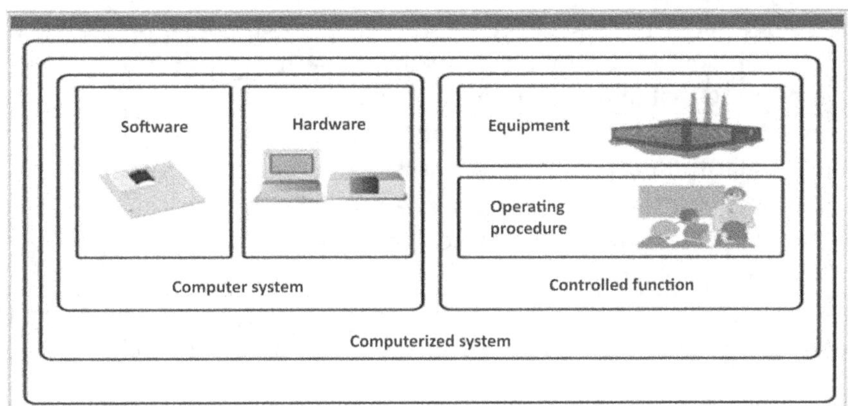

38.1 Key Aspects of Alarm Management Program

The key aspects of alarm management can be split into four classes:

- Procedure for handling of alarm management, react to activated alarm and define effective an alarm management life cycle.
- Those that deal with the alarm handling, e.g. operator interface, design, implementation, maintenance.
- Those that deal with the alarm contents, e.g. alarm definitions, response procedures, and alarm effectiveness/performance monitoring.
- Operator act as in given table

	Operator Must Act	FYL to Operator
Abnormal	Alarm	Alert
Expected	Prompt	Message

38.2 Regulators Expectation of Alarm Management System

A. FDA and Alarming: Six expectations for alarm management as per FDA Computerized Systems in Drug Establishments (2/1983) and General Principles of Software Validation; Final Guidance for Industry and FDA Staff, January 11, 2002

Sr. No.	Computerized Systems in Drug Establishments Guideline	Requirement Class
1	Documentation of alarm function. The condition that initiates the alarm must be documented. Any interlocks the alarm trigger must be documented	Alarm Contents
2	Documentation of alarm parameters/thresholds and their maintenance. The alarm set-points must be documented. (Typical values are trigger point and time delay).	
3	Determination and documentation of alarm response procedure.	
4	Definition of how alarms are recorded -in batch records, logs, or automatically by the control system, and maintenance of the history record. It is required to maintain alarm records for all alarms used that indicate out of range conditions	Alarm Handling
5	Documented design, validation, and maintenance, of the Operator Alarm Interface. (lights, alarm horn, control system Alarm Manager Interface). Maintenance requirements for the interface to ensure it continues to work as validated. E.g. verification of pilot lights.	
6	The alarm interface must be designed to ensure it captures all alarms and to ensure it provides just in time access to all alarms.	Alarm Handling

B. ISA-18.2 Alarm System Management Lifecycle

- **Philosophy:** Starting point of ASM provides guidance, capability of alarm control, plan and documentation.

- **Identification:** Iincludes activities like P&ID reviews, process hazard reviews, layer of protection analysis and environmental permits that identify potential alarms.

- **Rationalization:** tested against the criteria documented in the alarm management to justify that it meets the requirements such as consequence, response time, Operator action , limit, priority, classification, severity define groups of alarms with similar characteristics and common requirements for training, testing, documentation, or data retention.

- **Detailed Design:** designed to meet the requirements .Alarm design includes the basic alarm design, setting parameters like the alarm designed and or off-delay time, advanced alarm design, like using process or equipment state to automatically suppress an alarm, and HMI design, displaying the alarm to the Operator so that they can effectively detect, diagnose, and respond to it.

- **Implementation**: Putting the alarms into operation which includes the activities of training, testing, and commissioning.

- **Operation:** An alarm performs its function of notifying the Operator of the presence of an abnormal situation.

- **Maintenance:** In the maintenance stage the alarm does not perform its function of indicating the need for the Operator to take action as per procedure to remove an alarm from service and return an alarm to service.

- **Monitoring and Assessment**: It encompasses data gathered from the operation and maintenance stages. Assessment is the comparison of the alarm system performance against the stated performance.

- **Management of Change**: The management of change stage of the alarm lifecycle includes the activity of authorization for all changes to the alarm system, including the addition of alarms, changes to alarms, and the deletion of alarms. Once the change is approved, the modified alarm is treated as identified and processed through the stages of rationalization, detailed design and implementation again.

- **Audit**: The periodic review of the alarm system. The goal is to maintain the integrity of the alarm system throughout its lifecycle and to identify areas of improvement. The alarm philosophy document may need to be modified to reflect any changes resulting from the audit process.
  - (a) Verify alarm configuration settings against your design on a periodic basis. Be sure to distinguish permanent changes with those made automatically
  - (b) Control Alarm Set point and Priority Changes

(c) Follow procedures, update alarm rationalization and design documents, and identify any other alarms or functions effected.

(d) Return unapproved changes to their approved configuration state

(e) Monitor Alarm Shelving and Suppression: (List of suppressed alarms, Logged comments associated with suppressed alarms, Accumulated time each alarm was suppressed and Number of times each alarm was suppressed

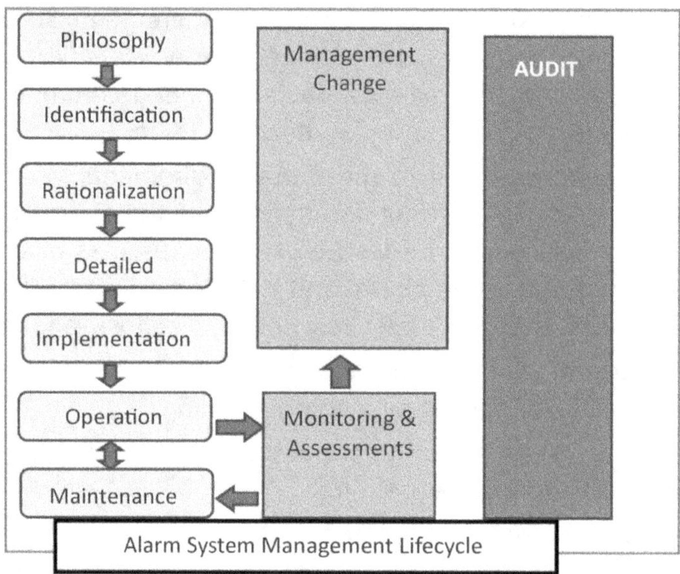

38.3 Element of a Good Alarm on Risk Assessment in Pharmaceutical Industry

Alarming requirements should be collected using formal approaches such as risk assessment and element for good alarm management shall be defined.

Element of a good alarm definition include documentation of:

(a) Alarm function, the potential cause(s) that initiate the alarm. alarm interlocks

(b) Alarm type /set point/parameters/thresholds,

(c) Escalation procedure

(d) Allowable response time.

(e) Alarm management shall appropriately fitted to the organizational process and environmental factors.

38.4 Categorization of Alarm

Alarms can be categorized into three groups:

(a) Safety Alarms

(b) Process controls alarms

(c) Equipment alarms.

- Within each group, alarm criticality will vary according to function. All alarms directly, or indirectly, involved with the process should be individually assessed for their GMP criticality to the process. Alarm prioritization is used to indicate relative criticality between alarms. Alarm prioritization requires expert knowledge about the alarm condition and alarm response.

- Based on risk assessment ICH Q9 and ISPE guidelines, a model was developed by merging concepts from Instrumentation Management and Hazards Risk Assessment. The following three categories are suggested:

GMP Criticality	Description
1	Product Quality Critical Alarm (Exception Alarm): An alarm whose failure may have a direct impact on product quality.
2	Process / System Alarms: An alarm whose failure may affect the process or system performance but does not directly impact product quality.
3	Non- Critical Alarms: Alarms whose failure has no impact on product quality, systems or the environment

38.5 Principle of Alarm Management System

The key principles of Alarm Management include defining the alarms, classifying the alarms, and developing alarm response procedures. The use of GMP classification and risk assessment allows for implementation of exception reporting. Pharmaceutical companies will be well served by implementing process efficiency alarms systems to prevent production discrepancies.

Chapter - 39

Handling of Objectionable Organisms- The Regulatory Perspective

Introduction

The topic of "objectionable organisms" is frequently found in regulatory observations and publications. An objectionable organism is one which can either cause illness or degrade the product thus making it less effective.

There is also now the expectation that the significance of other microorganisms recovered from the Microbial Limit Tests should be evaluated to see if they are objectionable to the product. The responsibility for determining whether a microorganism is objectionable belongs to the manufacturers and they must use a risk assessment based approach of the relevant factors to confirm this.

39.1 Regulatory Expectations

The FDA also addresses the need for non-sterile products to be free of objectionable organisms as set forth in 21 CFR 211.113 and 21 CFR 211.165. The factors that the USP, Pheur and FDA state for consideration when performing a risk based assessment into objectionable organisms are summarised below:

- The characteristics of the microorganism
- The number of microorganisms present
- The nature of the product as in whether or not the product supports growth
- Whether the product has adequate antimicrobial preservation
- The use of the product, due to the fact that hazard varies according to the route of administration via the eye, nose, or respiratory tract
- The method of application
- The presence of disease, wounds, or organ damage
- The intended recipient as risk may differ for neonates, infants, and the elderly
- The use of immunosuppressive agents and corticosteroids.

It is recommended that these risk based assessments are conducted by personnel with specialised training in microbiology and the interpretation of microbiological data.

Using the information provided from the USP and PhEur, one of the first factors that a manufacturer should consider is whether the organism is a known pathogen or opportunistic pathogen. A good place to start researching this is the FDA's "Bad Bug Book", as it provides more microorganisms which are capable of causing disease than those listed in the Pharmacopoeias.

39.2 Objectionable Microorganism Needs

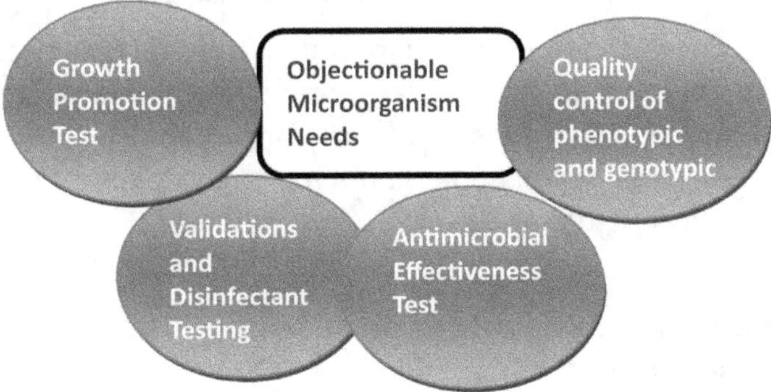

39.3 Microbiological Requirements

- SOP that describes the organisms that are "objectionable".

- Must have a plan for identifying microorganisms in addition to the —Specified organisms i.e (Non-selective media, Identification of any isolates ,Look at methods for food (helpful for oral products),Determine how you will screen for objectionable microorganisms (identify all isolates, identify if over an established Alert or Action Limit, identify total aerobic count, etc.)

- Absence of the organisms listed in the USP does not satisfy the requirement for the absence of objectionable microorganisms.

- There are more microorganisms than those listed in the USP that are capable of causing disease! (Reference: FDA's Bad Bug Book).

- Each pharmaceutical manufacturer must decide what microorganisms are considered to be objectionable in their products. Some microorganisms may not be present in your products due to the type of product (tablets and capsules vs. aqueous liquids)

- Currently, there is no published list of Objectionable Microorganisms—the —list‖ would be ever evolving and changing as new pathogens emerge, new organisms are associated with disease, and detection techniques improve. FDA-CFSAN—Bad Bug Book—helpful for oral products, lists food borne pathogens.

39.4 Detection of Objectionable Organism

A manufacturer can detect objectionable organisms in its product by:

- Performing tests for specified microorganisms as described in USP <62>.

- Identifying all suspicious colonies growing on agar plates used in microbial enumeration tests. To find out more about microbial enumeration tests, see USP <61>

- Performing a risk analysis to determine which organisms would be objectionable and then devising a plan for detecting the organisms.

Microorganisms, once isolated from a product, can be identified using phenotypic (biochemical) or genotypic methods. There are several microorganism identification systems which use phenotypic means to identify an organism. Genotypic testing can be done to the species or strain level. Some genotypic methods are PCR, 16S and 23S rRNA sequencing, and analytical rib typing.

39.5 What the Laboratories Can Do on Identification of Objectionable Organism?

If an objectionable microorganism is found during water testing, bioburden testing or environmental monitoring, the laboratory may want to consider preserving the microorganism for future investigations, disinfectant qualification studies, growth promotion tests, suitability tests and/or antimicrobial preservative tests. Laboratories can also take the following steps to ensure the safety of non-sterile products:

- Perform a risk assessment to determine which microorganisms could be harmful based on the route of administration, the user and the properties of the drug.
- Use enrichment and selective media to find objectionable organisms.
- Follow pharmacopeia guidelines for microbial enumeration and tests for specified microorganisms.
- Scan for potential problems by researching past recalls in similar products.
- Identify suspicious colonies found when performing total aerobic counts.
- Perform growth promotion testing on all batches of media used in aerobic plate counts and tests for specified and objectionable organisms.
- Validate all tests used for isolating and identifying organisms.
- Identify the organisms that are isolated from routine environmental monitoring and determine if these organisms could degrade the drug or be harmful to users of the drug.
- Perform disinfectant challenge testing to determine which disinfectant(s) to use at the manufacturing site.
- Routinely perform environmental monitoring. Identify the organisms that are isolated. Determine if these organisms could degrade the drug or be harmful to users of the drug.
- Perform disinfectant testing to determine which disinfectant(s) to use at the manufacturing
- Perform bioburden testing in order to discover any weaknesses in the manufacturing process.
- Keep current with trends in pharmaceutical microbiology by attending webinars and conferences and reading scholarly articles

39.6 Risk Assessment

- Presence of data supporting both objectionable and non-objectionable determinations
- Little information on the microorganism—gray area (a science-supported, risk-based approach may be needed.)
- The organism can be a pathogen under certain circumstances
- Scope for Risk Assessments (Raw materials including water, Non-Sterile Environmental Monitoring, Components—containers, closures, etc., Active Pharmaceutical Ingredients, Non-Sterile Finished Products and Dietary Supplements/Neutraceuticals).

- **Elements of a Science-Based Risk Assessment** (Identity of the microorganism, Number of microorganisms present in the product, route of administration, Nature of the product ,Method of application/ dosage form, Intended recipient/ patient population, Interference of the organism, Capability of the product to support growth or sustain the microorganisms , Inherent Product Characteristics-- pH, water activity, osmotic pressure, etc.

- The product interference of the organism with active ingredients, test methods, product stability or container closure system may need to be assessed.

- Product supports microbial growth, is the pH. If the pH of the product is in the same range for ideal growth of organisms, then this could cause microbial growth.

- Whether the product is anhydrous or water based can also have an effect on the ability of microorganisms to proliferate as, if there is sufficient water activity in the product, then this may support microbial growth. The USP chapter <1112> provides some representative examples of the water activity required to support the growth of different microorganisms. It is generally accepted that a water activity of 0.91 is required to support bacterial growth and a level of 0.7 or higher is needed to support fungal growth. Knowing the water activity of organisms of concern can then allow an assessment of the relative risk.

- A adequate antimicrobial preservation and whether the product contains antimicrobials, noting that antimicrobials are often only effective against some and not all organisms. If microorganisms proliferate then there is also a risk that the Total Aerobic Microbial Count and Total Yeast and Mould Count limits may be breached.

- The use of the product and method of application or dosage must also be examined.

Once You Have Completed the Risk Assessment…..

IF…

- You have confirmed the identification according to procedures you have in place.
- The patient population does not exclude those susceptible to the illness that this organism causes.
- The microorganism is known to cause illness

- Product route of administration is the same as the organism's route of infection(example: the bacterium causes illness via ingestion and your product is an oral product)
- The infective dose is low—It takes only a few cells to cause illness
- You cannot prove that the organism will NOT proliferate in your product

THEN....

It would be very risky to put the product on the market

"The product should be rejected."

Once the Assessment is Complete.....

- Be sure to document the references and resources you used in your assessment
- Make sure your references are acceptable (apply academic standards where unsure)
- Include qualifications of consultants (if used)
- Include special testing results used to support assessment

Reminders

- Cannot use assessments to release products that exceed established specifications. For example, if a product shows TAC results of 1,000 cfu/mL and the specification limit is 100 cfu/mL, the product exceeds specifications and cannot be released.
- A Risk Assessment should notbe performed for microorganisms that are known pathogens (E. coli, **Salmonella, Listeria, etc.)**

VS

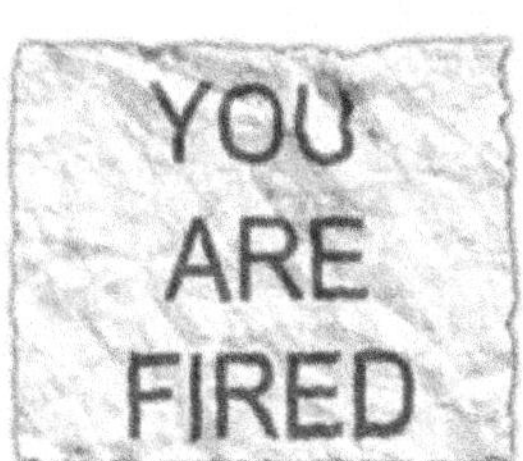

Making a Decision ....

- Based on the information you have gathered for the Risk Assessment, you should be able to make a decision and have the scientific data to justify the decision.
- Remember, the ultimate objective is to provide a safe and effective product for the consumer.

39.7 Resources for Assessing the Potential Harm

Several resources are available for assessing the potential harm a microorganism can cause. Information about the microorganism can be found in the U.S. Food and Drug

Administration Bad Bug Book, Manual of Clinical Microbiology, Bergey's Manual of Systematic Bacteriology, and in scientific articles found online. The National Institute of Health has classified human etiologic agents on the basis of hazard, click here to view the NIH Guidelines. Microorganisms are listed in Risk Groups 1-4, 4 being the most dangerous.

39.8 FDA Product Recalls due to Objectionable Microorganism given in the Table below

Organisms cited in FDA recalls		
Gram Negative Organisms		**Gram Positive Organisms**
Achromobacterxylosoxidans	Pseudomonas spinosa	Bacillus cereus and other Bacillus species
Burkholderiacepacia	Pseudomonas species	Staphylococcus aureus
Elizabethkingiameningoseptica	Salmonella species	Staphylococcus intermedius
Enterobactercloadae	Serratiafonticola	Streptococcus species
Enterobactergergoviae	Serratiamarcesans	
Escherichia coli	Stenotrophomonasmaltophilia	
Francisellatularensis	Pseudomonas fluorescens	
Klebsiellaoxytoca	Pseudomonas luteola	
Pseudomonas aeruginosa	Pseudomonas putida	
Yeasts and Molds		
Acremonium(mold)		Acremonium(mold)
Aspergillus species		Aspergillus species
Aspergillussydowii		Aspergillussydowii
Penicillium species		Penicillium species

39.9 Summary

The contaminants can lead to unpredictable pharmacological effects or super or Sub potency of the active drug substance and worse, they can be toxic. Setting up a program to identify objectionable organisms can be difficult. A great deal of research needs to be carried out. It is important to know why the organism is objectionable, as this information may be useful when performing investigations if this organism is found in your facility. Appropriate corrective actions be taken in the event that these organisms are found in the facility.

Chapter - 40

Quality by Design (QbD) Approach in the Product Life Cycle

Introduction

QbD is comprehensive understanding of the product and manufacturing process to consistently deliver the intended performance of the product throughout its life cycle.

"The pharmaceutical Quality by Design (QbD) is a systematic approach to development that begins with predefined objectives and emphasizes product and process understanding and process control, based on sound science and quality risk management. "

Principle of QbD Concepts is Zero defects and is based on the:

- Systematic approach to develop product on realisation by multivariate experiment
- Manufacturing process adjustable within the design space.
- Establish and maintain a state of control
- Focus on control strategy and robustness of the process to facilitate continual improvement
- Knowledge management and quality risk management based decisions
- Product ownership
- Continuous Improvement Cycle based on "PDCA Cycle": PLAN (Plan the steps), DO(Perform the steps), CHECK (Analyse the results) and ACT (Use the results)

40.1 Elements of QbD

Four main elements of QbD are:

- Assessment approaches which begin with mapping tools such as flow-down map, process map, Ishikawa diagram to evaluate their knowledge space and further risk management tools such as failure modes and effects analysis (FMEA).
- PAT tools including in-process monitoring and multivariate systems
- Mathematical and statistical tools which can be used in the planning, designing and analysing the experiment: statistical design of experiments (DoE).
- Continuous improvement tools which are implemented throughout process/product lifecycle to maintain the robust QbD construct

40.2 Steps Involved in QbD Development Process

- Begin with a target product profile that describes the use, safety and efficacy of the product
- Define a target product quality profile that will be used by formulators and process engineers as a quantitative surrogate for aspects of clinical safety and efficacy during product development
- Gather relevant prior knowledge about the drug substance, potential excipients and process operations into a knowledge space. Use risk assessment to prioritize knowledge gaps for further investigation
- Design a formulation and identify the critical material (quality) attributes of the final product that must be controlled to meet the target product quality profile.
- Design a manufacturing process to produce a final product having these critical material attributes.
- Identify the critical process parameters and input (raw) material attributes that must be controlled to achieve these critical material attributes of the final product. Use risk assessment to prioritize process parameters and material attributes for experimental verification. Combine prior knowledge with experiments to establish a design space or other representation of process understanding.
- Establish a control strategy for the entire process that may include input material controls, process controls and monitors, design spaces around individual or multiple unit operations, and/or final product tests. The control strategy should encompass expected changes in scale and can be guided by a risk assessment.
- Continually monitor and update the process to assure consistent quality.

40.3 The Principle Steps in QbD are

- ***Quality Target Product Profile (QTPP):*** A prospective summary of the quality characteristics of a drug product that ideally will be achieved to ensure the desired quality, taking into account safety and efficacy of the drug product.

- It relates to quality, safety and efficacy, considering e.g., the route of administration, dosage form, bioavailability, strength, closure container system, therapeutic moiety release and stability.
- ***Critical Quality Attribute (CQA):*** A physical, chemical, biological, or microbiological property or characteristic that should be within an appropriate limit, range, or distribution to ensure the desired product quality.
- ***Control Strategy:*** A planned set of controls, derived from current product and process understanding that ensures process performance and product quality. The controls can include parameters and attributes related to drug substance and drug product materials and components, facility and equipment operating conditions, in-process controls, finished product specifications, and the associated methods and frequency of monitoring and control.
- ***Critical Process Parameter (CPP):*** A process parameter whose variability has an impact on a critical quality attribute and therefore should be monitored or controlled to ensure the process produces the desired quality.
- ***Design Space:*** The relationship between the process inputs (material attributes and process parameters) and the critical quality attributes can be described in the design space.

  Selection of Variables, Describing a Design Space in a Submission, Unit Operation Design Space, Relationship of Design Space to Scale and Equipment, Design Space Versus Proven Acceptable Ranges, Design Space and Edge of Failure
- ***Continuous Improvements:*** Process performance can be monitored to ensure that it is working as anticipated to deliver product quality attributes as predicted by the design space. This monitoring could include trend analysis of the manufacturing process as additional experience is gained during routine manufacture.

40.4 QbD Documents

- ***Risk Assessment Report(s):*** Performed throughout QbD Process and Particularly important to process development
- ***Quality Target Product Profile (QTPP):*** Defines the desired product characteristics and sets development goals.
- ***Control Strategy Summary:*** Defines the process, its inputs and outputs, and how it is controlled.
- ***PPQ Report(s):*** Formal verification that the process Control Strategy has been defined appropriately and repeatedly produces the desired results.

- ***Continued Process Verification (CPV) Reports:*** Assuring that during routine commercial production, the process remains in a state **of control (FDA); involves feedback loops into the QbD "process" where intentional** process changes and/or observed variability is assessed for risk, characterized, re-validated, etc.

40.5 Process Analytical Technology (PAT) Frame Work

Before starting PAT one should answer following questions

- Why do PAT
- When to Apply
- Who benefits?
- Where does PAT begin and end?
- What to do with data?
- How to apply
- Is the PAT for Process Knowledge or Process Control?

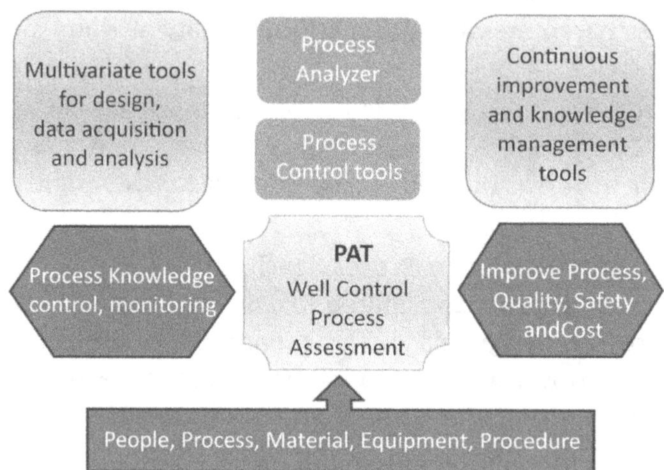

40.6 Use of PAT Tools in Continuous Manufacturing Process

There are many tools available that enables process understanding for scientific, risk-managed pharmaceutical Development, manufacture, and quality assurance. These tools, when used within a system, can provide effective and efficient means for acquiring information to facilitate process understanding, continuous improvement, and development of risk-mitigation strategies.

A. PAT Tools

- Multivariate tools for design, data acquisition and analysis
- Process analyzers

- Process control tools
- Continuous improvement and knowledge management tools

B. ***Risk-Based Approach:*** Within an established quality system and for a particular manufacturing process, one would expect an inverse relationship between the level of process understanding and the risk of producing a poor quality product. For processes that are well understood, opportunities exist to develop less restrictive regulatory approaches to manage change. Thus, a focus on process understanding can facilitate risk-based regulatory decisions and innovation.

C. ***Integrated Systems Approach:*** Bringing the development, manufacturing, QA, and information / knowledge management functions so closely together that these four areas should be coordinated in an integrated manner.

D. ***Real Time Release:*** Real time release is the ability to evaluate and ensure the acceptable quality of in-process and/or final product based on process data. Typically, the PAT component of real time release includes a valid combination of assessed material attributes and process controls.

E. ***Strategy for Implementation:*** PAT principles & tools should be introduced during the development phase. When using new measurement tools, such as on- or in-line process analyzers, certain data trends, intrinsic to a currently acceptable process, may be observed. Manufacturers should scientifically evaluate these data to determine how or if such trends affect quality and implementation of PAT tools.

40.7 PAT Regulatory Approach

- Improve the scientific basis for establishing regulatory specifications,
- Promote continuous improvement,
- Improve manufacturing while maintaining or improving the current level of product quality PAT implementation plans should be risk based. PAT can be implemented under the facility's own quality system. PAT implementation plans neither affect the current process nor require a change in specifications

40.8 Product Life Cycle

"Product life cycle means all phases in the life of a product from the initial development through marketing until the product's discontinuation".

Product and process knowledge should be managed from development through the commercial life of the product up to and including product discontinuation. Sources of knowledge include, but are not limited to prior knowledge, pharmaceutical development studies; technology transfer activities; process validation studies over the product lifecycle; manufacturing experience; innovation; continual improvement; and change management activities.

The product lifecycle includes the following technical activities for new and existing products:

- **Pharmaceutical Development**
  - Drug substance development;
  - Formulation development (including container/closure system);
  - Manufacture of investigational products;
  - Delivery system development (where relevant);
  - Manufacturing process development and scale-up;
  - Analytical method development.

- **Technology Transfer**
  - New product transfers during Development through Manufacturing;
  - Transfers within or between manufacturing and testing sites for marketed products.

- **Commercial Manufacturing**
  - Acquisition and control of materials;
  - Provision of facilities, utilities, and equipment;
  - Production (including packaging and labelling);
  - Quality control and assurance;
  - Release;
  - Storage;
  - Distribution (excluding wholesaler activities).

- **Product Discontinuation:**
  - Retention of documentation;
  - Sample retention;
  - Continued product assessment and reporting.

The implementation of QbD requires appropriate PAT to monitor CQAs and CPPs during the processing of pharmaceuticals products to ensure that the product meets the desired quality attributes and reduce the risk of the poor manufacturing process performance and substandard pharmaceutical reaching patient. Risk is minimized by reducing variations in the process, while accuracy, repeatability and reproducibility are increased. There is no signal way of implementing QbD in practices .The ICH guidance documents lay the ground work for a better understanding on how to achieve QbD and improve product and process understanding.

40.9 Definition

- ***Capability of a Process:*** Ability of a process to realise a product that will fulfil the requirements of that product. The concept of process capability can also be defined in statistical terms. (ISO 9000:2005).

- ***Continual Improvement:*** Recurring activity to increase the ability to fulfil requirements. (ISO 9000:2005).

- ***Control Strategy:*** A planned set of controls, derived from current product and process understanding that assures process performance and product quality. The controls can include parameters and attributes related to drug substance and drug product materials and components, facility and equipment operating conditions, in-process controls, finished product specifications, and the associated methods and frequency of monitoring and control.

- ***Design Space:*** The multidimensional combination and interaction of input variables (e.g., material attributes) and process parameters that have been demonstrated to provide assurance of quality.

- ***Enabler:*** A tool or process which provides the means to achieve an objective.

- ***Knowledge Management:*** Systematic approach to acquiring, analyzing, storing, and disseminating information related to products, manufacturing processes and components.

- ***State of Control:*** A condition in which the set of controls consistently provides assurance of continued process performance and product quality.

- ***Proven Acceptable Range:*** A characterized range of a process parameter for which operation within this range, while keeping other parameters constant, will result in producing a material meeting relevant quality criteria.

Chapter - 41

Validation Master Plan (VMP)

Introduction

Validation is an essential part of good manufacturing practices and Good Laboratory Practice. Validation of process and systems is fundamental to achieving the following goals,

(a) Quality, safety and efficacy must be designed and build into the product,

(b) Quality cannot be inspected or tested into the product.

(c) Each critical steps of the manufacturing process must be validated. Other steps in the process must be under control to maximize the probability that the finished product consistently and predictably meets all quality and design specification.

Validation is by design and manufacture can establish confidence that the manufactured products will consistently meets their product specifications.

Documentation associates with validation include:

(a) Validation master Plan (VMP)

(b) Validation Protocols and reports

(c) Qualification Protocols and Reports

(d) Specifications

Significant changes to premises, facility, system, utility, equipment and instrument should be routed through change control and direct or indirect effect on the product should be qualify and validated.

41.1 Relation between Validation and Qualification

Qualification is part of the validation. Generally validation and Qualification are the same concept. The term qualification is normally used for equipment and instrumentation, utility and system and validation is used for the process. Qualification and validation cannot be considered once-off exercises, but it should be on -going programme. It is responsibility of the pharmaceutical company to define the responsibility of the pharmaceutical company to define the respective responsibilities of its personnel's and of external contractors in qualification and validation programme. This should form a part of Validation Master Plan.

41.2 Validation Master Plan

The principle of VMP is preparation and carefully planning of various steps in the process under authorized standardization working procedure. Validation is a characterized by:

(a) *A multidisciplinary approach:* validation requires the collaboration of experts of various disciplines such as technologisists, pharmacists, microbiologist, engineering, chemical analysis, expert on validation.

(b) *Time constraints:* Validation work is submitted to rigorous time schedule. These studies are always the last stage prior to taking new processes, facilities into routine operation.

(c) *Costs:* Validation studies are costly as they required time of highly specialized personnel and expensive technology.

A. Purpose

(i) The VMP should present an overview of entire validation operation, its organizational structure, its content and planning. The core of the VMP being the list of the items to be validated and the planning schedule.

(ii) A VMP is a document that summarises the firms overall philosophy, intentions and approach to be used for establishing performance adequacy.

(iii) All critical production process be validated,

(iv) Validation studies are conducted in accordance with predefined protocols. Written reports summarizing recorded results and conclusions are prepared , evaluated, approved and maintained,

(v) Change to production process, operating parameters, equipment or materials that may affect product quality and /or the responsibility of the process are also to be validated prior to implementation.

B. Scope

(i) All validation activities relating technical operations, relevant to product and process controls with a firm should be included in a VMP. This includes qualification of critical manufacturing and control equipment.

(ii) The VMP should comprise all Qualification, requalification, Validation and Revalidations.

(iii) In case of large projects like the constructions of new facilities, often best approach is to create a separate VMP. (In such situation the VMP should be part of the total project management.)

(iv) VMP should help to understand the qualification and validation programme, responsibilities of team member and qualification and validation schedules.

(v) Validation should be performed in accordance with written protocols A report on the outcome of the validation should be produced.

(vi) Validation should be done over a period of time, e.g. at least three consecutive batches (full production scale) should be validated, to demonstrate consistency. Worst case situation should be considered.

(vii) Manufacturer should plan validation in a manner that will ensure regulatory compliance and ensuring that product quality, safety and consistency are not compromised.

C. VMP flow

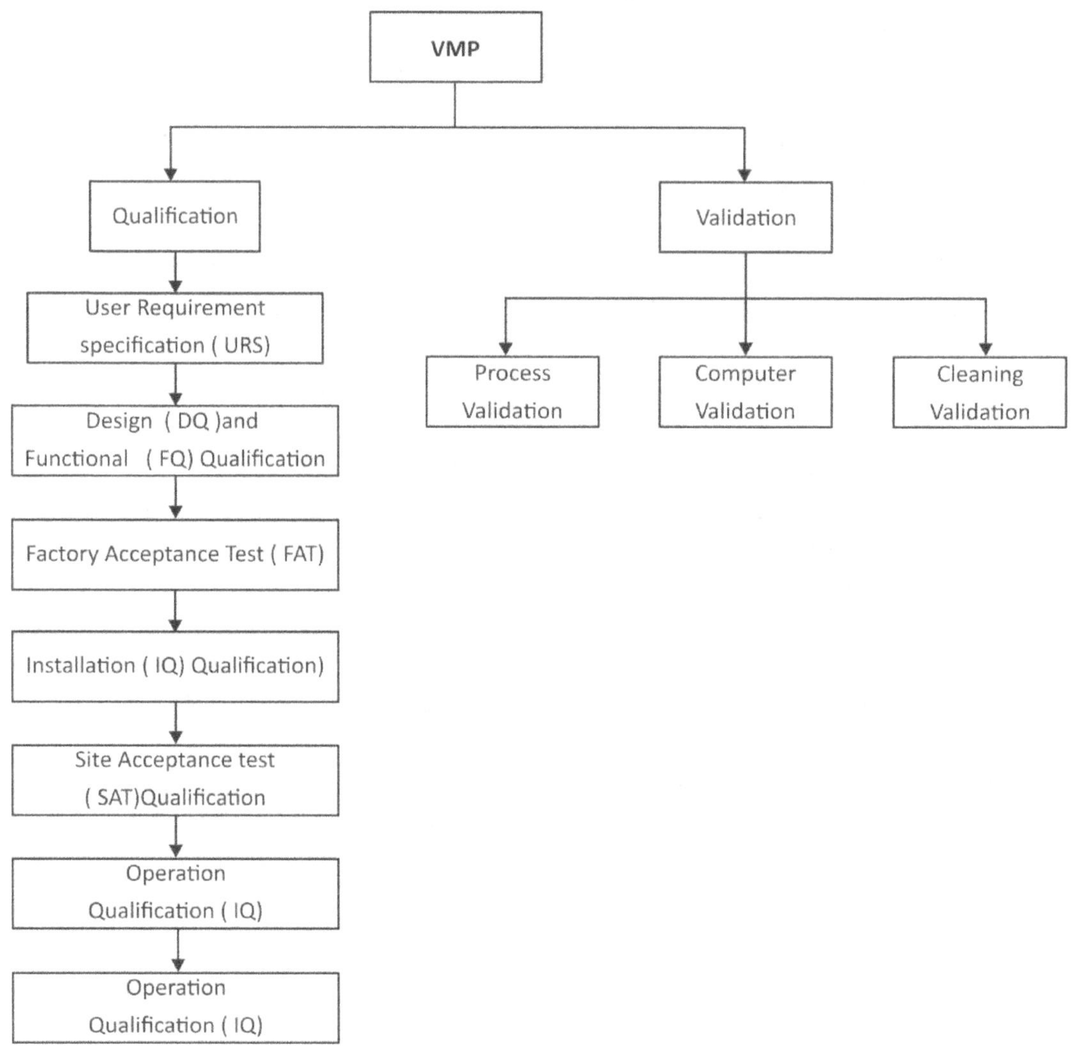

Responsibilities of Validation and Qualification team

The organizational structure of all validation activities, personnel responsibilities for VMP, protocol of individual validation projects, validation work, report and document preparation and control, approval / authorization of validation protocol and reports in all stages of validation process tracking system for reference and review, training needs in support of validation.

D. Plant / Process/ Product description

Provides a cross reference to other documents .A rationale for the inclusion or exclusion of validations, for the validation approach, the extent of validation and any challenge and "worst case" situation should included. Consideration can be given to the grouping of products / process for the purpose of validating "worst case" situations. Where "worst case" situations cannot be simulated the rationale for the groupings made should be defined.

E. Specific process considerations

Characteristics / requirements of plant / process etc. that are critical for yielding a quality product and need extra attention may be briefly outlined here.

F. List of products /process /Systems to be validated

All validation activities comprised in the VMP should be summarized and complied in a matrix format. Such matrix should provide an overview and contain:

- All items covered by the VMP that are subjected to validation describing the extent of validation required (i.e. IQ, OQ and/PQ).It should include validation of analytical techniques which are to be used in determining the validation status of other processes or systems.
- The validation approach, i.e. prospective , retrospective or concurrent,
- The re-validation activities,
- Actual status and future planning.

G. Acceptance criteria

A general statement on key acceptance criteria for the items should be made.

H. Documentation format:

The format to be used for protocol and reports should be described.

41.3 Validation Protocol/Report

The validation protocol should be numbered, signed and dated, and should contain as a minimum the following information's:

- Objectives , Scope of coverage of the validation study ,
- Validation team , their qualifications and responsibilities
- Justification for validation'

- Risk Assessment,
- Type of validation: Prospective, concurrent, Retrospective, revalidation,
- Number of batches to be validated,
- Approved mater formula, standard operating procedure and methods,
- A list of all equipment to be used with calibration, qualification, requalification and preventive maintenance details.
- Outcome of IQ , OQ for critical equipment,
- Critical process parameters and their respective tolerance;
- Description of the processing steps: copy of the master documents for the product
- Sampling points, Stages of sampling, methods of sampling and sampling plans;
- Statistical tools to be used in the analysis of data,
- Forms and chart to be used for documenting results,
- Non-conference (Out of Specification /Out of Trend/Deviation)
- Change control
- Conclusion
- Summary
- Approval of study

41.4 Time Plans of Each Validation Project and Sub-Project

An estimate of staffing (including training needs), equipment and other specific requirements to complete the validation effort , a time plan of the project with detailed planning of mentioned matrix.

41.5 Validation, Sampling Schedule

- Process Validation Schedule
- Cleaning Validation Schedule
- Utility validation Schedule (Water, Air, Gases etc)
- Computer Validation

41.6 Definitions

1. ***Validation Master Plan:*** An approved written procedure and plan of objectives and actions stating how and when a company will achieve compliance with the GMP requirements regarding validation.
2. ***Validation:*** The documented act of demonstrating that any procedure, process and activity will consistently lead to expected results. Includes the qualification of system and equipment's.

3. **Validation Team:** A multiple-disciplinary team of personnel primarily responsible for conducting and or supervising validation studies. Such studies may be conducted by person (S) qualifies by training and experience in a relevant discipline.

4. **Pre-Determined Acceptance Criteria:** The criteria assigned, before undertaking testing, to allow evaluation of test results to demonstrate compliance with a test phase of delivery requirement.

5. **Validation Protocol:** A written plan stating how validation will be conducted, including test parameters, product characteristics, production equipment and decision points on what constitutes acceptable test results.

6. **Validation Report:** Documents reporting the validation activities, the validation data and the conclusions drawn.

7. **Process Qualification:** The phase of validation dealing with sampling and testing at various stages of the manufacturing process to ensure that product specifications are met.

8. **Process Validation:** Establishing documents evidence with high degree of assurance, that specific process will consistently produce a product meeting its predetermined specifications and quality characteristics .Process validation may take the form of Prospective, Concurrent or Retrospective validation and Process Qualification or re-validation.

9. **Process Re-validation:** Required when there is change in any critical process parameters, formulation, primary packing components, raw material fabricators , major equipment or premises, failure to meet product and process specifications in sequential batches would require process re-validation.

10. **Prospective Validation:** Conducted prior to the distribution of either a new product or a product made under modified production process, where the modifications are significant and may affect the product's characteristics. It is pre-planned scientific approach and includes the initial stages of formulation development, process development, setting of process specifications, completion of pilot runs, and transfer technology from scale up batches to commercial size batches, listing major process equipment and environmental control.

11. **Concurrent Validation:** A process where current production batches are used to monitor processing parameters. It gives assurance of the present batch being studied and offers limited assurance regarding consistency of quality from batch to batch.

12. **Retrospective Validation:** Conducted for a product already being marketed, and is based on extensive data accumulated over several lots and overtime. Retrospective validation may be used for older products which were not validated by the fabricators at the time that they were first marketed and which is now to be validated to conform to the requirements of regulation.

13. **Revalidation:** A repeat of the process validation to provide an assurance with change control procedures does not adversely affect process characteristics and product quality.

14. **Validation Certification:** Final document of approval of a validation or revalidation issued by the person in charge of the validation activity.

15. **Process Capability:** Studies conducted to identify the critical process parameters that yield a resultant quality and their acceptable ranges, based on the established ± 3 sigma deviations of the process, and under stressed conditions but when free of any assignable causes.

16. **Process Capability study:** A process capability study is a statistical method that compares process information (e.g X and S) to the upper and lower specification limits.

17. **Process capability index (CpK):** A process capability index CpK represent the true measure of process capability,

$$cpk = \frac{X - LSL}{3S}$$

$$cpk = \frac{USL - X}{3S}$$

Where LSL = Lower limit specification limit

USL = Upper limit specification limit

X = Mean

S = Standard Deviation

18. **Critical Process Parameter:** A parameter which if snot controlled will contribute to the variability of the end products.

19. **Critical Variable Study:** A study that serves to measure variables (parameters) critical to the satisfactory operation of a piece of equipment or plant and to assure their operation within monitored and controlled limits. Examples of variables would be pressure, temperature, flow rates , time etc.

20. **Measuring device:** A device used in monitoring or measuring process parameters.

21. **Cleaning Validation:** The documented act of demonstrating that cleaning procedures for equipment used in manufacturing and packing will reduced to an acceptable level all residue (Product and cleaning agent) and to demonstrate that routine cleaning and storage of equipment does not allow microbial proliferation.

22. **Worst case condition:** A conditions or set of conditions encompassing upper and /or lower processing limits and circumstances within standard operating procedures, which pose the greatest chance of product or process failure when compared to ideal condition. Such conditions do not necessarily induce product or process failure.

or

A condition or set of conditions including the superior and inferior limits of processing and the corresponding circumstances, in compliance with the specifications of the Standard Operating Procedures, that present the greatest possibilities of product or process defect when compared with the ideal conditions.

23. **Equipment Qualification:** Studies which establish with confidence that the process equipment and ancillary systems are capable of consistently operating within established limits and tolerances. The studies must include equipment specifications, installation qualification (IQ), and operational qualification (OQ) of all major equipment to be used in the manufacturer of commercial scale batches. Equipment qualification should simulate actual production condition, including "worst case" / stressed condition.

24. **User requirement Specification (USR):** A requirement specification that describes what the equipment or system is supposed to do, thus containing at least a set of criteria or conditions that have to be meeting.

25. **Design Qualification (DQ):** The documented verification of the user requirements for an equipment and its ancillary system or documented evidences that the premises, supporting systems, utilities, equipment and processes have been designed in accordance with the requirements of GMP.

26. **Functional Qualification (FQ):** The document that specifies complete information of required design, system, process, compound and other related information.

27. **Factory Acceptance Test (FAT):** Testing of equipment is conducted at the supplier end to determine the equipment is as per USR, DQ and PQ and meets the requirement of regulatory bodies.

28. **Installation Qualification (IQ):** The performance and documentation of tests to ensure that equipment (such as machines, measuring equipment) used in a manufacturing process, are appropriately selected, correctly installed and work in accordance with established specification.

29. **Site Acceptance Test (SAT):** An acceptance test at the customer end involving the supplier.

30. **Operation Qualification (OQ):** Documented verification that the system or sub-system performs as intended throughout all anticipated operating ranges.

31. **Performance Qualification (PQ):** Document evidence which provides a high degree of assurance that a specific process will consistently and gives responsibility within defined specifications and parameters for prolong periods.

32. **Verification:** The application of methods, procedures, tests and other evaluations, in addition to monitoring, to determine compliance with the GMP principles.

33. ***Action Limit:*** The action limit is reached when the acceptance criteria of a critical parameter have been exceeded .Results outside these limits will require specified action and investigation.

34. ***Alter limit:*** The alert limit is reached when the normal operating range of a critical parameter has been exceeded, indicating that corrective measures may need to be taken to prevent the action limit being reached.

35. ***Calibration:*** Group of operations that determine, under specific conditions, the relation between values indicated by a measuring instrument, system or values presented by a measuring material, compared to those obtained with a correspondent reference standard.

36. ***Normal Operating Limits:*** The range that the manufacturer selects as the acceptable values for a parameter during normal operations. This range must be within the operating range.

37. ***Operating Limits:*** The minimum and/or maximum values that will ensure that product and safety requirements are met.

38. ***Operating Range:*** Operating range is the range of validated critical parameters within which acceptable products can be manufactured.

39. ***Operating Conditions:*** This condition relates to carrying out room classification tests with the normal production process with equipment in operation, and the normal staff present in the room.

40. ***Piping and Instrument Diagrams (P&IDs):*** Engineering schematic drawing that provides details of the interrelationship of equipment, services, material flows , plant controls and alarms. The P & ID also provides the reference for each tag or label used for identification.

41. ***Change Control:*** A formal system by which qualified representatives of appropriates disciplines review proposed or actual changes that might affect a validation status. The intent is to determine the need for action that would ensure and documented that the system is maintained in validation state.

42. ***Minor Changes:*** Changes having no direct impact on final or in –process product quality.

43. ***Major Changes:*** Changes have direct impact on final or in –process product quality.

44. ***Method Validation:*** Method validation should be consider following analytical methods;

45. ***Accuracy:*** Accuracy is the degree of agreement of test results with the true value, or the closeness of the results obtained by the procedure to the true value. It is normally established on samples of material to be examined that have been prepared to quantitative accuracy. Accuracy should be established across the specified range of here analytical procedure.

46. ***Precision:*** Precision is the degree of agreement among individual results. The complete procedure should be applied repeatedly to separate, identical samples drawn from the same homogeneous batch of material. It should be measured by the scatter of individual results from the mean (good grouping) and expressed as the relative standard deviation (RSD).

47. ***Repeatability:*** Repeatability should be assessed using a minimum of nine determinations covering the specified range for the procedure e.g. three concentrations / three replicates each, or a minimum of six determinations at 100% of the test concentration.

48. ***Intermediate Precision:*** Intermediate precision expresses within-laboratory variations (usually on different days, different analysts and different equipment).If reproducibility is assessed; a measure of intermediate precision is not required.

49. ***Reproducibility:*** Reproducibility expresses *precision* between laboratories.

50. ***Robustness (or ruggedness):*** Robustness (or ruggedness) is the ability of the procedure to provide analytical results of acceptable accuracy and precision under a variety of conditions. The results from separate samples are influenced by changes in the operational or environmental conditions. Robustness should be considered during the development phase, and should show the reliability of an analysis when deliberate variations are made in method parameters.

51. ***Linearity:*** Linearity indicates the ability to produce results that are directly proportional to the concentration of the analyte in samples. A series of samples should be prepared in which the analyte concentrations span the claimed range of the procedure. If there is a linear relationship, test results should be evaluated by appropriate statistical methods. A minimum of five concentrations should be used.

52. ***Range:*** Range is an expression of the lowest and highest levels of analyte that have been demonstrated to be determinable for the product. The specified range is normally derived from linearity studies.

53. ***Specificity (selectivity):*** Specificity (selectivity) is the ability to measure unequivocally the desired analyte in the presence of components such as excipients and impurities that may also be expected to be present. An investigation of specificity should be conducted during the validation of identification tests, the determination of impurities and assay.

54. ***Detection limit (limit of detection):*** Detection limit (limit of detection) is the smallest quantity of an analyte that can be detected, and not necessarily determined, in a quantitative fashion. Approaches may include instrumental or non-instrumental procedures and could include those based on:

- Visual evaluation;
- Signal to noise ratio;

- Standard deviation of the response and the slope;
- Standard deviation of the blank; and
- Calibration curve.

55. ***Quantitation limit (limit of quantitation):*** Quantitation limit (limit of quantitation) is the lowest concentration of an analyte in a sample that may be determined with acceptable accuracy and precision. Approaches may include instrumental or non-instrumental procedures and could include those based on:

- Visual evaluation;
- Signal to noise ratio;
- Standard deviation of the response and the slope;
- Standard deviation of the blank; and
- Calibration curve

Chapter - 42

Audit in Laboratory Management

Introduction

The complexity of today's pharmaceutical market requires more efficient drug development and production throughout the product Lifecycle along with Data management to meets the quality requirements. IA is an important performance indicator used to identify the gaps and correct them on timely basis to meets the quality requirements. IA should formally design to identify and manage the risks to drive realistic continual improvement.

ISO 9001 refers to a "Plan-Do-Check-Act" methodology for addressing processes in a quality management system. This methodology can be applied two ways in regards to IAs. The audit itself may be considered a process in which one plans by developing an auditing procedure and audit schedule, does the audit, checks that the audit process worked properly, and then acts upon any observations of the audit process. Secondly, the auditing process can also be used as a part of the checking step in the Plan-Do-Check-Act methodology. The Act step is the corrective and preventive actions taken as part of a continuous improvement of the audited process. Audits should not be viewed negatively as a means of finding weaknesses or problems, but in a positive light by looking an opportunity for continuous improvement in operations. Of course audit results are only one piece of the total picture that management should consider when performing management reviews of operations, but they should serve as unbiased observations of opportunities for improvement.

42.1 Traditional Approach to IA

Traditional IA was a singular functional and location view that fed a static audit plan.

Traditionally IA were conducted as on periodic basis and percentage of audit recommendation was implemented by management was 0-40%. Auditing staff for IA was also responsible for day to day activity in the organization. So IA was given as second priority of work (Figure 1: Historical Annual Process Map).

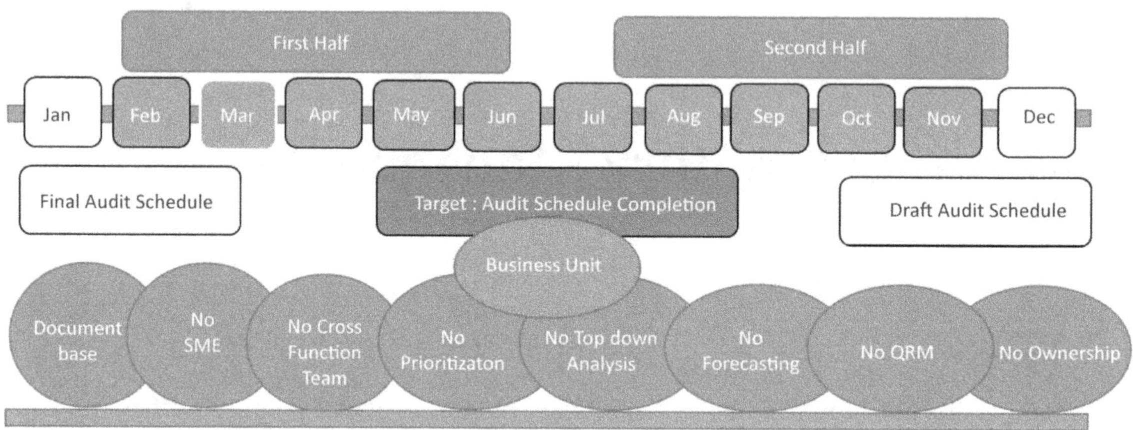

42.2 Forecasting Approach to IA

In 21 Century IA Principles & Techniques shall be depends on Preventative, Detective, Corrective, Directive and Compensating with IA methods such as Organizational Controls, Operational Controls, Personnel Controls-, Periodic Review Controls, and Facilities & Equipment Controls. Risk assessment tool will help the organization in continuous improvement, increased enterprise wide influence, End to end involvement in risk decisions, direct access to board or risk committees, Identification and mitigation of emerging risks.

Risk assessment compliance with help to meet regulatory expectation.

IA is a unique activity which involves a combination of audit approaches and techniques. These include interviews, document reviews, sampling, testing of controls, and analysis of transaction, processes and management information. The IA strategy play important role during IA activity such as planning, examination, evaluation and effectiveness of the system, audit conclusion, review, communication and follow-up.

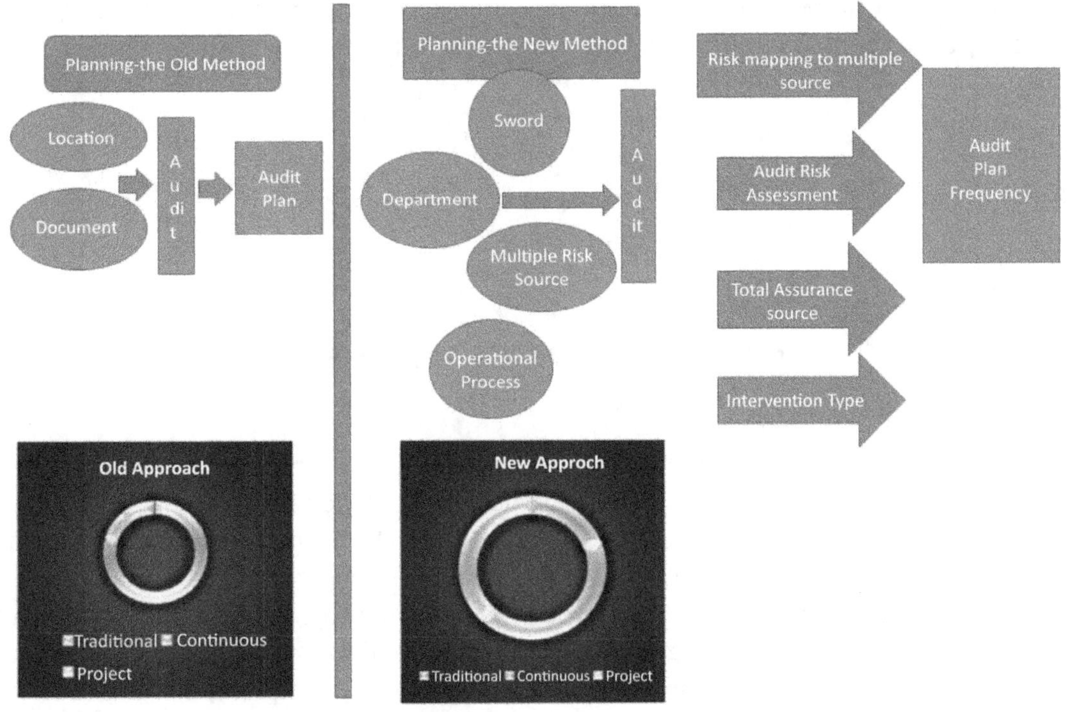

42.3 Benefits of QRM to IA Program

- Proactive management and Transparency of information
- Management override, Governance Risk control and Harmonies change management
- Effectiveness and efficiency of operations and programs.
- Compliance with laws, regulations, policies, procedures and contracts
- Facilitate risk-based regulatory oversight and optimization of resources for assessment and inspection
- Emphasize use of the control strategy as a key component of the regulatory commitment
- Help assure supply reliability by enabling strategic management
- Keeps the team focused on continuously considering and assessing risk

42.4 IA Approach to Organization

IA shall be conducted by different types of tools such as usage of flowchart, questionnaires, risk control matrixes, policy, procedural manual, vertical audit, horizontal audit, sampling based or system based. Approach of IA shall depend on the risk, loop falls and types of non-conformance [11]. System based or risk base approach can be selected for IA as mostly useful tool to identify the gap in the organization.

> 1. Quality System
>
> 2. Facilities and Equipment System
>
> 3. Production System
>
> 4. Materials System
>
> 5. Lab Control System

Figure 2 The Quality Six Systems Model.

I. System based Approach

The systems based IA program have the ability to assess whether each of the systems is in a state of control. The quality system audits consist of individual inspection of Quality System, Facilities and Equipment System, Materials System, Production System, Packaging and Labeling System, Laboratory System to correct the non-conformity. (Refer Figure 2: The Quality system module). System base IA can also be performed on fragile system identified during past audit. System review is a key component in any healthy quality system to ensure its continuing suitability, adequacy, and effectiveness. Under a quality system, senior managers are expected to conduct reviews of the whole quality system according to a planned schedule. Such a review typically includes both an assessment of the product as well as customer needs.

The review should consider at least the following:

- The appropriateness of the quality policy and its objectives
- The results of audits and other assessments
- Customer feedback, including complaints
- The analysis of data trending results
- The status of actions to prevent potential problems or their recurrence
- Any follow-up actions from previous management reviews
- Any changes in business practices or environment that may affect the quality system
- Whether product characteristics meet the customer's needs
- Science-based approaches

System based audit during IA can be approach as;

A. *Full Inspection Option :* Quality System + NLT 3 other systems

- initial establishment inspection
- previous inspection findings, non compliances

- significant changes since last inspection (New technologies, equipment's , facilities
- Follow up to non-conformities
- Revert to an abbreviated option with district concurrence

B. Abbreviated Inspection Option : Quality System + NMT 2 other systems

- good history
- no major changes to operations
- no pattern of recalls and problems
- When not using the full inspection option
- Surveillance inspections
- Adequate for routine coverage.
- Rotate system with the Abbreviated option-District will monitor

II. Linked Based -Process Approach

Linked **based** internal audit approach depends on six-system inspection model includes how the cGMP and the concepts of modern quality systems are linked. The new six-system inspection model shows the quality system and the five manufacturing systems. Under the Quality Systems Model, the Agency recommends that senior managers ensure that the quality system they design and implement provides clear organizational guidance and facilitates the systematic evaluation of issues. Refer Figure 4: Linked bases Approach.

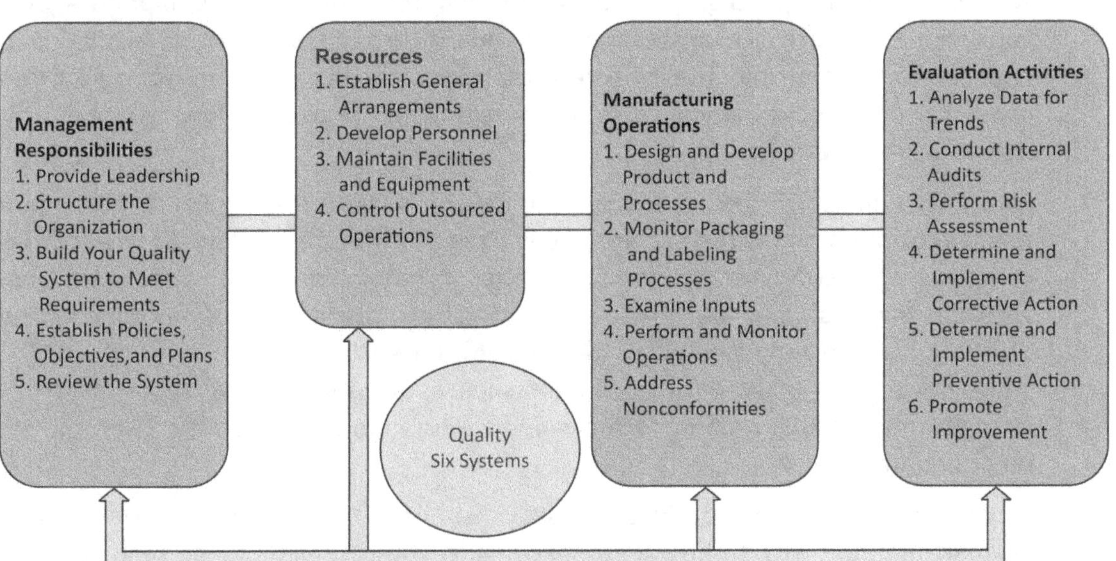

III. Risk Based Approach for IA program

New era for self-inspection by implementation of QRM. The concept of risk management is a major focus of the "Pharmaceutical cGMPs for the 21st Century" initiative. Risk management can guide the setting of specifications and process parameters. Risk assessment is also used in determining the need for discrepancy investigations and corrective actions and how to prioritize them. Adoption of risk assessment approach will promote innovation and continual improvement, and strengthen quality assurance and reliable supply of product. Risk-based approach can apply for assessing changes across the life cycle, identification of several gaps .IA with Quality risk management and knowledge management achieve product realization, maintain a state of control, and facilitate continual improvement. The purpose of risk based approach to IA program can be applied in two-fold as:

- To design IA as a quality risk management tool that can provide the objective evidence to the management about whether or not the current and potential risks to quality are effectively managed to acceptable levels. Management then can judge the effectiveness of process and functions within the quality system.

- Efficient inspection workload and resource management focusing on those areas within the quality system that present higher risk to the quality of medicinal product with aim of meeting the quality objective.

IV. Forensic Audit

Forensic audit is a process of investigation where or not fraud is happen against approved standards. Forensic Audits covers a broad spectrum of activities of organization. Main objective of forensic audit is to find out whether all system, process, data governance system meets the regulatory/procedure/policy/standards requirements.

Forensic audit shall be conducted to know data integrity in current good manufacturing practice (CGMP) for drugs. Forensic audit is conducted to verify the completeness, consistency, and accuracy of data. Complete, consistent, and accurate data should be attributable, legible, contemporaneously recorded, original or a true copy, and accurate (ALCOA). Forensic audit can be conducted on static documents (fixed-data document such as a paper record or an electronic image), and dynamic record (format allows interaction between the user and the record content).

Forensic audit many be conducted in ways, similar to the process of conducting a normal audit, but with some additional considerations shall be taken such as;

- Identifying the type of fraud that has been operating.
- how long it has been operating for,
- how the fraud was concealed for the duration

- identifying the fraudster(s) involved
- quantifying and gathering evidence
- Forensic audit can be done is two ways Reactive Forensic Audit and Proactive Forensic Audit
- **Reactive Forensic Auditing:** To investigate suspected fraud so as to prove the suspicions, and suspicions are proves by findings by evidence and present evidence in an acceptable format.
- **Proactive Forensic Auditing:** Forensic auditing in this sense could be viewed from different aspects depending on its application such as regulatory compliance

V. For-Cause-Audit

For-Cause-Audit is an audit conducted anything other than a routine internal audit. For cause audit is to investigate a specific problem such as Field Alert report, recall adverse events or other events. For-Cause-Audit is an in-depth examination of all components including but not limited to records, documents, interview etc.

Any above tools can be selected for internal audit to know the loop falls and take corrective and preventive action.

42.5 Planning

Risk management can be applied for IA planning The application of the risk management allows the estimation of the risk associated with areas within quality system and determine the scope, frequency, time, number of inspectors and allocation of inspectors to particular area based on SME, which help in risk-based inspection planning and better utilization of man power. During the risk assessment to IA schedule risk assessment process shall be done such as Figure.

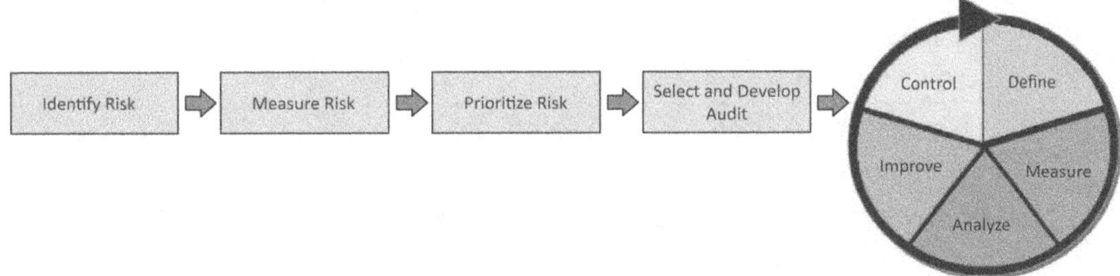

Quality risk Assessment. Planning of the IA shall be based on the prioritization and risk assed .Three primary methods can be selected for the annual audit plan.

- Cycle Approach
- Risk based approach
- Cycle-Based Risk Approach

Risk can be identified based on quality risk matrix such as the historical data of IAs, regulatory observation, and regulation guideline updating, warning letters or other factors as;

- Complexity of the site, manufacturing process, Batch Failure rate, Process Capability/Performance
- Critical areas
- Major changes in facility ,equipment & process
- Compliance status and history of unit
- Complaints and Recalls received
- the inherent risk of the drug manufactured,
- The inspection frequency and history of the establishment
- Laboratory failure investigation rate /Invalidated Out-of-Specification (OOS) Rate
- CAPA Effectiveness
- Warning letters
- Simple approach could be assigning a numerical descriptive value of 1 (low), 2(moderate) and 3(high) to established categories for the selection risk factors. The estimated values for risk factors could also be multiplied by significance-weighting factors to give a total. (Table 1- Approach to risk rating)

Table 1 Risk Rating level

Risk Rating level	Input from current inspection findings	Internal frequency	Approach
0	Serious triggering outside the inspection cycle	Immediate	Number of regulatory audits Results of previous IAs
1	Critical finding	Twice in year	Fragile system identified
2	> 6 major observation > 10 minor observation	Twice in year	
3	< 6 major observation < 10 minor observation	Yearly	
4	No Critical or Major finding or < 10 minor observation	Yearly	
5	No Critical or Major finding from current or previous inspection and < 6 other finding each.	Yearly	

Use of quality metrics will support to understanding of the inherent risk of manufacturing establishments and products. The collection of these data will help to direct our inspections. These metrics may provide a basis for IA to use improved risk based principles to determine the appropriate audit scheduling. Once all risk have been mapped to relevant audits, the audits are then ranked from highest to lowest base on audit score. The annual plan shall be chosen based on the percentage of "total risk". Understanding the top compliance risk is an important parameter in risk assessment audit scheduling. The compliance risk assessment will help the organization under standard the full range of its risks exposure, including likelihood that a risk event occur, the reason it may occur, and the potential severity of its impact. The risk associated can estimate by using different tools .Use of available tools is to understand processes, improve processes / products

42.6 Audit Sampling

Audit sampling plays important role to collect evidences to support observation and opinions. The type of sampling used and number of items selected should be based on the auditors understanding of the relative risk and exposures of the areas audited. Sampling involves some selection of certain sections of the data or population based on its quantitative or qualitative factors. Thus, the IA team must undertake statistically backed sample selection technique to ensure designing efficient samples, measuring sufficient evidence and evaluating results with objectivity.

Allowable Risk is calculated as: AR=IR x CR x DR
- AR= the allowable audit risk that a material misstatement might remain undetected
- IR= Inherent risk, the risk of a material misstatement in an assertion, assuming there were no related controls.
- CR= Control risk, the risk that a material misstatement that could occur in an assertion will not be prevented or detected on a timely basis by internal control.
- DR = Detection risk, the risk that the auditors' procedures will fail to detect a material misstatement if it exists.
- The sample size selected affects the level of sampling risk of the sample. Every increase in sample size reduces the sampling risk and allowance for sampling risk. The sample size is generally directly proportionate to the characteristics of the population and an increase in the population leads to an increase in the sample size. The different methodologies available for statistically based selection of sample size such as Random Sampling, Consecutive Sampling(snapshot: first or last), Consecutive Sampling (Systematic sampling), Haphazard Sampling (The Haphazard Sampling method means selecting items on an arbitrary basis, but without any),Block Sampling (Time period) and Stratification Sampling (dividing the population homogenously).

42.7 Audit Tools

Horizontal or vertical IA can be conducted to check the organization state of control. Vertical audits covers activities within given function or physical area, whereas, horizontal audit address a single contract, project or product Techniques such as interviewing, examine the records, observation, drawback and Pitfall can be used.

42.8 Audit Evidence and Analysis Phase

During IA sufficient, relevant, reliable and useful information shall be collected· Different types of evidence can be obtained during the course of IA such as physical, testimonial, documentary or analytical evidence to support audit observation. Sampling tools such as statically or probability, Attribute, Variable and judgment can be used during IA .For audit analysis root cause model for IA can be used.(Figure 4 :Root cause Model for IA).

42.9 Audit CLOSE-OUT and FOLLOW-UP

At the conclusion of the audit, there should be a close-out meeting to discuss the observations. Ensure that all parties have a clear understanding of the observations and any commitments to corrective actions. Timeframe should be provided for when the audit report will be made available and when responses to observations will be provided. Schedules for completing corrective actions should be developed. Depending on the observations made and the commitments to corrective actions, a follow-up audit may be necessary. If a decision is made to not conduct a follow-up audit, the corrective actions should be reviewed in any future audits. Documentation of the rationale for not conducting a follow-up audit can be useful for future audits; unless the audit procedure clearly defines when follow-up audits are or are not required. Preventive as well as corrective actions should be taken to address unfavorable observations. Preventive actions are focused on improving the quality system so that the same unfavorable observations are not repeated.

42.10 Good Laboratory Checklist for Internal Audit

Name of the Company	
Address of the Company	
Contact Person	
Type of Inspection	
Date(s) of Inspection	
Name of the Inspector	

GLP Checklist			
Sr. No.	**Principles of Good Laboratory Practice**	**Yes /No**	**Comments/Remarks**
1.	**TEST FACILITY ORGANIZATION AND PERSONNEL**		
1.1	*Test Facility Management's Responsibilities*		
1.1.1	Does the Test Facility Management ensure that Principles of Good Laboratory Practice are complied with, in its test facility?		
1.1.2 (a)	Does a statement exist to identify the individual(s) within a test facility who full fill the responsibilities of management as defined by these Principles of Good Laboratory Practice?		
(b)	Are there a sufficient number of Qualified personnel, appropriate facilities, equipment, and materials Available for the timely and proper conduct		
(c)	Does the management maintain proper record of the qualifications, training, experience and job description for each professional and technical individual?		
(d)	Does the management ensure that personnel clearly understand the functions they are to perform and, where necessary, provide training for these functions?		

Table *Contd...*

GLP Checklist			
Sr. No.	Principles of Good Laboratory Practice	Yes /No	Comments/Remarks
1.	**TEST FACILITY ORGANIZATION AND PERSONNEL**		
(e)	Are appropriate and technically valid Standard Operating Procedures (SOPs) established and followed?		
	Does the management approve all the original and revised SOPs?		
(f)	Is there a Quality Assurance Programme with designated personnel?		
	Does the management assure that the quality assurance responsibility is being performed in accordance with Principles of GLP?		
(g)	Does the management ensure that for each study, an individual with the appropriate qualifications, training, and experience is appointed as Study Director before the initiation of the study?		
	Does a procedure exist for replacement of a Study Director and is this procedure documented?		
(h)	Does the management ensure that in the event of a multi-site study, if needed, a Principal Investigator is designated who is appropriately trained, qualified and experienced to supervise the delegated phase(s) of the study?		
	Does a procedure exist for replacement of a Principal Investigator and is this procedure documented?		
(i)	Does the management ensure that there is a documented approval of the study plan by the Study Director?		
(j)	Does the management ensure that the Study Director makes the approved study plan available to the Quality Assurance personnel?		

Table Contd...

GLP Checklist			
Sr. No.	**Principles of Good Laboratory Practice**	**Yes /No**	**Comments/Remarks**
1.	**TEST FACILITY ORGANIZATION AND PERSONNEL**		
(k)	Does the management ensure the maintenance of a historical file of all SOPs?		
(l)	Does the management ensure that an individual is identified as responsible for the management of the archive(s)?		
(m)	Does the management ensure that a master schedule is maintained?		
(n)	Does the management ensure that test facility supplies meet requirements appropriate to their use in a study?		
(o)	Does the management ensure that for a multi-site study, clear lines of communication exist between the Study Director, Principal Investigator(s), the Quality Assurance Programme(s) and study personnel?		
(p)	Does the management ensure that test and reference items are appropriately characterized?		
(q)	Has the management established procedures to ensure that computerized systems are suitable for their intended purpose, and are validated, operated and maintained in accordance with the Principles of Good Laboratory Practice?		
1.1.3	When a phase(s) of a study is conducted at a test site, does the test site management have the responsibilies as defined above, with the exceptions of 1.1.2(g), (l), (j) and (o)?		
1.2	*Study Director's Responsibilities*		
1.2.2 (a)	Does the Study Director approve the study plan and any amendments to the plan by dated signature?		

GLP Checklist			
Sr. No.	**Principles of Good Laboratory Practice**	**Yes /No**	**Comments/Remarks**
1.	**TEST FACILITY ORGANIZATION AND PERSONNEL**		
(b)	Does the Study Director ensure that the Quality Assurance personnel have a copy of the study plan and any amendments in a timely manner?		
	Does the Study Director communicate effectively with the Quality Assurance personnel as required during the conduct of the study?		
(c)	Does the Study Director ensure that study plans and amendments and SOP's are available to study personnel?		
(d)	Does the Study Director ensure that the study plan and the final report for a multi-study identify and define the role of any Principal Investigator(s) and any test facilities and test sites involved in the conduct of the study?		
(e)	Does the Study Director ensure that the procedures specified in the study plan are followed?		
	Does the Study Director assess and document the impact of any deviations from the study plan on the quality and integrity of the study, and take appropriate corrective action if necessary?		
	Does the Study Director acknowledge deviations from the SOP's during the conduct of the study?		
(f)	Does the Study Director ensure that all raw data generated are fully documented and recorded?		
(g)	Does the Study Director ensure that computerized systems used in the study have been validated?		

Table Contd...

GLP Checklist			
Sr. No.	**Principles of Good Laboratory Practice**	**Yes /No**	**Comments/Remarks**
1.	**TEST FACILITY ORGANIZATION AND PERSONNEL**		
(h)	Does the Study Director sign and date the final report to indicate acceptance of responsibility for the validity of the data and to indicate the extent to which the study complies with the Principles of Good Laboratory Practice?		
(i)	Does the Study Director ensure that after completion (including termination) of the study, the study plan, the final report, raw data and supporting material are archived?		
1.3	*Principal Investigator's Responsibilities*		
1.3.1	Does the Principal Investigator ensure that the delegated phases of the study are conducted in accordance with the applicable Principles of Good Laboratory Practice?		
1.4	*Study Personnel's Responsibilities*		
1.4.1	Are all personnel involved in the conduct of the study knowledgeable in those parts of the Principles of Good Laboratory Practice, which are applicable to their involvement in the study?		
1.4.2	Do the Study Personnel have access to the study plan and appropriate SOP's applicable to their involvement in the study?		
	Do they comply with the instructions given in these documents?		
	Do they document and communicate all deviations from these instructions directly to the study director, and/or if appropriate, the Principal Investigator?		

Table *Contd...*

GLP Checklist			
Sr. No.	**Principles of Good Laboratory Practice**	**Yes /No**	**Comments/Remarks**
1.	**TEST FACILITY ORGANIZATION AND PERSONNEL**		
1.4.3	Do the Study Personnel record raw data promptly and accurately and in compliance with the Principles of Good Laboratory Practice?		
	Are they responsible for the quality of their data?		
1.4.4	Do the Study Personnel exercise health precautions to minimize risk to themselves and to ensure the integrity of the study?		
	Do they communicate to the appropriate person any relevant known health or medical condition in order that they can be excluded from operations that may affect the study?		
2	QUALITY ASSURANCE PROGRAMME		
2.1	*General*		
2.1.1	Does a documented Quality Assurance Programme exist in the test facility to ensure that studies performed are in compliance with the Principles of Good Laboratory Practice?		
2.1.2	Is the Quality Assurance Programme carried out by an individual(s) designated by, and directly responsible to management and who is familiar with the test procedures?		
2.1.3	Is this individual(s) free of involvement in the conduct of the study being assured?		
2.2	*Responsibilities of the Quality Assurance Personnel*		
2.2.1 (a)	Do the QA personnel maintain copies of all approved study plans and SOP's in use in the test facility and have access to an up-to-date copy of the master schedule?		

Table *Contd...*

GLP Checklist			
Sr. No.	**Principles of Good Laboratory Practice**	**Yes /No**	**Comments/Remarks**
1.	**TEST FACILITY ORGANIZATION AND PERSONNEL**		
(b)	Do the QA personnel verify that the study plan contains the information required for compliance with the Principles of Good Laboratory Practice, and is this verification documented?		
(c)	Do the QA personnel conduct inspections to determine if all studies are conducted in accordance with the Principles of Good Laboratory Practice?		
	Do they determine through inspections that study plans and SOP's have been made available to study personnel and are being followed?		
	Are records of such inspections (study-based inspectors, facility-based inspections, and process-based inspections) retained?		
(d)	Do the QA personnel inspect the final reports to confirm that the methods, procedures, and observations are accurately and completely described, and that the reported results accurately and completely reflect the raw data of the studies?		
(e)	Do the QA personnel promptly report any inspection results in writing to the management and to the Study Director, and to the Principal Investigator(s) and the respective management, when applicable?		

Table *Contd…*

GLP Checklist			
Sr. No.	**Principles of Good Laboratory Practice**	**Yes /No**	**Comments/Remarks**
1.	**TEST FACILITY ORGANIZATION AND PERSONNEL**		
(f)	Do the QA personnel prepare and sign a statement, to be included with the final report, which specifies types of inspections and their dates, including the phase(s) of the study inspected, and the dates inspection results were reported to management and the Study Director and Principal Investigator(s), if applicable? (This statement also serves to confirm that the final report reflects the raw data).		
3.	FACILITIES		
3.1	*General*		
3.1.1	Is the test facility of suitable size, construction and locations to meet the requirements of the study and to minimize disturbance that would interfere with the validity of the study?		
3.1.2	Does the design of the test facility provide an adequate degree of separation of the different activities to assure the proper conduct of the study?		
3.2	*Test system Facilities*		
3.2.1	Does the test facility have a sufficient number of rooms or areas to assure the isolation of test systems and the isolation of individual projects, involving substances or organisms known to be or suspected of being biohazardous?		
3.2.2	Are suitable rooms or areas available for the diagnosis, treatment and control of diseases, in order to ensure that there is no unacceptable degree of deterioration of test systems?		

Table *Contd...*

	GLP Checklist		
Sr. No.	**Principles of Good Laboratory Practice**	**Yes /No**	**Comments/Remarks**
1.	**TEST FACILITY ORGANIZATION AND PERSONNEL**		
3.2.3	Are there storage rooms or areas as needed for supplies and equipment?		
	Are these separated from rooms or areas housing the test systems and do they provide adequate protection against infestation, contamination and/or deterioration?		
3.3	*Facilities for Handling Test and Reference Items*		
3.3.1	To prevent contamination or mix-ups, are there separate rooms or areas for receipt and storage of the test and reference items, and for the mixing of the test items with a vehicle?		
3.3.2	Are the storage rooms or areas separate from rooms or areas containing the test systems? Are they adequate to preserve identity, concentration, purity and stability, and to ensure safe storage for hazardous substances?		
3.4	*Archive Facilities*		
	Are archive facilities provided for the secure storage and retrieval of study plans, raw data, final reports, samples of test items and specimens? Does the archives design and conditions protect the contents from untimely deterioration?		
3.5	*Waste Disposal*		
	Is the handling and disposal of wastes carried out in such a way that the integrity of studies is not jeopardized? (This includes provision for appropriate collection, storage and disposal facilities, and decontamination and transportation procedures)		

Table *Contd...*

GLP Checklist			
Sr. No.	**Principles of Good Laboratory Practice**	**Yes /No**	**Comments/Remarks**
1.	**TEST FACILITY ORGANIZATION AND PERSONNEL**		
4.	APPARATUS, MATERIALS AND REAGENTS		
4.1	Is the apparatus (including validated computerized systems) used for the generation, storage, and retrieval of data, and for controlling environmental factors relevant to the study, suitably located and of appropriate design and adequate capacity?		
4.2	Is the apparatus used in a study periodically inspected, cleaned, maintained and calibrated according to SOP's?		
	Are records of these activities maintained?		
	Is calibration, where appropriate, traceable to national or international standards of measurement?		
4.3	Do the apparatus and materials used in a study interfere adversely with the test systems?		
4.4	Are chemicals, reagents and solutions labeled to indicate identity (with concentration if appropriate), expiry date and specific storage instructions?		
	Is information concerning source, preparation date and stability available? (The expiry date may be extended on the basis of documented evaluation or analysis)		
5.	TEST SYSTEMS		
5.1	*Physical/Chemical*		
5.1.1	Is the apparatus used for the generation of physical/chemical data suitably located and of appropriate design and adequate capacity?		

GLP Checklist			
Sr. No.	**Principles of Good Laboratory Practice**	**Yes /No**	**Comments/Remarks**
1.	**TEST FACILITY ORGANIZATION AND PERSONNEL**		
5.1.2	Is the integrity of the physical/chemical test systems ensured?		
5.2	*Biological*		
5.2.1	Are proper conditions established and maintained for the storage, housing, handling and care of biological test systems, in order to ensure the quality of the data?		
5.2.2	Are newly received animal and plant test systems isolated until their health status has been evaluated?		
	If any unusual mortality or morbidity occurs, is this lot excluded from use in the studies, and, when appropriate, humanely destroyed?		
	At the experimental starting date of a study, are test systems checked to ensure that they are free of any disease or condition that might interfere with the purpose or conduct of the study?		
	Are test systems that become diseased or injured during the course of the study isolated and treated, if necessary, to maintain the integrity of the study?		
	Is all diagnosis and treatment of any disease before or during a study recorded?		
5.2.3	Are records maintained of source, date of arrival, and arrival conditions of test systems?		
5.2.4	Are biological test systems acclimatized to the test environment for an adequate period before the first administration or application of the test or reference item?		

Table *Contd...*

GLP Checklist			
Sr. No.	**Principles of Good Laboratory Practice**	**Yes /No**	**Comments/Remarks**
1.	**TEST FACILITY ORGANIZATION AND PERSONNEL**		
5.2.5	Does all the information needed to properly identify the test systems appear on their housing or containers?		
	Is appropriate identification given, wherever possible, to individual test systems that are to be removed from their housing or containers during the conduct of the study?		
5.2.6	During use, is the housing or containers for test systems cleaned and sanitized at appropriate intervals?		
	Is any material that comes into contact with the test system checked for being free of contaminants at levels that would interfere with the study?		
	Is bedding for animals changed as required by sound husbandry practice?		
	Is the use of pest control agents documented?		
5.2.7	Are test systems used in field studies located so as to avoid interference in the study from spray drift and from past usage of pesticides?		
6.	TEST AND REFERENCE ITEMS		
6.1	*Receipt, Handling, Sampling and Storage*		
6.1.1	Are records maintained for test item and reference item characterization, date of receipt, expiry date, and quantities received and used in studies?		
6.1.2	Are handling, sampling and storage procedures identified in order that the homogeneity and stability are assured to the degree possible and contamination or mix-up is precluded?		

Table *Contd...*

GLP Checklist			
Sr. No.	Principles of Good Laboratory Practice	Yes /No	Comments/Remarks
1.	**TEST FACILITY ORGANIZATION AND PERSONNEL**		
6.1.3	Do storage container(s) carry identification information, expiry date, and specific storage instructions?		
6.2	*Characterizations*		
6.2.1	Is each test and reference system appropriately identified e.g. code, CAS number (Chemical Abstract Service Registry Number), name, and biological parameters?		
6.2.2	For each study, is the identity, including batch number, purity, composition, concentrations, or other characteristics to appropriately define each batch of the test or reference items, known?		
6.2.3	In cases where the sponsor supplies the test item, is there a mechanism developed in cooperation between the sponsor and the test facility, to verify the identity of the test item subject to the study?		
6.2.4	Is the stability of test and reference items under storage and test conditions known for all studies?		
6.2.5	If the test item is administered or applied in a vehicle, is the homogeneity, concentration and stability of the test item in that vehicle determined? (For test items used in field studies (e.g. tank mixes), these may be determined through separate laboratory experiments)		
6.2.6	Is a sample for analytical purposes taken from each batch of test items retained for all studies except short-term studies?		

Table *Contd...*

<table>
<tr><td colspan="4" align="center">GLP Checklist</td></tr>
<tr><td>Sr. No.</td><td>Principles of Good Laboratory Practice</td><td>Yes /No</td><td>Comments/Remarks</td></tr>
<tr><td align="center">1.</td><td align="center">TEST FACILITY ORGANIZATION
AND PERSONNEL</td><td></td><td></td></tr>
<tr><td align="center">7.</td><td>STANDARD OPERATING PROCEDURES</td><td></td><td></td></tr>
<tr><td align="center">7.1</td><td>Does the test facility have management approved written SOP's intended to ensure the quality and integrity of the data generated by that test facility?</td><td></td><td></td></tr>
<tr><td></td><td>Are revisions to SOPs approved by the test facility management?</td><td></td><td></td></tr>
<tr><td align="center">7.2</td><td>Does each separate test facility unit or area have current SOP's relevant to the activities being performed therein available immediately? (Published text books, analytical methods, articles and manuals may be used as supplements to these SOPs)</td><td></td><td></td></tr>
<tr><td align="center">7.3</td><td>Are deviations from SOPs related to the study documented, and acknowledged by the Study Director and the Principal Investigator(s), as applicable?</td><td></td><td></td></tr>
<tr><td align="center">7.4</td><td>Are SOP's available for, but not limited to, the following categories of test facility activities (the details given under each heading are to be considered as illustrative examples?):</td><td></td><td></td></tr>
<tr><td align="center">7.4.1</td><td>Test and Reference Items</td><td></td><td></td></tr>
<tr><td></td><td>Receipt, identification, labelling, handling, sampling and storage.</td><td></td><td></td></tr>
<tr><td align="center">7.4.2 (a)</td><td>Apparatus, Materials and Reagents
Apparatus: Use, maintenance, cleaning and calibration</td><td></td><td></td></tr>
<tr><td align="center">(b)</td><td>Computerized systems: Validation, operation, maintenance, security, changes control and back up.</td><td></td><td></td></tr>
<tr><td align="center">(c)</td><td>Materials, Reagents and Solutions Preparation and labelling.</td><td></td><td></td></tr>
</table>

Table Contd…

GLP Checklist			
Sr. No.	**Principles of Good Laboratory Practice**	**Yes /No**	**Comments/Remarks**
1.	**TEST FACILITY ORGANIZATION AND PERSONNEL**		
7.4.3	*Record-keeping, Reporting, Storage and Retrieval:* Coding of studies, data collection, preparation of reports, indexing systems, handling of data including the use of computerized systems.		
7.4.4	*Test System* (where appropriate)		
(a)	Room preparations and environmental room conditions for the test system.		
(b)	Procedures for receipt, transfer, proper placement, characterization, identification and care of the test system.		
(c)	Test system preparation, observations and examinations, before, during, and at the conclusion of the study.		
(d)	Handling of test system individuals found moribund or dead during the study.		
(e)	Collection, identification and handling of specimens including necropsy and histopathology.		
(f)	Siting and placement of test systems in test plots.		
7.4.5	Quality Assurance Procedures Operation of Quality Assurance personnel in planning, scheduling, performing, documenting and reporting inspections.		
8.	PERFORMANCE OF THE STUDY		
8.1	*Study Plan*		
8.1.1	Does a written plan exist prior to the initiation of each study? Is the study plan approved by dated signature of the Study Director and verified for GLP compliance by QA personnel as specified in Section 2.2.1(b)? Is the study plan approved by the test facility management and the		

	GLP Checklist		
Sr. No.	**Principles of Good Laboratory Practice**	**Yes /No**	**Comments/Remarks**
1.	**TEST FACILITY ORGANIZATION AND PERSONNEL**		
	sponsor, if required by national regulation or legislation in the country where the study is being performed?		
8.1.2 (a)	Are amendments to the study plan justified and approved by dated signatures of the Study Director and maintained with the study plan?		
(b)	Are deviations from the study plan described, explained, acknowledged and dated in a timely fashion by the Study Director and/or Principal Investigator(s) and maintained with the study raw data?		
8.1.3	For short-term studies, is the general study plan accompanied by a study specific supplement? (Optional)		
8.2	*Content of the Study Plan* Does the study plan contain at least the following information?		
8.2.1	Identification of the Study, the Test Item and Reference Item		
(a)	A descriptive title		
(b)	A statement, which reveals the nature and purpose of the study.		
(c)	Identification of the test item by code or name (IUPAC; CAS number, biological parameters, etc.)		
(d)	The reference item to be used.		
8.2.2	Information Concerning the Sponsor and Test Facility		
(a)	Name and address of the sponsor.		
(b)	Name and address of any test facilities and test sites involved.		
(c)	Name and address of the Study Director.		

Table *Contd...*

GLP Checklist			
Sr. No.	**Principles of Good Laboratory Practice**	**Yes /No**	**Comments/Remarks**
1.	**TEST FACILITY ORGANIZATION AND PERSONNEL**		
(d)	Name and address of the Principal Investigator(s) and the phases of the study delegated by the Study Director and under the responsibility of the Principal Investigator(s).		
8.2.3	Dates		
(a)	The date of approval of the study plan by signature of the Study Director.		
(b)	The proposed experimental starting and completion dates.		
8.2.4	Test Methods Reference to the OECD Guideline or other test guideline or method to be used.		
8.2.5	Issues (where applicable)		
(a)	The justification for selection of the test system.		
(b)	Characterization of the test system, such as the species, strain, substrain, source of supply, number, body weight range, sex, age, and other pertinent information.		
(c)	The method of administration and the reason for its choice.		
(d)	The dose levels and/or concentration(s), frequency, and duration of administration/application.		
(e)	Detailed information on the experimental design, including a description of the chronological procedure of the study, all methods, materials and conditions, type and frequency of analysis, measurements, observations and examinations to be performed, and statistical methods to be used (if any).		
8.2.6	Records A list of records to be retained		

Table Contd...

GLP Checklist			
Sr. No.	**Principles of Good Laboratory Practice**	**Yes /No**	**Comments/Remarks**
1.	**TEST FACILITY ORGANIZATION AND PERSONNEL**		
8.3	*Conduct of the Study*		
8.3.1	Is a unique identification given to each study, and do all items concerning that study carry this identification?		
	Are specimens from the study identified to confirm their origin? (Such identification should enable traceability, as appropriate for the specimen and study).		
8.3.2	Is the study conducted in accordance with the study plan?		
8.3.3	Are all data generated during the conduct of the study recorded directly, promptly, accurately, and legibly by the individual entering the data, and have all entries been signed or initialed and dated?		
8.3.4	Are any changes in the raw data made so as not to obscure the previous entry and have the reasons for the changes been indicated, and dated and signed?		
8.3.5	Is data generated as a direct computer input identified at the time of data input by the individual(s) responsible for direct data entries?		
	Does computerized system design provide for the retention of full audit trails to show all changes to the data without obscuring the original data? (It should be possible to associate all changes to data with the persons having made these changes, for example, by use of timed and dated (electronic) signatures).		
	Are the reasons for the changes given?		

Table *Contd...*

GLP Checklist			
Sr. No.	**Principles of Good Laboratory Practice**	**Yes /No**	**Comments/Remarks**
1.	**TEST FACILITY ORGANIZATION AND PERSONNEL**		
9.	REPORTING OF STUDY RESULTS		
9.1	*General*		
9.1.1	Is a final report prepared for each study? (In the case of short-term studies, a standardized final report accompanied by a study specific extension may be prepared).		
9.1.2	Are reports of Principal Investigators or scientists involved in the study signed and dated by them?		
9.1.3	Is the final report signed and dated by the Study Director to indicate acceptance of responsibility for the validity of the data?		
	Is the extent of compliance with GLP principles indicated?		
9.1.4	Are corrections and additions to a final report made in the form of amendments?		
	Do such amendments clearly specify the reason(s) for the corrections or additions and are they signed and dated by the Study Director?		
9.1.5	(Reformatting of the final report to comply with the submission requirements of a national registration or regulatory authority does not constitute a correction, addition or amendment to the final report).		
9.2	*Content of the Final Report*		
	Does the final report contain at least the following information:		
9.2.1	Identification of the Study, the Test Item and Reference Item		
(a)	A descriptive title.		
(b)	Identification of the test item by code or name (IUPAC, CAS number, biological parameters, etc).		

Table Contd…

GLP Checklist			
Sr. No.	**Principles of Good Laboratory Practice**	**Yes /No**	**Comments/Remarks**
1.	**TEST FACILITY ORGANIZATION AND PERSONNEL**		
(c)	Identification of the reference item by name.		
(d)	Characterization of the test item, including purity, stability and homogeneity.		
9.2.2	Information concerning the Sponsor and the Test Facility		
(a)	Name and address of the sponsor..		
(b)	Name and address of any test facilities and test sites involved.		
(c)	Name and address of the Study Director.		
(d)	Name and address of the Principal Investigator(s) and the phase(s) of the study delegated, if applicable.		
(e)	Name and address of scientists having contributed reports to the final report		
9.2.3	Dates Experimental starting and completion dates.		
9.2.4	Statement A QA Programme statement listing the types of inspections made and their dates, including the phase(s) inspected, and the dates any inspection results were reported to management and to the Study Director and Principal Investigator(s), if applicable. (This statement would also serve to confirm that the final report reflects the raw data).		
9.2.5	Description of Materials and Test Methods		
(a)	Description of methods and materials used.		
(b)	Reference to OECD Test Guideline or other test guideline or method.		

Table *Contd...*

GLP Checklist			
Sr. No.	**Principles of Good Laboratory Practice**	**Yes /No**	**Comments/Remarks**
1.	**TEST FACILITY ORGANIZATION AND PERSONNEL**		
9.2.6	Results		
(a)	A summary of results.		
(b)	All information and data required by the study plan.		
(c)	A presentation of the results, including calculations and determinations of statistical significance.		
(d)	An evaluation and discussion of the results, and, where appropriate, conclusions.		
9.2.7	Storage The location(s) where the study plan, samples of test and reference items, specimens, raw data and the final report are to be stored.		
10.	Storage and retention of records and materials		
10.1	Have the following been retained in the archives?		
(a)	The study plan, raw data, samples of test and reference items, specimens, and the final report of each study.		
(b)	Records of all inspections performed by the QA Programme, as well as master schedules.		
(c)	Records of qualifications, training, experience and job descriptions of personnel.		
(d)	Records and reports of the maintenance and calibration of apparatus.		
(e)	Validation documentation for computerized systems.		
(f)	The historical file of all SOP's.		
(g)	Environmental monitoring records.		

Table *Contd...*

GLP Checklist			
Sr. No.	**Principles of Good Laboratory Practice**	**Yes /No**	**Comments/Remarks**
1.	**TEST FACILITY ORGANIZATION AND PERSONNEL**		
10.2	Is the material contained in the archives indexed so as to facilitate orderly storage and retrieval?		
10.3	Is access to the archives limited to only personnel authorized by the management?		
	Is movement of material in and out of the archives properly recorded?		
10.4	Is there a documented procedure for the transfer of material stored in the test facility archives to the archives of the sponsor(s) of the study(s), in case the test facility go out of business and has no legal successor?		

Chapter - 43

Laboratory Environment Condition and Monitoring

Introduction

The quality control (QC) laboratory plays a critical role in pharmaceutical production, for both in-process and finished product testing. Environmental monitoring is designed to understand the natural environment and protect it from any negative outcomes of human activity. The laboratory shall have space allocated for the performance of its work that is designed to ensure the quality, safety and efficacy of the service provided to the users and the health and safety of laboratory personnel, patients and visitors. The laboratory shall evaluate and determine the sufficiency and adequacy of the space allocated for the performance of the work.

Appropriate an environmental monitoring programme should be in place which covers, for example, use of air settlement plates and surface swabbing, temperature and pressure differentials. Alert and action limits should be defined. Trending of environmental monitoring results shall be carried out.

Environmental Monitoring: (E/M) is a program designed to demonstrate the control of viable (living microorganisms) and non-viable particles in critical areas. It includes the **monitoring** of personnel, air and area surfaces for microbial contamination

Establish an Environmental Monitoring Programme is important:
- Identifying weaknesses in contamination control systems
- Identifying locations which will provide "early warning" signals of loss of control
- Preparing useful environmental monitoring SOPs
- Keeping manageable records
- Temperature and humidity monitoring

43.1 Laboratory Environmental Conditions

The laboratory must have adequate space that is properly organized so that the quality of work and the safety of staff, patients, customers and visitors is not compromised. Measures must be taken to ensure good housekeeping (general tidiness, cleanliness, hygiene, freedom from rodents and insects), and maintain all work areas well. Laboratory section leaders should arrange equipment and work stations to ensure efficient and convenient workflow.

- The laboratory must have an appropriate bio safety environment and facilities to safely handle microorganisms belonging to different bio risk levels as per the mandate of the laboratory.

- Laboratories must be provided with appropriate utilities including clean running water, lighting (natural and artificial), ventilation, electric outlets, back-up power (if required), drainage systems that comply with environmental regulations, and sanitation facilities for patients and staff.

- Where primary sample collection is carried out, consideration must be given to patient access (including patients with disabilities), comfort and privacy. Separate rooms should be available for sample collection and blood donor activities.

- Potentially hazardous activities must be carried out in a separate area to prevent cross-contamination and reduce potential safety risks to all staff and visitors. Examples include: TB bacteriology, handling and examination of high-risk samples, nucleic acid amplifications, and controlled environments for large computer systems and some high-capacity analysers.

- Adequate storage space with the right conditions, including refrigerators and freezers, must be available and protection from light, damp, dust, insects and vermin ensured to maintain the integrity of samples, slides, histology blocks, histology samples, retained microorganisms, documents, manuals, equipment, reagents and other supplies, records and results. Storage areas must be adequately secured to prevent unauthorized access.

- Disposal of all infectious waste including sharps must be managed safely and effectively according to waste management regulations. The laboratory must use separate waste disposal systems for infectious and non-infectious waste.

43.2 Laboratory Temperature and Humidity Recording Program

Laboratory trained person shall ensure that all the air conditioners of the respective areas switch "ON". Laboratory person calibration date of instrument. Laboratory person shall note down the temp & humidity twice in a day. The specific time for noting temperature shall be 8.00 to 9.30 am & 12.00 to 13.30 Hrs. If any abnormalities are observed during the recording, laboratory person shall inform to Head. Record the abnormalities investigate and take appropriate action.

Temperature and Relative Humidity by Digital hygrometer

- Record the previous day minimum and maximum Temperature and Relative Humidity.
- Recording of previous day minimum and maximum Temperature & Relative Humidity shall be performed at the time of observing daily initial reading.
- Record the minimum and maximum Temperature and Relative Humidity by pressing "Memory" key. Press one time memory key to record max. Temperature & max. Relative Humidity of previous day and again press memory key to record min. Temperature & min. Relative Humidity.
- Delete the minimum and maximum reading from the memory of hygrometer that is required to leave memory space to record the next day min. and max. reading.
- For resetting the minimum & maximum value, press "RESET" button provided back side of the digital hygrometer with the help of pin.
- Temperature and relative humidity is important from the point of view of-
- To avoid degradation of products.
- To ensure comfortable and hygienic working of workmen as per GMP norms.
- To avoid damage to hard gelatine capsule shells.
- Ensure the thermometer/ temperature indicator is calibrated.
- Room Temperature: The temperature prevailing in a working area.
- Cool: The temperature between 8°C to 25°C.
- Cold: The temperature between 8°C to 15°C.
- Deep Freezer: The temperature between 2°C to 8°C.
- Humidity: NMT 55 ±5%

43.3 Monitoring of Differential Pressure in Different Areas

- Enter in the respective area and ensure that all doors are tightly closed.
- Ensure that the HVAC is in running condition.
- Open the door of respective area and ensure zero (0) set, while door is opened.
- Take the magnehelic gauge reading in the reference of positive and negative pressure of rooms.
- Record the value.

43.4 Microbiological Environmental Monitoring

Microbiological environmental monitoring involves the collection of data relating to the numbers of microorganisms present in a clean room or clean zone. These microorganisms are recovered from surfaces, air, and people. Nonviable particle counting, a physical test, is often included within the program because this function has often resided with the microbiology department to perform and due to the theoretical relationship between high numbers of nonviable particles and viable counts.

The main aim of microbiological environmental monitoring is to assess the monitoring of trends over time and the detection of an upward or downward movement, within clean areas.

The viable count aspect of environmental monitoring consists enumerating the numbers of microorganisms present in a clean room by collection results by using the following sample types:

- passive air-sampling: settle plates;
- active air-sampling: volumetric air-sampler;
- surface samples: contact (RODAC) plates;
- surface samples: swabs;
- finger plates;
- plates of sleeves/gowns

Microbiological environmental monitoring is a key part of the assessment of pharmaceutical manufacturing facilities. Environmental monitoring data indicates if clean rooms are operating correctly. The microbiologist must be familiar with parameters include temperature, humidity, pressure, room air changes, and air-flow pattern. Environmental monitoring is divided into viable and nonviable monitoring. Viable monitoring is the examination of microorganisms (bacteria and fungi) located within the manufacturing environment. The most important element for a microbiologist to establish is the overall monitoring program. This involves the consideration of several factors, which include:

- The monitoring methods (to assess air, surface, and personnel contamination);
- Culture media (such as general purpose media or fungal specific media) and whether a disinfectant neutralizer is required;
- Incubation times and temperatures;
- Sampling procedures;
- Sample locations within the clean room;
- Frequency of monitoring;
- Room conditions for the sampling (at rest or in operation);
- Who undertakes the monitoring (production or quality control (QC)) staff;
- Establishing alert and action levels;
- Guidance on dealing with out of limits results;
- Characterization of the micro flora.

These various considerations should be incorporated into a policy and into a plan. From this, local standard operating procedures (SOPs) should be generated. Environmental monitoring programs should adapt to changes to the pharmaceutical manufacturing environment and, therefore, the program should be regularly reviewed.

There are inherent weaknesses with environmental monitoring programs. Environmental monitoring data only provides an indication of the background environment of a clean room at the time it as used. A single result may or may not be representative of the longer-term room conditions. Furthermore, the methods are very insensitive, and they cannot be "validated" in the way that an analytical method can be. In addition, the frequency of samples and locations in which the samples are taken may or may not indicate the actual level of contamination risk, and there is often no direct correlation between number of organisms found and product contamination risk

43.5 Air and Surface Environment Monitoring

Air sampling using a "spore trap" sampler. Considerable information can be obtained about the environment from which the sample was taken. For example, determining the types and concentrations of mold spores and other particles in the sample.

Surface sampling using a "tape-lift" method. This method can determine what molds are growing on a surface, if mold is present at sub-visible concentrations, or if mold spores have been settling out of the air in an unusual quantity.

The main purpose of air and surface mold sampling as part of our assessment is to determine things that the naked eye is *not* capable of discerning. Among them, whether...

- mold particles or spores have been released from areas of visible growth into the general environment so that occupants are being exposed.
- mold growth is present at sub-visible concentrations. This is especially important in cases where the source of moisture was elevated humidity rather than liquid water.
- mold contaminants have spread beyond areas known to be affected into adjacent spaces. Also, whether it appears they have been entrained through the air conditioning systems.
- mold settling on surfaces indicates an ongoing issue. This is especially useful because air sampling provides only a "snapshot in time" of mold concentrations. Surface samples from horizontal surfaces often provide some indication of the extent to which mold contaminants have been released into the air over an extended time.

43.6 Difference between Viable Particle and Non-Viable Particle

A. Viable Monitoring or Viable Particle Monitoring

- Capable of working successfully.
- Living microorganism.
- Able to germinate.
- Capable of surviving or living successfully
- A viable particle is a particle that contains one or more living microorganisms. These can affect the sterility of the pharmaceutical product and generally range from ~0.2µm to ~30µm in size
- Viable particles include bacteria, Fungi.
- Different types of monitoring method used for viable particle detecting, such like.
    (a) Passive Air Sampling (settle plate method)
    (b) Active Air Sampling (Volumetric Air Sampling)
    (c) Surface monitoring (swab sampling or Contact method)
    (d) Personnel monitoring (Contact method)
- Viable particles are not free floating they need a carrier to float from one area to another area.
    (a) Non-viable particles main source for movement of the viable particles
    (b) Viable particles (Bacteria and fungi) 0.5um to 5.0 um size. So this range of particles are main source of product contamination
    (c) Less than 0.5 um is critical of attachment of viable particle.
    (d) Greater than 5.0 um is large size of particle cannot move easily and settle down.

B. Non-Viable Particles Monitoring

- Does not contain living organism.
- Non-feasible (particle)
- Unable to survive independently.
- Manufacturing area, equipment, utilities and personnel are main source of particle contamination in aseptic area.
- Check the particles of 0.5 um to 5.0 size
- Electronic particle counter is used for particle counting.
- This instrument can't differentiate between viable and non-viable particles.
- Any particle detected by the sensor of this instrument it will count this particles.

- For example particles are generated through the movement in aseptic area so movement of the personnel should be controlled and rhythmic in aseptic area. Good manufacturing practices should be followed to control the particle contamination in aseptic area.

- As per EU GMP 1 m³ air should be sampled per location for particle monitoring. Particle contamination depends upon the activity in aseptic area. If there is no activity in aseptic area then particle contamination would be low and if activity is in progress in aseptic area then particle contamination would be increased. That's why guidelines has defined different limits for at rest and in operation conditions.

- In pharmaceuticals, online and continuous monitoring of particles are also performed in which particle counter gives results in form of print out during production hours. This is very important tool to have great control over the aseptic area during production hours.

- During monitoring any deviation from the specified limits could results in investigation. Below is the chart showing maximum number of permitted particles in different classes of area as per EU GMP guideline.

- So, to control classified area we need to control both viable and non-viable particles

Chapter - 44

Laboratory Quality Standards Management

Introduction

Establishing and maintaining laboratory quality standards are essential. They are important for several reasons, including ensuring the accuracy of test results, increase the confidence of patients, management the quality and traceability of results, procuring equipment; use of standard techniques and reagents; for sharing documentation; training programmes; quality assurance; meeting requirements for reimbursement for national insurance schemes; and compliance with national or international accreditation and licensing systems.

The internationally accepted standards of the International Standards Organization (ISO) 15189 and ISO 17025 have stringent requirements and can usually be met only by laboratories either at the national level or by specialized reference laboratories

44.1 International Laboratory Management Quality Standards

There are several internationally accepted standards applicable to laboratories (Table 1) and many of these have been developed by ISO. Standards ensure desirable characteristics of products and services such as quality, safety, reliability, efficiency and reproducibility. ISO standards provide a technical base for health, safety and conformity assessment. They are accepted everywhere in the world, are voluntary and ISO has no legal authority to enforce the implementation of its standards. For the health services, there is the ISO 9000 series, which relates to administrative procedures and, in 2004, ISO 15189 was officially launched. It includes both management as well as technical requirements for medical laboratories. Table 1 shows the different international standards applicable to laboratories.

Table 1 International standard applicable to laboratories

1.	ISO/IEC 17025	General requirements for the competence of testing and calibration laboratories
2.	ISO 15189	Medical laboratories – particular requirements for quality and competence
3.	ISO/IEC 17043	Conformity assessment – general requirements for proficiency testing
4.	ISO 13528	Statistical methods for use in proficiency testing by Inter laboratory comparison
5.	OECD GLP	OECD principles on good laboratory practice
6.	ISO Guide 34	General requirement for the competence of reference material producers
7.	ISO 8402	Quality management and quality assurance –vocabulary
8.	ISO 19011	Guidelines for quality and/or environmental management system auditing
9.	ISO 9001	Quality management systems – requirements

44.2 Steps Involved in Laboratory Quality Management System

Quality standards are an integral part of the quality system. They are designed to help laboratories meet regulatory requirements. A quality system can be developed in a step-wise manner as shown in Table 2.

Table 2 Development of the laboratory Quality standards

Quality policy	→	Mission statement
Quality plan	→	Implementation of policy
	→	
	→	
Quality Environment	→	Top to bottom management
Procedure	→	Development and application of SOPs
	→	
	→	
Morning andevaluation	→	Assessment of quality and correction
Life cycle management	→	Assessment start to retirement of process, system or equipment

44.3 ISO 15189 Laboratory Standards

While developing these Laboratory Quality Standards, an effort has been made to retain the essential components of the ISO 15189 laboratory standards.

These are divided into eleven sections:

1. Organization and management
2. Quality management system
3. Human resources (personnel)
4. Accommodation and environmental conditions
5. Laboratory safety
6. Laboratory equipment
7. Procurement and supplies management
8. Information management
9. Managing laboratory specimens
10. Customer service and resolution of complaints
11. Outbreak alert and laboratory network

44.4 Steps Involved in Implementing the Quality Standard in Laboratory

The steps will be of help to policy-makers as well as regulators in developing laboratory quality standards. It provides a simple approach to meet the minimum requirements set, and it is hoped that eventually laboratories can improve their systems and aspire to meet ISO 15189 in a logical and step-by-step manner.

Once laboratory quality standards have been developed, they are required to be approved by the appropriate authority. An implementation plan then needs to be drawn up with short-, medium- and long-term objectives and activities, implementing partners identified, and the necessary budgetary support provided.

The purpose of establishing laboratory quality standards is to ensure the accuracy of test results, increase the confidence of patients, clinicians and communities in the value of laboratory testing, and to inform patient management.

The laboratory head will need to take a leadership role and involve all staff in the process. Some changes are easy to implement and cost little, such as reorganization; other changes require moderate inputs and funding; and yet other changes are more expensive or more difficult to implement.

Start by making simple and easy-to-implement changes, for example:

- Introduce SOPs for particular procedures or activities one by one. This could be sample collection, including phlebotomy, or an SOP for the examination of a particular analyte.

- Make arrangements to conduct regular meetings with users of the service. This will have the benefit of keeping users informed of the efforts being made to improve the quality of the laboratory service.
- A checklist (Annex 1) may be used to establish a baseline of implementation of laboratory quality standards as well as monitoring the progress made.

44.5 Laboratory Management Error

All laboratory activities may be subject to errors, and studies have shown that errors in the laboratory can occur in all the phases of diagnostic procedures. Examples of errors that can occur in each phase are given below:

A. Pre-analytical Management phase
- Incorrect test request or test selection
- Incomplete laboratory request forms
- Incorrect specimen collection, labelling and transportation

B. Analytical Management phase
- Use of faulty equipment, improper use of equipment
- Use of substandard or expired reagents
- Incorrect reagent preparation and storage
- Incorrect technical procedures; non-adherence to standard operating procedures (SOPs) or internal quality control (IQC)

C. Post- analytical Management phase
- Inaccurate reporting and recording
- Inaccurate calculations, computation or transcription
- Return of results to the clinician too late to influence patient management
- Incorrect interpretation of results

44.6 Laboratory Management Phases

A. Pre-analytical Management phase

The quality of the final result is profoundly affected by the quality of the sample material. Failure to take steps to assure quality at this stage will result in poor-quality results from the laboratory, despite good quality analytical processes.

- Proper request forms must be used and contain information to correctly identify the source of the sample and the authorized person requesting the test. Request forms must contain the following:

(a) Identification,

   (b) Location/source of specimens;
   (c) Identity of the requesting person;
   (d) Type of sample;
   (e) Examinations (tests) required;
   (f) Time and date sample taken.

- Proper management of samples during collection, transport and storage must be ensured according to SOPs, which must address:
   (a) Collection of timed samples;
   (b) Type of sample container to be used for various laboratory tests;
   (c) volume of sample required; any special/necessary additives
   (d) Sample collection technique;
   (e) Correct labelling;
   (f) Special transport arrangements between the site of primary sampling and the laboratory;
   (g) Safe disposal of materials used to collect primary samples;
   (h) Procedures to be followed if the sample quality is suboptimal
   (i) or unsatisfactory;
   (j) Recording unusual physical characteristics

- All primary samples must be given a unique identifying number recorded with the date and time of receipt. Taking aliquots of samples and sub-sampling should be done using appropriate laboratory safety precautions.

- The laboratory must provide instructions and monitor transportation of samples to the laboratory within the correct time frame, at the correct temperature, and in the designated preservatives for the requested analysis to be performed.

- Triple packing should be used when shipping toxic or infectious substances or dangerous goods, according to current International Air Transport Association (IATA) regulations (Infectious Substances Shipping Guidelines, 2006). All persons involved in the shipping process should be trained in the correct procedures for packaging and transportation.

- There must be a written policy to deal with incorrectly identified samples received by the laboratory. Criteria must be developed and adopted for acceptance and rejection of specimens.

- Great care must be taken not to contaminate or cross-contaminate specimens.

B. Analytical Management phase

- Careful selection of the examination procedure is important and depends on the facilities, equipment and staff available, and the number of samples for examination.

- SOPs must be available for all analytical methods. The methods must be evaluated by the laboratory to ensure that they are suitable for the examinations requested. SOPs must be available in the appropriate language. An authorized work instructions may be made available at work stations.

- The laboratory must have an IQC system to verify that the intended quality of results is achieved for every batch of examinations. Action must be taken if there is non-compliance. This may mean that the results of a batch of tests are rejected if the QC results are outside the pre-set tolerance ranges. In these circumstances, the samples will need to be re-examined once the required corrective action has been taken.

C. Post-analytical Management phase

- Designated staff must review and authorize release of the test results.

- Laboratory results must be legible, without transcription mistakes, and preferably reported in System International (SI) units.

- Laboratory reports should include:
    - (a) Identification of the laboratory issuing the report;
    - (b) Requester's identification;
    - (c) Type of sample;
    - (d) Date and time of primary sample collection,
    - (e) Date and time of receipt by the laboratory;
    - (f) Date and time of reporting;
    - (g) Comments on the primary sample which might have a bearing on the interpretation of the result,
    - (h) Comments on the quality of the primary sample which might invalidate the result,
    - (i) Method of testing used;
    - (j) The results and units of measurement where appropriate;
    - (k) Identity and signature of the person releasing the report.

- There must be a documented procedure for reporting urgent results by telephone.

- Procedures must be in place for storage of samples post-examination to enable re-examination if required, for a specified time and for their eventual safe disposal.

D. Occurrence management Phase

- The laboratory must have a mechanism for staff to document and report problems in laboratory operations.
- Appropriate correction action must be planned and implemented for the problems identified, reported and reviewed.

44.7 Checklist for Laboratory Quality Standards Management

Checklist for laboratory quality standards					
Name of Laboratory : Name of Auditor : Audit Date :					
Sr. No	**Quality System**	**Y**	**P**	**N**	**Remark**
Instructions: For each item, please circle either Y=Yes, P=Partial or N=No.					
A.	**General information**				
1.	Is there a laboratory organizational chart that describes the internal management and supervisory arrangements in the laboratory?				
2.	Is there an adequate number of staff with appropriate qualifications to operate the laboratory?				
3.	Is a member of the laboratory staff appointed as the Quality Manager? Does he/she meet regularly with the Laboratory Head to discuss quality issues?				
4.	a. Is a member of the laboratory staff appointed as Safety Officer? b. Does he/she meet regularly with the Laboratory Head to discuss safety issues?				
5.	Does the management provide a safe and adequate working environment and facility?				
6.	Does the management provide essential equipment and ensure its functionality?				
7.	Does the management provide adequate supplies to ensure continuity of service?				
8.	Is there an effective system in place for documentation and record-keeping?				
9.	Is there an effective quality management system in place?				
10.	Is there adequate communication between the laboratory and all its users?				

Table Contd...

	Checklist for laboratory quality standards				
Sr. No	**Quality System**	**Y**	**P**	**N**	**Remark**
B.	**Quality management system (QMS)**				
11.	Does the Quality Manager have delegated responsibility to oversee compliance with the quality management system?				
12.	Does the laboratory have a Quality Manual which documents all policies and procedures currently in use in the laboratory?				
13.	Does the laboratory have a document control system in place? Are obsolete documents removed and marked "obsolete"?				
14.	Are all documents reviewed at least annually?				
15.	Is there an internal quality control (IQC) system in place that controls every batch of examinations?				
16.	Is the IQC performed, documented and reviewed prior to release of results?				
17.	Are the results of these IQC samples recorded and acted on if found to be outside the acceptable range?				
18.	Are QC results monitored statistically?				
19.	Are deviations followed by timely troubleshooting and corrective action?				
20.	Does the laboratory participate in an external quality assessment scheme (EQAS)?				
21.	Are the results of the EQAS recorded and discussed with staff to resolve any possible discrepancies?				
22.	Are all operational procedures in the Quality Manual reviewed continuously to identify issues of potential non-compliance?				
23.	Is there an annual review of the QMS which is properly documented?				
24.	Are regular audits carried out in each section of the laboratory at the intervals prescribed in the Quality Manual?				
25.	Are the recommendations and/or corrective actions from these audits documented, discussed with staff and an action plan developed with clear timelines and documented follow up?				

Name of Laboratory :   *Name of Auditor :*   *Audit Date :*

Table *Contd...*

Checklist for laboratory quality standards				
Name of Laboratory : Name of Auditor : Audit Date :				

Sr. No	Quality System	Y	P	N	Remark
C.	**Human resources (personnel)**				
26.	Is the Laboratory Head a person with appropriate qualifications and training for the position as per the job description?				
27.	Are there sufficient numbers of staff who are appropriately trained to carry out all the tasks in the laboratory?				
28.	Is each section of the laboratory led by a person who is adequately qualified, trained and competent in that discipline?				
29.	Are lines of authority and responsibility clearly defined for all laboratory staff, including the designation of a supervisor and deputies for all key functions?				
30.	Do all staff have written job descriptions?				
31.	Are appropriate and regular continuing professional development (CPD) programmes available to all staff?				
32.	Is there a performance appraisal system in place for every staff member, which includes an annual competency assessment?				
33.	Are there regular (at least monthly) staff meetings between the Laboratory Head anal staff to disseminate information and discuss problems?				
D.	**Environmental conditions**				
34.	Does the laboratory have bio safety facilities corresponding to the bio risk hazard of the pathogens it is mandated to handle?				
35.	Are the patient and testing areas of the laboratory, and incompatible testing activities, effectively separated from one another?				
36.	Is the work area kept clean and tidy for efficient operation, including cleaning and disinfecting of workbench areas on a daily basis?				
37.	Are adequate utilities provided (water, ventilation, electricity [including back-up power sources such as UPS]) for the laboratory to operate effectively?				

Table *Contd...*

Checklist for laboratory quality standards					
Name of Laboratory :	Name of Auditor :		Audit Date :		
Sr. No	**Quality System**	**Y**	**P**	**N**	**Remark**
38.	Are there separate rooms designated for sample collection?				
39.	Is there adequate storage space available, with the correct conditions, to ensure the integrity of all archived samples and supplies?				
40.	Are sufficient waste disposal facilities available and is waste separated into infectious and non-infectious waste, with infectious waste autoclaved, incinerated or buried?				
E.	**Laboratory safety**				
41.	Are all staff aware of the *Laboratory bio safety manual* and have they received training in its procedures?				
42.	Is the laboratory properly secured from unauthorized access with appropriate signage?				
43.	Are standard operating procedures (SOPs) available at all workstations for safe handling of all samples?				
44.	Are the work and storage areas of the laboratory free of staff food items?				
45.	Are all samples stored separately from reagents and products in the laboratory refrigerators and freezers?				
46.	Are "sharps" handled and disposed of properly in "sharps" containers?				
47.	Are all staff provided with appropriate personal protective equipment (PPE) ?				
48.	Are there adequate hand washing facilities available in the laboratory, including soap and towels?				
49.	Is there an SOP to address spillage or leakage of chemicals, specimens, etc. including breakages in centrifuges?				
50.	Are adequate first aid materials and facilities available?				
51.	Are all accidents and incidents recorded and reported as per national regulations?				
F.	**Laboratory equipment**				
52.	Does the equipment provided meet the minimum standards for the level of healthcare?				
53.	Is there a written policy that details the process of procuring equipment to ensure that it is appropriate and meets the requirements?				

Table *Contd…*

Checklist for laboratory quality standards				
Name of Laboratory : Name of Auditor : Audit Date :				
Sr. No	**Quality System**	**Y**	**P**	**N**	**Remark**
54.	Does the laboratory have a list of manufacturers and the contact persons for all equipment?				
55.	Is the equipment purchased from reputable suppliers who can provide appropriate maintenance, servicing and spare parts for the equipment?				
56.	Can the supplier assure continuity of supply of reagents and other commodities for the life of the equipment?				
57.	Has training on each piece of equipment been provided to a sufficient number of staff, including engineers, to ensure that it can be operated correctly at all times?				
58.	Is relevant equipment maintenance and service information readily available in the laboratory? a. Service contract information b. Contact details of service provider c. Performance and maintenance records d. Last date of service e. Next date of service				
59.	Is the maintenance schedule adhered to with records kept of all maintenance carried out?				
60.	Is all equipment uniquely identified?				
61.	Is there a mechanism for validation of equipment prior to use and after repair/maintenance?				
62.	Are current equipment inventory data available on all equipment in the laboratory? (a) Name of the equipment (b) Manufacturer (c) Condition received (new, used, reconditioned, donated) (d) Serial number (e) Date of purchase/acquisition (f) Date of entry into service				
63.	Is there a documented procedure for the decommissioning of redundant or non-functional equipment?				

Table Contd...

Sr. No	Quality System	Y	P	N	Remark
	Checklist for laboratory quality standards				
	Name of Laboratory : Name of Auditor : Audit Date :				
G.	**Procurement and supplies management**				
64.	Is there a policy that defines and documents the procedures for selection and purchase of consumables, reagents and other supplies?				
65.	Is there an effective inventory management system in place which ensures avoidance of "stock-outs" and details maximum and minimum stock levels of all items?				
66.	Are supply and reagent specifications periodically reviewed and approved suppliers identified?				
67.	Does the policy detail the procedures and criteria for inspection, acceptance/rejection, and storage of consumable materials?				
68.	Does the policy ensure that supplies are used on a "first-in first-out" basis?				
69.	Is there adequate storage available to ensure that all supplies are held at the correct environmental conditions? a. Is the storage area well organized and free of clutter? b. Are set places labelled for all inventory items? c. Are hazardous chemicals stored appropriately? d. Is adequate cold storage available? e. Is temperature monitoring conducted according to material safety data sheet (MSDS) instructions? f. Is storage in direct sunlight avoided? g. Is the storage area adequately ventilated? h. Is the storage area clean and free of dust and pests?				
70.	Is there a documented policy to ensure that when reagents are put into use, the batch number and date are recorded and they are never used beyond their expiry date?				
71.	Is there a mechanism for validation of supplies, especially diagnostic reagents?				
72.	Are expired products disposed of properly?				
H.	**Testing**				
73.	Does the laboratory have a written policy giving all the information and details required on the report form to ensure that the person who requested the tests is supplied with all relevant information regarding the result, including the normal reference range for the procedure?				

Table Contd...

Checklist for laboratory quality standards					
Name of Laboratory : **Name of Auditor :** **Audit Date :**					
Sr. No	**Quality System**	**Y**	**P**	**N**	**Remark**
74.	Is there a documented turnaround time for each test which has been agreed up on between the laboratory and those requesting the test?				
75.	Is there a documented list of critical/panic result levels for critical tests and details of how requesters should be notified when results fall within these levels?				
76.	Is there a written procedure for reporting results by telephone to ensure that results reach the requester?				
77.	Is there a documented policy that details what, how, where and for how long records should be retained?				
78.	Is there a documented policy that detail show long samples and subsamples should be stored?				
79.	Is there a documented policy that ensures the confidentiality of all records?				
I.	**Managing laboratory specimens**				
	I. Pre-analytical phase				
80.	Does the laboratory have a request form that provides all the details required to properly identify the source of the sample the requester, the test(s) required, and the date and time the sample was taken?				
81.	Are guidelines for source sample identification specimen collection, labelling and transport readily available to persons responsible for primary sample collection?				
82.	Are SOPs available on the correct collection procedures for all types of specimens sent to the laboratory?				
83.	Do these instructions include recommendations on safe and timely transport of specimens to the laboratory?				
84.	Is there a written policy to deal with incorrectly identified or incorrect specimens received in the laboratory?				
85.	Is each primary sample given a unique accession number along with the date and time of receipt?				
86.	Is there a written procedure for preparing aliquots of samples, which ensures the accurate transfer of patient information?				
87.	Is there a procedure to ensure that samples once registered, and if necessary made into a liqots, are delivered to the testing laboratory in a timely manner?				

Table Contd...

Checklist for laboratory quality standards					
Name of Laboratory :	Name of Auditor :		Audit Date :		
Sr. No	**Quality System**	**Y**	**P**	**N**	**Remark**
88.	Are there written SOPs which are in accordance with current International Air Transport Association (IATA) regulations for the packing and shipping of samples to external laboratories?				
89.	Are specimens packaged appropriately using a triple packaging system and transported to referral laboratories within acceptable timeframes?				
90.	Are there adequate supplies of the correct shipping containers?				
91.	Do some of the staff responsible for shipping have current IATA shippers' certificates?				
92.	Are referred specimens tracked properly, using a logbook or tracking form to ensure that there is no loss of specimens?				
93.	Are procedures in place to ensure the receipt of results from referral laboratories?				
	II. Analytical phase				
94.	Are there SOPs for all analytical methods and are they available at the workbench?				
95.	Do these SOPs follow the protocol laid out in the documentation control policy of the Quality Manual?				
96.	Are there work instructions at the workbench and do these follow documentation control requirements?				
97.	Is there an IQC process in place to verify the accuracy of results on each batch that is analysed?				
98.	Is there a reagent logbook for the lot number and dates of opening, which reflects the verification of new lots?				
99.	Is action taken and documented, including rejection of patient test results, if there is noncompliance with IQC results?				
	III. Post-analytical phase				
100.	Are all test requests cross-checked with test results to ensure that all tests have been completed?				
101.	Is there a procedure whereby all results are reviewed and signed out by an authorized person?				

Table Contd...

Checklist for laboratory quality standards				
Name of Laboratory : **Name of Auditor :** **Audit Date :**				

Sr. No	Quality System	Y	P	N	Remark
102.	Is the laboratory report(s) in a standard format including the following (answer each item): (a) Is the testing laboratory clearly identified in the report? (b) Does the report contain the name, number, address. (c) Is the name of the person requesting the test indicated on the report? (d) Is the type of sample received and the test requested included in the report? (e) Are the date and time of specimen collection, receipt of specimen and release of report indicated? (f) Does the report indicate reference ranges for each test where applicable? (g) Is there space for interpretation of results, where applicable? (h) Does the result contain the name of the person authorizing release of the report?				
103.	Is there a policy for communicating to the requester about specimens unsuitable for processing, specimens where testing is delayed, corrected reports and critical results?				

Chapter - 45

Laboratory Instrument Calibration Program

Introduction

Every pharmaceutical company should never compromise on, it is the quality of their products. Even a tiny variance can lead to life-threatening situations for the users. To ensure the high quality of the end products, all pharma companies must ensure that all of their equipment is well calibrated. Calibration ensures the integrity, accuracy and reliability of measurement data for equipment and instruments used in the laboratory. Calibration gives confidence in

- Determine the accuracy, precision, reliability and deviation of the measurements produced by all the instruments

- Ensure that the readings of equipment or instruments are consistent with other measurements and display the correct readings every single time.

- To establish the reliability of the instrument being used and whether it can be trusted to deliver repeatable results each time.

- Calibration is one of the major parts of pharmaceutical manufacturing standards and quality assurance because of these reasons:

- With regular use and time, instruments and devices used in the industry will undergo damage, which causes a shift in the measurements.

- Regulatory inspection will help in tracking the shifts and rectifying through calibration.

When the instrument does not perform within the acceptable range of error, its quality will be greatly affected. Calibration guarantees the quality is not affected by constant errors. This is crucial for processes where the quality of measurement is directly related to the quality of the product.

Certain measurements will be affected by ambient condition such as temperature, pressure and humidity. A change in the surrounding atmospheric conditions will cause a drift in the measurement, and regular calibration will keep this drift within acceptable limits.

Periodic calibration will ensure the measurements and outputs achieved are accurate at all times without affecting the quality of the final product.

If an instrument is not calibrated for a long time, the following issues can occur:

- **Indefinite quality:** The measurements will become faulty, and discrepancies will be seen, which will reflect on the final product. This will result in questionable quality of the products.

- **Protection issues:** In the pharmaceutical industry, the quality of the final product holds extreme importance because it can affect thousands of lives. Non-calibrated instruments will cause wrong measurements of raw materials when producing medicines or instruments of sensitive nature such as thermometers, which can cause serious safety threats to the final users.

- **Mistreatment of resources:** Non-calibrated instruments will waste time and resources, which prove expensive in the long run.

- **Stoppage increases:** When the final products show deviance from the required quality, shutting down the process to conduct a complete recalibration is required. This results in increased downtime. Regular calibration will eliminate that and help in detecting warning signs before they affect the final product or cause significant damage.

- **Risk of lawsuit:** When customers realize the product they received does not match up to the expected quality or if they suffer seriously due to the low quality of the pharmaceutical products, they will return the product and ask for a full refund or press legal charges due to the damage caused. This will affect reputation and increase costs. It might even call for a product recall, which is also expensive.

45.1 Laboratory Instrument Calibration Authorized Agencies

All measurement standards used to calibrate measuring devices should be traceable to a national standard of measurement. This can be done:

- Directly, through purchase of pre-calibrated certified standards. These shall be supported by calibration documents or certificate from the supplier stating the date, accuracy (assigned value and units of measure), traceability and conditions under which the results were obtained. These standards shall be re-calibrated at pre-determined intervals.

- Indirectly, by preparation of an internal working standard calibrated against a certified standard. Such standards shall be supported by internal test reports and any other supporting documentation.

- Where no recognized external standard exists, an internal standard may be prepared and calibrated, provided a written procedure is prepared and a rationale for assigning values, accuracy and units is established.

- Some of the calibration standards include MIL-STD-45662A, A2LA Accreditations and ANSI/NCSL Z540-1, with each one having their own set of requirements for establishing and maintaining a calibration system. These standards help in controlling the accuracy of the measuring and test equipment, and measurement standards used to assure that the services delivered are satisfactory.

- The International Laboratory Accreditation Cooperation (ILAC) has published the ILAC P10:01,23 which describes who should perform calibration of equipment and reference standards. The options include, but are not limited to, calibration laboratories accredited under ISO 17025:2005 by an accreditation body (AB) covered by the ILAC Arrangement, as well as National Metrology Institutes (NMI) whose services can be found in the BIPM key comparison database (KCDB).

45.2 Method used for the Laboratory Instrument Calibration

There are different ways that are used to calibrate an instrument. These methods are chosen based on the desired results of the calibration and regulatory authorities' requirements, like FDA guidelines. Let us look at three such procedures:

I. ***Standard Calibration:*** This method is mostly preferred for calibrating instruments that are non-critical to quality or are not required for accreditation and license purposes. Even if they are required for licensing purposes, they have to be calibrated to be effective. Use traceable standards and document its performance.

II. ***Calibration with Data:*** Procedures for calibrations with data are similar to that of accredited calibration. The only exception being that these procedures are not accredited to the ISO standard. Moreover, they are not accompanied by data on measurement uncertainties.

III. ***ISO 17025 Accredited Calibration:*** This has to be the strictest method of calibration. Generally, it requires a measurement report which has the details of the measurements that are made against a standard of 'as found' (before calibration is started) and 'as left' (once the calibration is completed). Every single standard and measurement used for the procedure has to be traceable to an acceptable national or international organization for standards. If the calibration is done by a calibration service provider, they must issue a certificate.

45.3 Setting of Laboratory Instrument Calibration Limits

Calibration is concerned with the measurement of values and their comparison with acceptable limits of standards, resulting in adjustment or correction, if necessary.

Compare calibration results with established limits of accuracy for the measuring device. If the device being calibrated does not fall within the limits, then readjust and recalibrate it until it falls within pre-established limits. If not, remove from use.

The establishment of limits should be based on a combination of:

- Those specified at the time of purchase
- recommendations from the manufacturer
- Limits established in reference standards.
- The acceptable limits required for satisfactory calibration of each instrument should be identified or referenced in the relevant procedure.

45.4 Calibration Certification Requirement

The calibration certificate is required by the end-users of the products that are tested with the concerned instrument

Every calibration certificate has a unique serial number which associates one calibration with one instrument. Most of them would cite the basis of the calibration procedure too. The format of the certificate may vary from one geographical location to another.

Calibration certificate must include:

- Dates and environmental conditions at the time of calibration
- Traceability statement
- Received Condition & Returned Condition
- In tolerance/meets all specifications
- Identification of the standards used during calibration. Associates specific traceable instruments with this certificate.
- Calibration procedure used, including revision level if applicable.
- Calibration interval and source of recommendation
- Contact information for inquiries about this Certificate

45.5 Instrument Calibration Procedure

Prepare documented procedures based on the instrument manufacturer's written instructions, and use these for calibration and performance checks of all measuring instruments. Calibration procedure should include the following:

- A list of equipment to which the procedure is applicable

- Calibration points, environmental requirements and special conditions
- Limits of accuracy
- List and identity of traceable standards
- Sequence of steps for calibration
- Instructions for recording data with reference to the relevant standard form.

45.6 Instrument Calibration Responsibility

It is the responsibility of the supervisor of the section to which the equipment belongs:

- To plan, schedule, organize and maintain records of the calibration programs for various equipment under their control;
- To ensure that equipment and instruments are continuously calibrated or removed from use;
- To train staff for performing calibration/performance checks.

45.7 Calibration Frequency Determination

Calibration mainly depends upon its tendency to drift from the true measurement and how it impacts the quality of the end product. Based on the information, one can design a calibration schedule for each instrument. The frequency mainly depends on

- *Criticality of the measurement:* Make sure to calibrate the more critical locations more frequently than the less critical ones.
- *Stability history:* Check how often calibration was required in the past. This will provide a good framework to plan and schedule calibration.
- *Workload and operating conditions:* Instruments with too much workload or extreme operating conditions call for more frequent calibration than others.
- *Before extremely important measurements:* Calibration is a good idea before a particularly important measurement has to be taken.
- *Sudden unforeseen events:* If an electric fault, external conditions that are out of one's control such as drastic weather and pressure changes, a fall or any other impact occur, conduct a calibration process immediately.
- *Discrepancies suspected in the final product:* Any discrepancy or shift from acceptable performance criteria is an indication to calibrate the instrument. Slight variations are normal, but anything that crosses an acceptable limit means calibration is needed immediately.

  The interval between calibrations can vary as Weekly, Monthly or bi-monthly, quarterly, semi-annually or annually. Risk assessment should be carried out while defining the frequency

45.8 Laboratory Calibration Schedules

- Purchase each new piece of equipment or instrument according to specifications.
- Enter new equipment in an asset register prior to use.
- Ask the supplier prior to delivery or after installation to calibrate new equipment and provide a certificate of calibration.
- Maintain calibration/maintenance schedules for all equipment.
- The schedules of calibration or performance checks should be based on: manufacturer's recommendations, Reference standards and recalibration of the measuring devices based on recommended time intervals.

45.9 Instrument Labelling and Documentation

All calibrated equipment should be labelled with the following information:

- Date of last calibration
- Signature of the person who performed the calibration
- Date when the next calibration is due.
- Identify and label equipment that has passed its due date of calibration until it is recalibrated.
- Maintain complete records for the calibration and performance checks of all equipment and instruments. Calibration and performance check test records should include (where appropriate)
- Asset register number
- Instrument serial number
- Limits for calibration
- Date of calibration/performance check
- Due date of next calibration
- Any details of adjustment* or repair
- Results of the calibration*/performance check
- Statement of compliance or details of non-compliance and action taken
- Signature/initials of the person performing the calibration/performance check.

* It is important to record the results of calibration before and after any adjustment.

Maintain calibration and performance check records for five years

Conduct a review if any measuring device is found to be out of calibration and requires adjustment. Take corrective action where appropriate.

If the item can be adjusted back into calibration, it may continue to be used. If the item cannot be adjusted back into calibration, it must not be used until the problem is corrected. Under these circumstances, attach an identifying label stating that the item is under repair and is not to be used.

The supervisor must assess the likely impact of the inaccuracy of the affected measurement on the quality of the current product and that produced since the last satisfactory calibration. Factors influencing the degree of risk include the following:

- Critical nature of the measurement
- Sensitivity of quality control testing
- History of product records and performance checks.
- Additional quality control testing may be instituted to determine whether quality has been compromised. Where it is likely that quality has been compromised, this shall be communicated to senior management and documented in a report

45.10 Relocation of Laboratory Instruments

Recalibrate equipment (especially those that are non-portable) that have been relocated. The manufacturer's recommendations on the need for recalibration shall be sought when relocating non-portable instruments.

45.11 External Calibration Contractors

Make an agreement with the contractors to supply written reports of calibrations, which should include the following:

- Use of standards and references traceable to national standards
- Certification/licensing by the equipment manufacturer, if available
- Checking of all certificates or reports supplied by approved external laboratories on receipt.

45.12 Definitions

- **Calibration**

  A set of operations which, under special conditions, establishes the relationship between the values indicated by measuring instruments and standards.

- **Performance checks**

  The routine checking of the performance of an instrument to verify that it has remained within the specified range of accuracy and precision.

- **Accuracy**

  The closeness of agreement between the result of a measure and the true value of measurement. Calibration is used to determine the accuracy of an instrument.

- **Precision (repeatability)**

  The closeness of agreement between the results of successive measurements of a defined procedure under prescribed conditions.

- **Measurement standard**

  A measuring instrument or material that physically defines a unit of measurement or value of a quantity. Measurement standards used for calibration should be traceable to the SI units of standard measurements.

Chapter - 46

Laboratory Safety Management Program

Introduction

Pharmaceutical laboratories are potentially dangerous places to work in due to chemical, electrical, physical, mechanical, and biological or radiation hazards. Those at risk include laboratory staff, customers and visitors entering the laboratory environment; so it is important for all laboratory staff to recognize the potential dangers and reduce risk to a minimum.

The level of risk depends on the activities within the laboratory, and the types and sources of material entering the laboratory. Occupational injuries and illnesses may result from bad practices, ignorance, inexperience and failure to follow established procedures.

Laboratory safety is governed by numerous local, state and federal regulations. The Occupational Exposure to Hazardous Chemicals in Laboratories standard (29 CFR 1910.1450) was created specifically for non-production laboratories. Additional OSHA standards provide rules that protect personnel, including those that who in laboratories chemical hazards as well as biological, physical and safety hazards.

Occupational Safety and Health Act of 1970

"To assure safe and healthful working conditions for working men and women; by authorizing n for cement of the standards developed under the Act; by assisting and encouraging the States in their efforts to assure safe and healthful working conditions; by providing for research, information, education, and training in the field of occupational safety and health."

The purpose of the Laboratory standard is to ensure that personnel in non-production laboratories are informed about the hazards of chemicals in their workplace and are protected from chemical exposures exceeding allowable and as specified in other substance-specific health standards. The Laboratory standard achieves this protection by establishing safe work practices in laboratories to implement a Chemical Hygiene Plan (CHP).

Safety management Program of laboratory standard consists of five major elements:

- Hazard identification;
- Chemical Hygiene Plan;
- Information and training;
- Exposure monitoring; and
- Medical consultation and examinations

Each laboratory covered by the Laboratory standard must appoint a Chemical Hygiene Officer (CHO) to develop and implement a Chemical Hygiene Plan. The CHO is responsible for duties such as monitoring processes, procuring chemicals, helping project directors upgrade facilities, and advising administrator son improved chemical hygiene policies and practices. A worker designated as the CHO must be qualified, by training or experience, to provide technical guidance in developing and implementing the provisions of the CHP.

46.1 Laboratory Safety Standard-Roles and Responsibilities

The following are the National Research Council's recommendations concerning the responsibilities of

Various individuals for chemical hygiene in laboratories.

A. Safety Chief Executive Officer

- Bears ultimate responsibility for chemical hygiene within the facility.
- Provides continuing support for institutional chemical hygiene.

B. Safety Chemical Hygiene Officer

- Develops and implements appropriate chemical hygiene policies and practices.
- Monitors procurement, use, and disposal of chemicals used in the lab.
- Ensures that appropriate audits are maintained.
- Helps project directors develop precautions and adequate facilities.
- Knows the current legal requirements concerning regulated substances.
- Seeks ways to improve the chemical hygiene program.

C. Safety Laboratory Supervisors

- Have overall responsibility for chemical hygiene in the laboratory.
- Ensure that laboratory personnel know and follow the chemical hygiene rules.
- Ensure that protective equipment is available and in working order.
- Ensure that appropriate training has been provided.

- Provide regular, formal chemical hygiene and housekeeping inspections, including routine
- Inspections of emergency equipment.
- Know the current legal requirements concerning regulated substances.
- Determine the required levels of PPE and equipment.
- Ensure that facilities and training for use of any material being ordered are adequate.

D.　Safety Laboratory Personnel

- Plan and conduct each operation in accord with the facility's chemical hygiene procedures, including use of PPE and engineering controls, as appropriate.
- Develop good personal chemical hygiene habits.
- Report all accidents and potential chemical exposures immediately.

E.　Safety Officer

- Giving safety advice;
- Establishing the safety policy, in consultation with the laboratory head;
- Administering the safety policy;
- Assisting in the design and maintenance of the safety programme;
- Orientating and training all staff in the elements of the safety programme;
- Organizing membership of the hospital or institution safety committee, as appropriate;
- Submitting regular reports on the safety status to the laboratory head;
- Maintaining accident records;
- Investigating all laboratory accidents;
- Documenting regular safety inspections;
- Establishing a waste management programme addressing the disposal of all waste
- Ensuring that all staff comply with policies, rules and procedures

46.2 Requirement of Safety Department

- Safety rules must be established to reduce risks to staff, customers and visitors. Staff must comply with these rules and ensure that customers and visitors are sufficiently briefed to ensure safety.
- SOPs must be prepared to ensure safe handling of laboratory equipment.
- SOPs must be developed to ensure the safe handling of all samples and procedures such as phlebotomy, sample transport, sub sampling, analytical

procedures, storage and disposal of samples. They should be available at all work stations and provided to appropriate staff.

- SOPs for safe handling of all referred samples must be available and provided to appropriate staff. A list of diseases of national and international concern that require emergency action must be available in the laboratory.

- All staff handling patient samples and other biological materials must wear appropriate personal protective equipment (PPE). These must be removed before leaving the laboratory or undertaking clerical work. Hands must be washed immediately after removing the protection and before leaving the laboratory.

- SOPs must be available in the event of a spillage/leakage of biological, chemical or radiochemical materials or patient samples, including when containers are broken in a centrifuge.

- First-aid materials and facilities must be readily available to deal with accidents. All accidents, however small, or accidents that might have occurred ("near misses"), must be recorded and reported as per national regulations.

- The employing authority, through the laboratory head, is responsible for ensuring adequate protection of laboratory personnel to avoid occupational hazards.

46.3 Safety Rules for Laboratories

- Eating, drinking, smoking and applying cosmetics are prohibited in the laboratory.
- Pipe ting by mouth is prohibited.
- Appropriate protective clothing must be worn at all times in the laboratory, and gloves should be worn when required.
- The laboratory must be kept clean and tidy, and should contain only those items necessary for the work carried out.
- All work surfaces must be appropriately decontaminated at the end of each working day and immediately after any spillage.
- All staff must wash their hands when leaving the laboratory.
- Care must be taken to avoid the formation of aerosols or splashing of materials.
- All contaminated waste or reusable materials must be appropriately decontaminated before disposal or reuse.
- Access to the laboratory must be restricted to authorized personnel only.
- All incidents or accidents must be reported immediately and appropriate action taken to prevent further occurrences.
- All staff working in the laboratory must be adequately trained, both in the duties they perform and in all safety aspects of work.
- All waste must be appropriately marked before disposal.

- The disinfectant used must be appropriate and its efficacy must be ensured.

" Safety rules are mandatory for all staff."

46.4 Laboratory Personnel Information and Training

Laboratory personnel must be provided with information and training relevant to the hazards of the chemicals present in their laboratory. The training must be provided at the time of initial assignment to a laboratory and prior to assignments involving new exposure situations.

Training must include the following:

- Methods and observations used to detect the presence or release of a hazardous chemical.
- These may include employer monitoring, continuous monitoring devices, and familiarity with
- the appearance and od or of the chemicals;
- The physical and health hazards of chemicals in the laboratory work area;
- The measures that personnel can take to protect themselves from these hazards, including protective equipment, appropriate work practices, and emergency procedures;
- Applicable details of the employer's written Chemical Hygiene Plan;
- Retraining, if necessary.

A. Personal protective equipment (PPE) Training

Employers must train personnel to use the appropriate personal protective equipment (PPE)

- face shield or safety goggles;
- safety gloves;
- Long-sleeved shirts, lab coats, aprons.
- Eye protection is required at all times
- a full face shield

B. Fires Plan work Training

Employers should ensure that personnel are trained to do the following in order to prevent fires Plan work. Have a written emergency plan for your space and/or operation.

- Minimize materials. Have present in the immediate work area and use only the minimum quantities necessary for work in progress. Not only does this minimize fire risk, it reduces costs and waste.

- Observe proper housekeeping. Keep work areas uncluttered, and clean frequently. Put unneeded materials back in storage promptly. Keep aisles, doors, and access to emergency equipment unobstructed at all times.
- Observe restrictions on equipment (i.e., keeping solvents only in an explosion-proof refrigerator).
- Keep barriers in place (shields, hood doors, lab doors).
- Wear proper clothing and personal protective equipment.
- Avoid working alone.
- Store solvents properly in approved flammable liquid storage cabinets.
- Shut door behind you when evacuating.
- Limit open flames use to under fume hoods and only when constantly attended.
- Keep combustibles away from open flames.
- Do not heat solvents using hot plates
- **Remember the "RACE" rule in case of a fire.**
    - **R= Rescue/remove all occupants**
    - **A= Activate the alarm system**
    - **C= Confine the fire by closing doors**
    - **E= Evacuate/Extinguish**

C. Emergency procedures Training

Employers should ensure that personnel are trained in the following emergency procedures

- Know what to do. You tend to do under stress what you have practiced or pre-planned. Therefore, planning, practice and drills are essential.
- Know where things are: The nearest fire extinguisher, fire alarm box, exit(s), telephone, emergency Shower/eyewash and first-aid kit, etc.
- Be aware that emergencies are rarely "clean" and will often involve more than one type of problem. For example, an explosion may generate medical, fire, and contamination emergencies simultaneously.
- Train personnel and exercise the emergency plan.
- Learn to use the emergency equipment provided.

D. PASS rule for fire extinguishers training

Employers should train personnel to remember the **"PASS"** rule for fire extinguishers

PASS summarizes the operation of a fire extinguisher.

- **P – Pull the pin**
- **A – Aim extinguisher nozzle at the base of the fire**

- S – Squeeze the trigger while holding the extinguisher upright
- S – Sweep the extinguisher from side to side; cover the fire with the spray

E. Clothing fire Training

Employers should train personnel on appropriate procedures in the event of a clothing fire

- If the floor is not on fire, STOP, DROP and ROLL to extinguish the flames or use a fire blanket or a safety shower if not contraindicated (i.e., there are no chemicals or electricity involved).
- If a co-worker's clothing catches fire and he/she runs down the hallway in panic, tackle him/her and smother the flames as quickly as possible, using appropriate means that are available (e.g., fire blanket, fire extinguisher).

F. Lockout/Tagout training

"Lockout/Tag out" standard, requires the adoption and implementation of practices and procedures to shut down equipment, isolate it from its energy source(s), and prevent the release of potentially hazardous energy while maintenance and servicing activities are being performed. It contains minimum performance requirements, and definitive criteria for establishing an effective program for the control of hazardous energy. However, employers have the flexibility to develop Lockout/Tag out programs that are suitable for their respective facilities.

G. Trips, Slips and Falls training

Worker exposure to wet floors or spills and clutter can lead to slips/trips/falls and other possible injuries.

46.5 Material Safety Data Sheets (MSDSs)

Material Safety Data Sheets (MSDSs) for chemicals received by the laboratory must be supplied by the manufacturer, distributor, or importer and must be maintained and readily accessible to laboratory personnel. MSDSs are written or printed materials concerning a hazardous chemical. Employers must have an MSDS in the workplace for each hazardous chemical in use.

MSDS sheets must contain:

(a) Name of the chemical;
(b) Manufacturer's information;
(c) Hazardous ingredients/identity information;
(d) Physical/chemical characteristics;
(e) Fire and explosion hazard data;
(f) Reactivity data;

(g) Health hazard data;

(h) Precautions for safe handling and use; and

(i) Control measures.

46.6 Safety Hazards

Employers must assess tasks to identify potential worksite hazards and provide and ensure that personnel use appropriate personal protective equipment (PPE) as stated in the PPE standard, 29 CFR 1910.132. Employers must require personnel to use appropriate hand protection when hands are exposed to hazards such as sharp instruments and potential thermal burns. Examples of PPE which may be selected include using oven mitts when handling hot items, and steel mesh or cut-resistant gloves when handling or sorting sharp instruments as stated in the Hand Protection standard, 29 CFR 1910.138.

Chapter - 47

Roll Back Good Analytical Practices

Introduction

"A new type of thinking is essential if mankind is to survive and move toward high levels.- Albert Einstein. If we cannot break old pattern of thinking we cannot move to the higher standard level needed to solve the sustainability problem.

47.1 Analytical Technique Procedure

An analytical procedure must be shown to be fit for its intended purpose. It is useful to consider the entire lifecycle of an analytical procedure when approaching development of the procedure, i.e. its design, development, qualification, and continued verification. The current concepts of validation, verification, and transfer of procedures address portions of the lifecycle but do not consider them holistically. This approach is consistent with the concepts of Quality by Design (QbD) as described in ICH Q8 (R2), 9, 10, and 11.

The lifecycle approach for analytical procedure:

- Analytical target profile (ATP): Structure and Application throughout the Analytical Lifecycle and Analytical Control Strategy.
- Comparable to the Quality Target Product Profile (QTPP).

Stage 1: Procedure design, development and understanding

(a) Procedure design and development

(b) Procedure understanding

(c) Preparing for qualification

Stage 2 : Procedure Performance Qualification

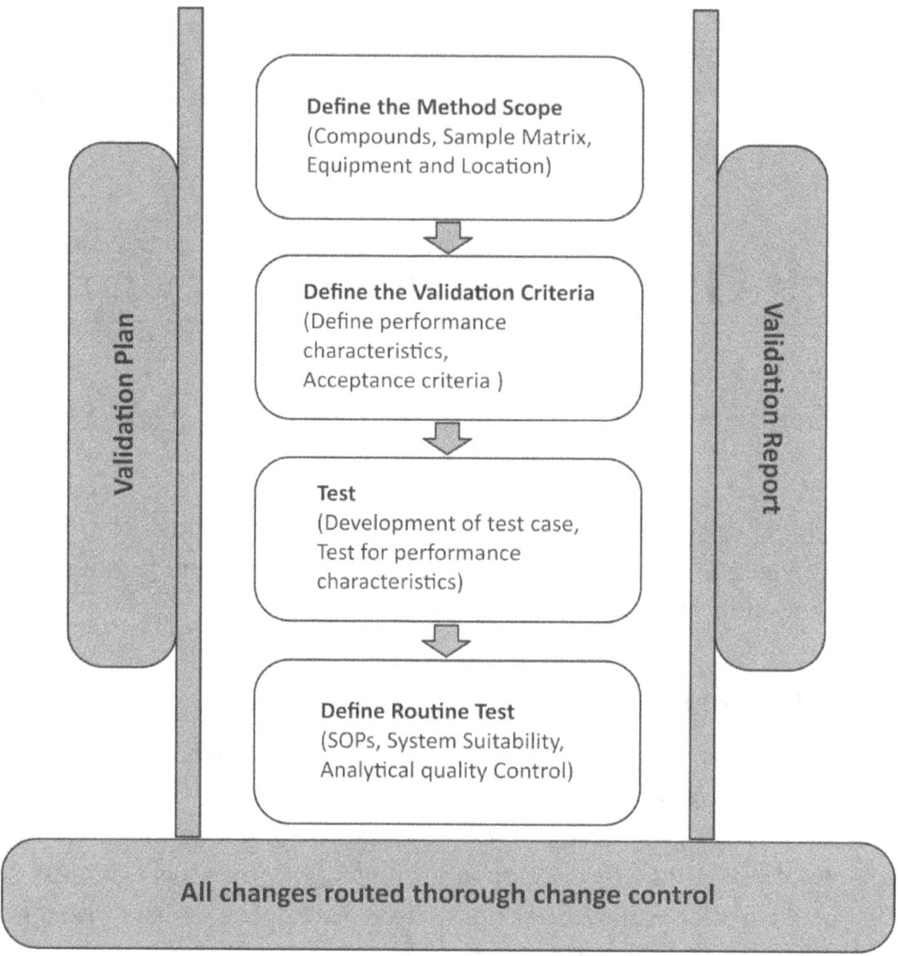

A. **Stage 3 :Implementation and continued procedure performance verification**
    (a) Routine monitoring
    (b) Analytical control strategy
    (c) Knowledge management
    (d) Change control

47.2 Laboratory Instrument Qualification

Laboratory equipment qualification, also referred to as analytical instrument qualification (AIQ), is an essential component of Good analytical practice. It provides assurance that the instrument will perform accurately and consistently in accordance with specification and user requirements. The ultimate objective for equipment qualification is the documented and demonstrated assurance that the unit will produce valid data. Qualification must be performed initially and on an going basis (no less than) annually. Master Qualification and Validation plan should be prepared and detailed the laboratory approach to ensuring equipment is adequately qualified and controlled.

47.3 Statistical Concepts

Appropriate statistical analysis should be applied, when necessary, to quantitative data reported. The methods of analysis, including justification and rationale, should be described fully. These descriptions should be sufficiently clear to permit independent calculation of the results presented.

47.4 Analysis Platinum Rules

- Before initiating the analysis, confirm that the required reagents, sufficient stocks of glassware and chemicals are available.
- Volumetric solutions, test solutions prepared and standardized.
- Check chemicals, reagents, solutions, indicators used are within the valid date.
- Check the availability of required instruments and equipment for analysis.
- Check that the instrument / Equipments are within the calibration period.
- Ensure related working standards are within the valid period.
- Wear all the necessary safety appliances during the analysis. i.e. eye-glasses, hand gloves, etc.
- Weighing shall be done using appropriate calibrated analytical balance.
- The weighing of samples for testing, unless and otherwise stated in method of analysis under individual test shall follow the method.
- The material is added to the tarred receiver, the amount is determined by difference, and then the whole amount is transferred quantitatively E.g. by using a solvent to the final receiver.
- While measuring volume using pipette or measuring jars, reading burette volume, making up solutions in volumetric flasks or using Nesslers cylinders analyst shall follow lower meniscus for colour less solutions and upper meniscus for coloured solutions.

47.5 Laboratory Data Integrity

Data integrity is the assurance that data records are accurate, complete, intact and maintained within their original context, including their relationship to other data records. This definition applies to data recorded in electronic and paper formats or a hybrid of both. A good way to understand data integrity is through an analogy from the legal world. A data record is similar to a contract. A contract is valid only if all the pages of the document are complete and legible, contain the required, authentic signatures and properly state the terms and conditions. In this sense, integrity denotes validity. Ensuring data integrity means protecting original data from accidental or intentional modification, falsification or even deletion, which is the key to reliable and trustworthy records that will withstand scrutiny during regulatory inspections

47.6 Laboratory Management Practice

- ***Establish and Follow Procedures:*** Develop basic procedures, for example, to receive, identify, assign, cue, test, report and dispose of samples.

- ***Maintain Your Proficiency:*** Analysts must have the education, training and experience, acquired through formal education or on-the-job training, sufficient to perform assigned analytic duties.

- ***Validate Methods:*** Method validation needs and techniques will change as the group using a particular method changes

- ***Use Traceable Standard Reference Materials:*** Reference material uses include validating methods that help ensure accurate data from individual test runs, calibrating instruments and assessing analyst proficiency

- ***Run in Duplicate:*** The purpose of duplicate (sometimes triplicate) testing is to add to the confidence that the test run has produced good data for the test object.

- ***Raw data:*** Any laboratory worksheets, records, memorandum, notes that are the result of original observations and activities and are necessary for the reconstruction and evaluation of-the report of that study."

- ***Keep Original Data:*** Whether data is first recorded in electronic/digital form, in a notebook or on the closest piece of scrap paper, keep it.

- ***Assign Instruments and Equipment to Analysts:*** Scientific instruments are temperamental tools; they need individual attention.

- ***Use Control Charts:*** Control charts are excellent tools for several uses, including those already noted. A control chart enables a laboratory to track the results of a reference material and/or control sample at the end of each test run. It gives the laboratory a snapshot of test run quality and a picture of the quality of the laboratory's results for that particular test over time.

- ***Document Everything and Maintain Good Records:*** To return to the golden rule of, "If you didn't document it, you didn't do it," organized records benefit a test organization. An ordered records system can be a prima facie indication to customers, auditors, government and legal authorities, and others that the organization follows its procedures. Records provide a fount of information for training new staff members to perform the stable of methods of the laboratory

47.7 Risk-Based Approach

Risk-based approach on the need for revalidation of existing analytical methods may need to be considered when the manufacturing process changes during the product's life cycle.

Knowledge gained during the will help to assess the method performance.

Over the life cycle of a product, new information and risk assessments (e.g., a better understanding of product critical quality attributes or awareness of a new impurity) may warrant the development and validation of a new or alternative analytical method. New technologies may allow for greater understanding and/or confidence when ensuring product quality.

47.8 Important Element in Quality Control

A. **Quality assurance**
   - Ensuring Suppliers Quality
   - Self-Inspection
   - Change Control
   - Trending
   - Risk Management

B. **Documentation system**
   - Genera information system
   - Laboratory Documentation
   - Data Traceability
   - Electronic Documentation/Computerised Systems

C. **Personnel**
   - General
   - Training

D. **Premises and Equipment**
   - Premises
   - Equipment.
   - Equipment Validation
   - Cleaning/Sanitation
   - Maintenance

E. **Materials and Supplies**
   - Materials
   - Water/Water system

- Sampling and Samples..
- Sampling
- Samples
- Personnel for Sampling

F. Testing
- Testing General
- Testing of Raw Materials
- In Process Controls (IPC)
- Testing of Intermediates
- Testing of Final Products
- Stability Testing
- Validation of Test Methods

G. Results and Release of Test Results
- Handling of Test results
- Failures - Out of Specification Test Results (OOS)
- Failures - Re-testing and Re-sampling
- Test results release/ analytical reports /certification

7.9 Laboratories Contents and its Verification

Sr. No	Important area	Content	Verification and important question
1.0	**General information of Laboratory :**		
1.1	**General information**	• Name of the establishment • Physical address, phone No., FAX • No. Email • Postal address	• Who is the contact person (name, phone No, email)
1.2	**Test activities**	• QC laboratory status and activities on site	• Licensed by a competent national authority • Regularly inspected by a competent national • Authority QC laboratory status and activities on site
1.3	**Activities contracted out (Contract testing)**	• Name(s)/ address/addresses) of the company / companies • Type of activities, written contract	• Licensed by a competent national authority • Evaluation / Re-evaluation of the contract laboratory by the customer (contract giver)
2.0	**Quality Assurance :**		
2.1	**General**	• QA system description; definition of the quality policy and legal conditions • Organisation chart /QA staff • Functionality of QA	• Document available? • Key personnel; reporting lines, responsibilities and release criteria clearly defined? • Review period (procedures, processes)
2.2	**Suppliers quality ensuring**	• Suppliers approvals, contracting, Purchasing control/vendors evaluation	• Policy for supplier's quality assessment defined? • Audits, qualification/ evaluation made?
2.3	**Self-inspection**	• Self inspection / audit system and performance	• How and by whom performed? How reported? • How are corrective measures implemented? • Schedule available and is adhered to?
2.4	**Trending**	• Results/OOS-results	• Do you assess trends? How and by whom are trends evaluated? SOP exists?
2.5	**Change control**	• System, responsibilities, follow up • Actions	• How are changes documented, managed, controlled?
2.6	**Risk Management**	• Risk management method/approach	• Are all critical parameters included? • How is this related to validation process?

Table *Contd...*

Sr. No	Important area	Content	Verification and important question
3.0	DOCUMENTATION :		
3.1	**General information**	• System description (preparation, revision. distribution, archiving) of documentation (change control) • Handling of copies from controlled documents • Syntax of documents (electronic or paper)	• Defined in writing (format, numbering system, approval criteria, distribution, return, interval for revision etc.). • How the process of archiving is managed (location, protection)? • SOP comprises the authorisation for copying, identification of copies from official and controlled documents?
3.2	**Laboratory documentation**	• Specifications (SPECs)	• Specifications are consistent with the information currently held in the dossier?
		• SOPs	• Exists for, sampling, testing, equipment handling and other laboratory processes? Are they complete? Where are previous versions archived? • Standard form introduced?
		• Log books	• What is the form and content of the personnel analytical notebooks, worksheets, general lab notebooks? • Exists for equipment, calibration, maintenance, standards, sample receipt etc.? • Standard form includes complete data (lab staff identity, dedication, data to be recorded etc)?
		• Raw data /e.g. chromatograms, spectra, results), out prints	• What is your definition of raw data? • Recorded/attached directly into relevant laboratory notebooks data (no scrap or loose paper)
3.3	**Data traceability**	• Procedure, record on receipt and usage of materials, standards • Sample tracking • Analytical raw data traceability	• How is traceability ensured? How the system of identification defined is (e.g. how is traceability of working standards to primary standards ensured)? • How the (identified) history of sample is recorded (receipt log, storage conditions, handling, security, safety data etc.)
3.4	**Electronic documentation/**	• No difference in requirements on documentation whether in paper form or electronic form.	

Sr. No	Important area	Content	Verification and important question
	computerised systems		
4.0	PERSONNEL :		
4.1	General	• Number of employees (total);specified to positions and different testing • QC Manager and deputy (other key personnel)	• The number is adequate? What is the annual average staff turnover? • Is there a document specifying qualifications, experiences, duties, responsibilities and staff presence on site?
4.2	Training	• system, programme/plan • Training on special reasons • Documentation on training • Evaluation of training effectiveness (evaluation and re-evaluation)	• System is described? Who is responsible? Who trains (trainer's qualification)? • Is a competent list available and update regularly? • What specific training is given e.g. maintenance, cleaning staff etc.)? • SOP/records on training exist? Was programme fulfilled? • What is the interval for training re-evaluation?
5.0	EMISES :		
5.1	PREMISES	• Location of the QC laboratories • Facility design, rooms separation (e.g. clean and dirty, different testing activities). • Temperature, humidity, ventilation and recording systems/alarms. • Storage areas (e.g. for documents, for samples etc.) • Labelling	• Are QC labs separated from production areas? • Where type of testing is carried out e.g. chemical, biological (microbiological) testing? Is the laboratory equipment located in appropriate area (e.g. clean equipment in clean room)? • How is the system of ventilation / humidification / temperature designed? Is it monitored continuously? Is this system separated for QC area from other areas? System of alarms for critical equipment exists? • How are the documents/sensitive instruments protected? Are storage conditions monitored? Where is defective equipment stored? • Is there clearly indicated dedication of rooms, areas (directions), status? • Are dedicated rooms/laboratories clearly identified?

Table *Contd...*

Sr. No	Important area	Content	Verification and important question
5.2		• Areas / Sterility testing area	• How is the design and fittings of the area? Where are sterility tests carried out? In isolator? How is assessed protection against microbiological contamination during aseptic operations? Appropriate instructions for access into the critical areas exist?
5.3	**PREMISES**	• Areas for positive control tests and fertility testing	• Fertility testing performed? Where? • Is there area for positive tests/fertility testing separate from areas where product is tested?
5.4		• Air supply and ventilation system in microbiological laboratory	• Is the system properly designed? Show me the schematic drawing. Is the ventilation system separated from other areas? What are the pressure differentials (e.g. airlock-test room)? Are there visual alarms
5.5		• Area of preparation	• Where are prepared materials for aseptic operations? Where are prepared culture media for testing? Is there any segregated area/room for manipulation with culture media?
5.6		• Washing room	• How and where are decontaminated materials from microbiological testing? Is there some segregation of washing area to the clean and non-clean part?
6.0	**Equipment :**		
6.1	**Equipment**	• Instrumentation • Assembly (DQ,IQ,OQ,PQ) • Calibration • Labelling • Log books • Cleaning	• Brief description of major equipment available? • Relevant validation documents, SOP(s) for line with qualified assembly and documentation of all possible configurations available? Appropriate environmental conditions clearly stated? • Calibration procedures defined in writing? Documentation (e.g. records) available? Intervals for calibration defined? Calibration status indicated? • Equipment status indicated? • Exist for every major equipment? Are data complete • SOP(s) exists? Records available (require for critical equipment (as applicable)?

Sr. No	Important area	Content	Verification and important question?
6.2		• Isolators used for the sterility testing	• IQ, OQ, PQ made? Show me the results (report)! Show me the results of leak test in general (same as LAF)!
6.3		• Incubators	• Show me the document involving the temperature mapping! Calibration of instrument for measurements of humidity was made? How is tested CO2 (calibration)?
6.4	**Equipment**	• Autoclave	• Show us the results of validation (cold points, cycle's number, stacking)! How do you maintain autoclave? What is the quality of steam (quality SPECS)? What kinds of control test are performed in the steam?
6.5		• Sterilisation by autoclave	• What equipment is sterilised? How is equipment sterilised? Raw data available (cycle/ temperature records)? How are goods handled, which have been run a failed cycle (SOP)?
6.6		• HEPA Filters(validation/maintenance)	• What ways do you have to ensure the integrity of filters (HEPA)? What is the frequency of their replacement? How do you ensure there is no leakage after replacement
6.7		• Colony counter unit • Particle counter • Microscope	• Calibration made? How? • Show me the document on qualification! • Which type used? Was qualified?
6.8	**Equipment validation**	• Qualification • Design qualification (DQ) • Installation qualifications(IQ) • Operational qualifications (OQ) • Performance qualifications (PQ)	• IQ, OQ, PQ was carried out prior to first use? • Who approves? Fully documented (including possible involvement of suppliers and/or third-party)? • Included intervals for revalidation? Requirements and specification for delivered equipment (URS) exists? Relevant checks were made? • Details on specifications and acceptance criteria are provided? Operators have been trained? • Sufficient details of procedures, materials and certified reference materials are available? Results are recorded in a manner amenable to establishing

Table *Contd...*

Sr. No	Important area	Content	Verification and important question
			trends? Results are checked and evaluated by supervisor or delegate? Raw data are consistent with data in summary report? Results are within acceptance criteria applied?
6.9	**Cleaning sanitation**	• Cleaning/sanitation system	• Validation was carried out? Relevant documents available? What are the limits for equipment cleaning? • Which equipment, glassware and other containers are used for cleaning/sanitation? • Who cleans? What are the intervals for cleaning areas specified? SOP?
6.10	**Maintenance**	• System • Preventive maintenance • Documentation	• How maintenance is programmed, performed and documented? Are the critical systems, areas, equipment included? Which work is contracted? • Are the regular/extraordinary maintenance programs available? Who "releases" equipment after maintenance/ repair for laboratory performance? • Is there a schedule including time frames available? Is there inventory of items to be included into the maintenance system? Is there system for the formal acceptance of equipment back into service (and vice versa)?
7.0	**MATERIALS AND SUPPLIES :**		
7.1	**Materials**	• Laboratory reagents, standards	• List(s) of materials available? The suppliers are listed, assessed? What are requirements of identification tests? How is labelling (e.g. date of receipt)? Testing • Kits used? How new lots are traced back to the previous lots?
7.2		• Reference substances)RM	• How are RM handled, labelled, stored (expiration)? Primary standards are available? Which? Are the secondary RM are acceptable? Traceability to official standards assessed? Working standards prepared? How used? Is there an SOP for the in-house

Sr. No	Important area	Content	Verification and important question
			calibration of reference materials? How expiry date and potency value are assigned to each reference material characterised in –house?
7.3		• Handling of highly toxic, hazardous and sensitising materials, poisons	• How are these materials handled (stored)? Are there relevant instructions available? Which measures are introduced to avoid cross contamination? SOP for waste disposal available?
7.4	**Testing materials**	• Settle plates	• What media type is used (Bacteria &yeasts moulds suitable)? Have the plates been irradiated (zero results!)? What is exposure time and how is it calculated (dryness!)?
7.5		• Culture media(kind, purpose)	• Is there tested each batch (growth promotion, selectivity, sterility)? Is there an agreement available on shipment (plates) of prepared media? How is shipment validation performed? How do you guarantee that shipment conditions are kept constant?
		• Culture collection/ Reference standards	• How do you store the reagents and strains used for the identification? Show me the inventory! Are identifications of strains carried out on arrival? Expiration indicated on labels?
7.6	**Protective garments**	• Preparation	• How are gowns washed / sterilised?
		• Use in performance	• Which protective garments are used by operators at sterility testing? Are they suitable for intended use? Are the laboratory coats located in appropriate manner? Instructions for use exist? Training of operators was made? What standard for micro assays is used?
7.7	**Water and water systems**	• Water system/quality	• Is the laboratory water system described? Howis the laboratory water prepared? Water quality is defined (SPECS)? What is the quality of water used for microbiological testing)?
		• Water sampling	• Where, when and how is your laboratory water sampled (SOP)

Table *Contd…*

Sr. No	Important area	Content	Verification and important question
		• Water testing	• What kind of quality testing is done for your water used for different types of analyses?
8.0	**SAMPLING AND SAMPLES**		
8.1	**Sampling**	• General policy	• Show me the description of sampling system (authorisation, statistics application, sampling tools/areas)! What is the number of samples taken and justification for reduced sampling?
8.2		• Sampling	• Sampling performed how? By whom? SOP(s) for sampling available? Includes the details on containers, labelling, equipment cleaning etc.?)
8.3	**Sample**	• Place of sampling for raw materials	• Is there separate sampling area or area in stores? How is the risk of cross contamination / bacterial contamination prevented?
8.4		• Starting/packaging materials sampling	• What the representative amount of sample is as defined in the SOP?
8.5		• IPC's sampling /Intermediates sampling	• System of IPC's/intermediates sampling described (SOP)?
8.6		• System of air sampling	• What type of air sampler is used and why?
8.7		• System of water sampling	• What is the sample volume/testing time, media used, transfer time (SOPs)? Is equipment calibrated (protocol)? What disinfection procedure is used?
8.8		• Procedures/records	• Documentation record
8.9		• Re-sampling	• What sampling techniques including equipment used?
8.10		• Retained samples	• Show me the store for retained samples of raw materials and final products! SOP (time period, number specified?) exists?
8.11	**Samples**	• Samples tracking	• The amount, time period, storage conditions defined? How long samples are stored prior testing? Show me the documentation!
8.12		• Handling of samples	• How are the composite samples blended? How are (labelled, transferred, registered, distributed)

Sr. No	Important area	Content	Verification and important question
			samples for testing and contract testing handled (if applicable)? SOP(s) available? Proper accountability of samples assessed? Are there used some contract facilities? Responsibilities defined?
8.13	**Personnel for sampling**	• Staff	• Specifically trained?
9.0	**TESTING :**		
9.1	**Testing general**	• QC system • Flow sheets • Methods • Contract testing • Re-testing	• Written document available? Which types of testing performed (e.g. microbiological, immunological, chemical etc.)? • Specifying important steps? • Which methods are used for testing: standardised (e.g. Pharmacopeia), modified or developed "in house? • Which analyses are performed on contract?
9.2	**Testing of raw materials**	• System • Methods	• Procedure(s) available? Comply with marketing authorisation? Which is extending of (all raw material tested, full testing made)? Identity testing made? Which materials are released on base of supplier's certificate? • Approved/validated? Acceptance limits specified?
9.3	**Testing in process, controls (IPC)**	• Testing methods/equipment • Testing in the processing areas(laboratory)	• Procedures available? Approved, validated? By whom? Parameters/limits comply with SPECs (or to values specified in processing documents)? • Who prepares and controls quality of reagents and standards used? Who controls the equipment and quality of testing? Personnel (operators) trained? Where is documented?
9.4	**Testing of intermediates**	• System • Sampling	• What is the testing strategy (extend, methods, parameters and limits used)? Which results are transferred into the final product protocol? • Who takes samples for this testing?

Table *Contd...*

Sr. No	Important area	Content	Verification and important question
9.5	**Testing of final products**	• System • Sampling • Samples handling	• In which stages of processing are taken samples for final product testing? What kind of control is performed on final packages? • What is the sampling plan (which norm is used).How do you ensure the representatively of samples per batch? • How you handle the rests of final produce samples (e.g. large volume containers)?
9.6	**Microbiological testing (product)**	• Incubation • Growth promotion • Positive controls • Negative controls • Bio burden • Micro-organism – identification	• Show us the limits of incubation! What is the incubation time and temperature? What is the frequency of observation of sample during incubation? • Do you have records (registration/incubation)? Which micro-organisms used? • The procedure for sub-cultures available? Which controls do you have? How often do you perform positive controls and why? • Do you perform negative product controls? Show me SOP (number of containers/samples)! • Limits defined/used in production/in process controls? Does bio burden IPC´S show the worst case conditions? • Which system do you use to identify? If certain system is chosen VITEC- how is validated?
9.7	**Microbiological testing Environmental**	• System of performance • Sampling location • Swabs • Settle plates • Contact plates • Limits for microbiological testing	• How is monitoring performed at rest/in operation? What is the frequency? How is testing verification performed? Is the monitoring involved in validation plan? Are there available documents on pre-qualification/re-qualification? • Location/sampling sites selected? How is worst-case determined? Water system/area included? Which mode/method is used (SOP)? How are deviations handled? Validation made? • Which types of swabs/solutions are used? Which swabbing techniques are used?

Sr. No	Important area	Content	Verification and important question
			<ul><li>Which media used? What is the method, time exposure, surface area/limits and recovery rate?</li><li>The recovery plate for the surfaces validated: How?</li><li>Vendor verification was made? Preparation/expiry date specified? Sterility test of contact plates performed? Incubation infrequency described (SOP)? Identification of organisms made? What is the time lapse from fumigation to taking sample?</li><li>Limits defined for raw materials/finished products, water, air quality, equipment within processing and testing areas, for personnel ,storage areas, and detergents (cleaning validation)?Limits justified? How?</li><li>What are your action/alert limits? What is the procedure if the limits overshoot? Note: The particulate matter (with respect to situation "in site") should be controlled in addition.</li></ul>
9.8	**Stability testing**	<ul><li>System</li><li>On going</li><li>Premises/equipment</li></ul>	<ul><li>Approach/policy described? Matrixing/bracketing applied? Performed in place or contracted? Full testing made? Critical parameters defined?</li><li>What is the program (intervals, number of batches/products defined)? Analytical methods are suitable? What is the extend of testing incise of changes? Which measures are taken in case if OOS results were tested?</li><li>Appropriate storage stations available? Dedicated, validated, labelled?</li><li>Thermometers and humidity meters calibrated? Is there continual monitoring of temperature and humidity? How are stored light sensitive materials?</li><li>Alarms system exists/described (log book)?</li></ul>
9.9	**Validation of test methods**	<ul><li>Policy</li><li>Validation process</li><li>Validation data</li></ul>	<ul><li>Method validation is part of VMP? General SOP on method validation available? Validation report formally approved? Who approves?</li></ul>

Table *Contd...*

Sr. No	Important area	Content	Verification and important question

		• Method transfer	• Validation purposes specify? Validation completed and documented in each protocol for parameters defined in ICH:
			  ○ precision (System and method)
			  ○ intermediate precision
			  ○ Accuracy,
			  ○ Specificity
			  ○ Reproducibility
			  ○ linearity (range),
			  ○ limit of detection,
			  ○ limit of quantisation,
			  ○ robustness (including solution stability and filter compatibility)
			• Documented in each SOP or protocol:
			  ○ Acceptance criterion for each parameter defined and met,
			  ○ System suitability test procedure has been developed,
			  ○ Acceptance criterion for each system suitability parameter defined and met.
			• Raw data stored
			• Is there a SOP on method transfer?

9.10	**Handling of test results**	• Transfer of raw data • Laboratory Management System(LIMS) • Summary of raw data • Evaluation of test results • Trending	• How are testing results (raw data) transferred into the summary protocol? Are analytical data reviewed by responsible person? How? • Is the system validated? Training of operating personnel was carried out? Is the access authorise and controlled? How? Security of results ensured? What is your change control system? • Who writes the final protocol? How and where are the raw data • Who is responsible for comments and evaluation of the results (QC manager)?

Sr. No	Important area	Content	Verification and important question
9.11	Failures - Out of Specification (OOS) test results	• System/OOS • Laboratory errors (operator, equipment) • Process/Procedure related errors • Evaluation of OOS results • Test results invalidation • Corrective action	• Is there a SOP for OOS result investigation? • How is laboratory investigation and formal investigation beyond the lab performed? • What is your reporting procedure? QA is Involved? • What is the procedure on decision related to OOS? Reasons defined? • How are invalidated test results? Who can invalidate the testing results? • How a corrective action implemented?
9.12	Failures- Retesting and Resembling	• Company's procedure(re-testing programme, • criteria for re-sampling˘)	• How often can a retest be performed? How many times could be testing repeated (testing into compliance)? • What are criteria for re-sampling (e.g. if the sample was not representative)?
9.13	Release of test results/ analytical reports/ certification	• Process release of test results • Feedback to batch release • Preparation of Analytical (summary)Report • Preparation and release of • Certificate of analysis	• SOP available? • Who's responsible for review, decisions, conclusions and formal release of batch? How much is taken into the consideration the validity of analytical results? • Who prepares and approves Summary Report (QC manager)? • Who approves Certificate of analyses? QA involved?
9.14	Chemical testing	• Procedures in place • Reagents preparation • Volumetric glassware • Volumetric solutions • Indicators for titration • Water bath	• Up dated? Valid? • Date of preparation and factors indicated on label comply with relevant method? • What is the level of volumetric glassware (e.g. pipettes calibration)? • How solutions are labelled (indicated date of preparation)? What is the accuracy, stability, storage conditions defined (SOP)? • Show me your system for housekeeping of indicators! • What is the interval for titration indicators change? • Scale of thermometer/calibration level corresponds to the parameters specified in relevant methods?

Table *Contd…*

Sr. No	Important area	Content	Verification and important question
9.15	**Physical and physical chemical testing**	<ul><li>Titrations</li><li>Conduct metric and pH measurements</li><li>Refractometry</li><li>Relative density testing</li><li>Polarimetry</li><li>Viscosity testing</li></ul>	<ul><li>Performed visual or using instruments? If visual how is personnel trained and tested?</li><li>Under which conditions is sample solution measured? Temperature adjusted? If not how is result calculated?</li><li>Temperature controlled/adjusted? Controlled adjusted?</li><li>Temperature controlled/adjusted?</li><li>RM used? What is traceability to official standards? How many readings made?</li></ul>
10.0	**Qualification for some laboratory apparatuses**		
10.1		<ul><li>Balances</li></ul>	<ul><li>Are balances calibrated prior to use by suitable masses</li><li>Are masses certified? Is certificate available? How often is certification performed?</li><li>Show me your in-house calibration programme? Is there a SOP? What are the acceptance criteria and how is it linked to the external calibration results?</li><li>How often balances are calibrated? Are calibration certificates available?</li></ul>
10.2		<ul><li>pH meters</li></ul>	<ul><li>How and when is calibration performed? (daily ,before use)?</li><li>Calibration buffers are relevant for pH range measured in laboratory?</li></ul>
10.3		<ul><li>Conduct meter</li></ul>	<ul><li>Which calibration material is used for determination of cell constant?</li></ul>
10.4		<ul><li>Titrator (KF determination only)</li></ul>	<ul><li>Which material is used for the calibration of the titrator?</li></ul>
10.5		<ul><li>Reference thermometers</li></ul>	<ul><li>Are the working thermometers compared with acetified thermometer within relevant range?</li><li>How often the certified thermometer is sent for Calibration? Are records available?</li></ul>
10.6		<ul><li>Melting point apparatus Refract meter Polari meter</li></ul>	<ul><li>How each instrument is calibrated (SOP)?</li></ul>

Sr.No	Important area	Content	Verification and important question
10.7		• Disintegration	• Is instrument calibrated? • Is thermometer for water bath suitable and calibrated?
10.8		• Dissolution	• Show me the documentation on physical calibration (shaft wobble, level, spindle speed, vibration, vessel temperature)! • Which materials are used for chemical calibration of the system (USP calibrator)?
10.9		• IR spectrophotometers	• Verification of wave number scale has been carried out? Accuracy, resolution, base line flatness controlled for the instrument?
10.10		• Atomic absorption (AA)	• Performance monitored? Linearity and trends assessed? How (by choosing frequently analysed elements or another element such as cooper)?
10.11		• HPLC/GC	• Is there well defined system suitability tests? • Acceptance criteria have been defined?
10.12		• Water testing TOC equipment	• Which material is used for calibration (RM)?

Chapter - 48

Laboratory Entry-Exit Procedure

Introduction

A laboratory is a facility that provides controlled conditions in which scientific or technological research, experiments, and measurement may be performed. Access to the laboratory should be restricted to authorized personnel.

Personnel should be made aware of:

- The appropriate entry and exit procedures including gowning;
- The intended use of a particular area;
- The restrictions imposed on working within such areas;
- The reasons for imposing such restrictions; and
- The appropriate containment levels.

48.1 Entry Procedure in Pharmaceutical Laboratory

Enter in the change room and put off the street shoe, keep them in the shoe rack. Disinfect both the hand. Remove personal belongings (wristwatch, jewellery, sticker binds) and hand over to the attendant for safekeeping. Then take a clean cap and wear over the head all hairs are covered. Wear beard mask (if applicable). Take clean shoe covers or shoes from "clean shoe cover" section provided in the cupboard. Wear the shoe covers while crossing over the bench ensuring that the feet do not touch the floor.

Disinfect both the hands with hand disinfectant solution. Take clean linen, apron or disposable apron from clean linen section provided and wear it. Ensure that the attire is satisfactory by looking in the mirror. Enter the corridor. Nose mask should be used while entering in core area.

48.2 Exit Procedure in Pharmaceutical Laboratory

Enter the visitors change room. Sit on the step over bench and remove the shoe cover while crossing over the bench by holding the elastic and rolling in such a way that the external potion of shoe cover goes inside. Hold this shoe cover over the next shoe cover to get one roll of both the shoe cover goes inside. Hold this shoe cover the next shoe cover to get one roll of both of the both the shoe covers.

Dispose of the "used caps and shoe covers" section provided in the cupboard. Remove cap and put it into the "used caps and shoe covers" section provided in the cupboard. Remove the disposable aprons or linen and put it into the "used linen" section provided in the cupboard. Remove the disposable apron or linen and put it into the "used linen" section provided in the cupboard. Disinfect both the hands with hand disinfectant solution. Collect the personal belongings. Exit the change room ensuring that the door is closed properly.

References

1. Australian GMP Guidelines
   - Questions & answers on the code of good manufacturing practice for medicinal products
   - Technical Guidance on the Interpretation of Manufacturing Standards for Supplier Qualification

2. Canadian GMP Guidelines
   - Annex 2 to the Current Edition of the Good Manufacturing Practices Guidelines Schedule D Drugs (Biological Drugs)(GUI-0027)
   - Consultation: Draft Documents for Drug Good Manufacturing Practices Inspection Program (7 August2009)
   - Consultation on Good Manufacturing Practices-Inspection Program Review (26 January 2011)
   - Drug Good Manufacturing Practices (GMP) and Establishment Licensing (EL) Enforcement Directive(POL-0004)
   - GMP Inspection Policy for Canadian Drug Establishments (POL-0011)
   - Good Manufacturing Practices - Audit Report Form(FRM-0211)
   - Good Manufacturing Practices - Audit Report Form(FRM-0211) Instructions
   - Good Manufacturing Practices - Foreign Site Submission Form (FRM-0212)
   - Good Manufacturing Practices - Request for an Inspection of a Foreign Site Form (FRM-0213)
   - Good Manufacturing Practices - Foreign Site Inspection Services Agreement Form (FRM-0214)
   - Good Manufacturing Practices (GMP) for Schedule D Drugs, Part 2, Human Blood and Blood Components
   - Good Manufacturing Practices (GMP) Guidelines(4 March 2011)
   - Guidance Document - Annex 13 to the Current Edition of the Good Manufacturing Practices Guidelines Drugs
   - Used in Clinical Trials (GUI-0036)
   - Guidance on Evidence to Demonstrate Drug GMP Compliance of Foreign Sites (GUI-0080)Cover Letter

- GUIDE-0023: Risk Classification of GMP Observations,2003 edition
- Summary Report: Stakeholder Consultations on the Good Manufacturing Practices (GMP) Inspection Program Review (26 January 2011)
- Veterinary Drugs Annex to Current Edition of the Good Manufacturing Practices Guidelines

3. European Union GMP Guidelines
   - EudraLex - Volume 4: Good manufacturing practice (GMP)Guidelines
   - Q & A: Good Manufacturing Practice (GMP)

4. Japanese GMP Guidelines
   - GMP Compliance Inspection concerning Pharmaceuticals(including APIs)

5. US FDA GMP Guidelines
   - Center For Drug Evaluation and Research
   - Center For Veterinary Medicine
   - Center For Biologics Evaluation and Research
   - Center For Device and Radiological Health

6. World Health Organization Guidelines
   - GMP Questions and Answers
   - Quality Assurance of Pharmaceuticals - A Compendium of Guidelines and Related Materials

7. Australia - Therapeutic Goods Administration
   - Australian codes of good manufacturing practice – current status
   - Australian Code of Good Manufacturing Practice for Medicinal Products (16 August 2002)
   - Australian Code of GMP for Human Blood and Tissues (24August 2000)

8. Canada - Health Canada
   - Canadian Good Manufacturing Practices for Drugs (Part C Division 2 of Food and Drug Regulations)
   - Canadian GMP Resources

9. China
   - Regulations for Implementation of the Drug Administration Law of the People's Republic of China

10. European Union - European Medicines Agency
    - EudraLex Volume 4 - Good Manufacturing Practice Guidelines
    - Directive 2003/94/EC for medicinal products for human use and investigational medicinal products for human use
    - Historical Documents

11. India - Central Drug Standard Control Organization
   - Schedule M - Good Manufacturing Practices and Requirements of Premises, Plant and Equipment For Pharmaceutical Products

12. Japan - Pharmaceuticals and Medical Devices Agency
   - Ministerial Ordinance on Standards for Manufacturing Control and Quality Control for Drugs and Quasi-drugs(Tentative Translation VER. 09092005) [GMP] (24 December 2004)

13. United States - Food and Drug Administration GMP Regulations
   - 21 CFR Part 4 - Current Good Manufacturing Practice Requirements for Combination Products (As of 1 April2013)
   - 21 CFR Part 210 - Current Good Manufacturing Practice in Manufacturing, Processing, Packing, or Holding of Drugs(As of 1 April 2013)
   - 21 CFR Part 211 - Current Good Manufacturing Practice For Finished Pharmaceuticals- (As of 1 April 2013) Historical preambles announcing changes and comments regarding21 CFR Parts 210 and 211.
   - 21 CFR Part 212 Current Good Manufacturing Practice for Positron Emission Tomography Drugs - (As of 1 April 2013)
   - 21CFR Part 110 - Current Good Manufacturing Practice in Manufacturing, Packing, or Holding Human Food (As of 1April 2013) Historical preambles announcing changes and comments regarding 21 CFR Part 110.
   - 21 CFR Part 606 - Current Good Manufacturing Practice For Blood and Blood Components (As of 1 April 2013)
   - Historical preambles announcing changes and comments regarding 21 CFR Part 606
   - 21 CFR Part 820 - Quality System Regulation (As of 1 April2013) Historical preambles announcing changes and comments regarding 21 CFR Part 820.
   - 21 CFR Part 111 - Current Good Manufacturing Practice in Manufacturing, Packaging, Labeling, or Holding
   - Operations for Dietary Supplements (As of 1 April 2013) Historical preambles announcing changes and comments regarding 21 CFR Part 111.

14. World Health Organization
   - WHO Good Manufacturing Practices

15. Directive 2001/83/EC of the European Parliament and of the Council of 6 November 2001 on the community code relating to medicinal products for human ..use. Consolidated version. December 2008.

16. Regulation (EC) No 726/2004 of the European Parliament and of the Council of 31 March 2004 laying down Community procedures for the authorization and ..supervision of

medicinal products for human and veterinary use and establishing a European Medicines Agency. Consolidated version. July 2009.

17. Notice to applicants and regulatory guidelines medicinal products for human use. Volume 2B - Presentation and content of the dossier. Incorporating the ..Common Technical Document (CTD). May 2008.

18. Notice to applicants and regulatory guidelines medicinal products for human use. Volume 2A - Procedures for marketing authorization. Chapter 1 - Marketing Authorisation. November 2005.

19. Notice to applicants and regulatory guidelines medicinal products for human use. Volume 2A - Procedures for marketing authorization. Chapter 2 - Mutual ..Recognition. February 2007.

20. Notice to applicants and regulatory guidelines medicinal products for human use. Volume 2A - Procedures for marketing authorization. Chapter 3 - Community ..Referral. September 2007.

21. Notice to applicants and regulatory guidelines medicinal products for human use. Volume 2A - Procedures for marketing authorization. Chapter 4 - Centralized ..Procedure. April 2006.

22. Pharmacopoeia

23. USP Pharmacopoeia

24. EP Pharmacopoeia

25. BP Pharmacopoeia

26. International Pharmacopeia